Business, Education, Health Sciences and Nursing, Law, Medicine, Political and Policy Sciences, and Public Affairs

Business, Education, Health Sciences and Nursing, Law, Medicine, Political and Policy Sciences, and Public Affairs

LONDON AND NEW YORK

Published 2015 by River Publishers
River Publishers
Alsbjergvej 10, 9260 Gistrup, Denmark
www.riverpublishers.com

Distributed exclusively by Routledge
4 Park Square, Milton Park, Abingdon, Oxon OX14 4RN
605 Third Avenue, New York, NY 10017, USA

Business, Education, Health Sciences and Nursing, Law, Medicine, Political and Policy Sciences, and Public Affairs.

© 2015 River Publishers. All rights reserved. No part of this publication may be reproduced, stored in a retrieval systems, or transmitted in any form or by any means, mechanical, photocopying, recording or otherwise, without prior written permission of the publishers.

Routledge is an imprint of the Taylor & Francis Group, an informa business

ISBN 978-87-93379-01-5 (print)

While every effort is made to provide dependable information, the publisher, authors, and editors cannot be held responsible for any errors or omissions.

Contents

Introduction	xix
Biographies	xxi
Abilene Christian University	1
Alfred State College	1
Allan Hancock College	2
Alliant International University	3
Alliant International University - LA	4
Angelo State University	5
Anne Arundel Community College	5
Appalachian State University	6
Arkansas State University	6
Arkansas Tech University	7
Armstrong Atlantic State University	8
Athens State University	12
Auburn University	13
Austin Peay State University	15
Baker College of Muskegon	15
Barry University (All)	15
Baruch College	17
Bay Community College	19
Bentley University	19
Binghamton University (SUNY)	23
Blinn College	26
Bloomsburg University	26
Bluegrass Community & Technical College	26
Borough of Manhattan Community College	27
Bowie State University	28
Bowling Green State University	29
Bradley University	29
Brenau University	30
Brigham Young University - Idaho	30

Bronx Community College 32
Brookdale Community College 32
Brooklyn College 33
Broward College (All Campuses) 33
Buffalo State College 36
Cabarrus College of Health Sciences 36
Cabrini College 37
California Lutheran University 38
California State Polytechnic University 40
California State University Bakersfield (CSUB) 40
California State University Channel Islands 42
California State University Chico 42
California State University Fresno (Fresno State) 44
California State University Fullerton 45
California State University Long Beach 53
California State University San Bernardino 55
California State University San Marcos 55
Cambridge College 55
Cape Cod Community College 56
Carolina Law 57
Carthage College 57
Central Connecticut State University 58
Central New Mexico Community College 59
Century College 60
Cerritos College 60
Chabot College 61
Chadron State College 61
Chaffey College (All Campuses) 62
Chaminade University 63
Champlain College 64
Chandler-Gilbert Community College 65
Chapman University 66
Chattanooga State Community College 69
Chestnut Hill College 70
Chicago State University 70
City College of New York 71
Claremont Graduate University 71
Clayton State University 72
Clemson University 73

Cleveland State University	74
College of Charleston	77
College of Lake County	78
College of New Rochelle	78
College of Southern Maryland	79
College of Southern Nevada	80
College of Staten Island (CUNY - Staten Island)	82
College of the Redwoods	86
Collin College	87
Colorado State University	87
Columbus State Community College	90
Community College of Baltimore County	91
Contra Costa College - San Pablo	91
Cosumnes River College	92
Culinary Institute of America	92
CUNY Queens College	93
Cuyahoga Community College	96
Dalton State College	97
Daytona State College	97
De Anza College	97
Delta College	98
Delta State University	99
DePaul University	99
DeVry University	100
Diablo Valley College	103
Dominican University of California	103
Drake University	104
Drexel University	105
Duquesne University	106
Durham Technical Community College	109
East Carolina University	109
East Stroudsburg University	109
Eastern Illinois University	110
Eastern Kentucky University	111
Eastern Michigan University	126
Eastern New Mexico University	130
Eastern University	131
Eastfield College	131
Edgewood College	132

Edinboro University of Pennsylvania 134
Edison State College 134
El Camino College 135
El Camino College Compton Center 135
El Centro College 136
Elizabethtown College 136
Eureka College 137
Fairfield University 137
Fayetteville State University 139
Ferris State University 140
Fitchburg State University 141
Florida Agricultural and Mechanical University 142
Florida Atlantic University 143
Florida Atlantic University - Treasure Coast 144
Florida Coastal School of Law 145
Florida Gulf Coast University 145
Florida International University 147
Florida State College at Jacksonville 150
Florida State University 151
Foothill College 152
Franciscan University of Steubenville 153
Franklin University 155
Frostburg State University 158
Fullerton College 159
Galveston College 160
George Mason University 161
George Washington University 162
George Washington University Law Center 167
Georgetown University Law Center 171
Georgia College 173
Georgia Gwinnett College 173
Georgia Highlands College 174
Georgia Perimeter College - Decatur 175
Georgia State University 175
Georgian Court University 181
Gettysburg College 181
Gonzaga University 182
Goodwin College 183
Governors State University 183

Graduate Center - CUNY 184
Grand Valley State University 184
Grinnell College 185
Hampton University 186
Harper College 186
Harrisburg Area Community College 187
Hartnell College 188
Harvard University 188
Henry Ford Community College 191
Hofstra University School of Law 192
Horry-Georgetown Technical College 193
Hostos Community College 193
Houston Community College (All Campuses) 193
Howard University 195
Hudson County Community College 196
Hudson Valley Community College 196
Illinois Central College 196
Illinois Valley Community College 197
Indiana State University 197
Indiana University Bloomington 201
Indiana University of Pennsylvania 202
Indiana University South Bend 203
Indiana University-Purdue University Fort Wayne 204
Irvine Valley College 205
Ivy Tech Community College: Indianapolis 206
J. Sargeant Reynolds Community College 206
Jacksonville State University 207
Jacksonville University 210
James Madison University 211
Jefferson College 212
John Cabot University 212
John Jay College of Criminal Justice 213
John Marshall Law School 214
Johns Hopkins University 218
Kansas State University 218
Kennesaw State University 221
Kent State University 225
Kutztown University of Pennsylvania 227
La Salle University 228

Lake Forest College 231
Lake Superior State University 231
Lamar State College at Orange 231
Lamar University 232
Lancaster Bible College 232
Lane Community College 232
Lansing Community College 233
Lehman College 233
Lewis and Clark College 234
Liberty University 234
Lock Haven University 235
Loma Linda University 236
Lone Star College (All) 236
Long Beach City College 237
Long Island University - C.W. Post 238
Lorain County Community College 239
Los Angeles Mission College 240
Los Angeles Southwest College 240
Los Angeles Valley College 240
Louisiana Tech University 241
Loyola Marymount University 241
Loyola University 243
Loyola University Maryland 245
Macomb Community College (All Campuses) 248
Madonna University 249
Marist College 249
Marquette University 250
Marshall University 251
Massachusetts College of Pharmacy & Health Science 251
McNeese State University 252
Medgar Evers College 252
Mercy College 252
Meridian Community College 253
Messiah College 253
Metropolitan Community College 254
Metropolitan State University 254
Miami Dade College (All) 255
Michigan State University 259
Middle Tennessee State University 260

Middlesex Community College 260
Midlands Technical College 260
Midstate College 261
Millersville University 262
Minnesota State University 263
Miramar College 263
Mission College 263
Mississippi State University 264
Mississippi University for Women 265
Missouri Southern State University 266
Missouri State University 266
Missouri University of Science and Technology 269
Mohawk Valley Community College 270
Molloy College 271
Monmouth University 272
Monroe Community College 273
Montclair State University 274
Montgomery County Community College (All) 277
Mott Community College 278
Mount Saint Mary College 278
Mt. San Antonio College 278
Muskegon Community College 281
Nashville State Community College 281
National University 281
National-Louis University 282
Neumann University 283
New Jersey City University 284
New Mexico State University 284
New York City College of Technology 286
New York College of Podiatric Medicine 286
New York Law School 287
New York University 289
New York University School of Law 289
Norfolk State University 290
North Carolina State University 291
North Central Texas College 292
North Lake College (All Campuses) 293
Northampton Community College 293
Northeast Iowa Community College 293

Northeastern Illinois University 294
Northeastern State University 295
Northern Arizona University 295
Northern Illinois University 297
Northern Virginia Community College 297
Northern Virginia Community College - Medical 298
Northwest Missouri State University 298
Northwest Vista College 299
Nova Southeastern University 299
Oakland University 301
Ocean County College 301
Ohio State University 301
Ohio University 306
Ohio University Southern 307
Oklahoma State University: Oklahoma City 308
Old Dominion University 309
Oral Roberts University 313
Orange Coast College 315
Orange County Community College (SUNY) 315
Oregon State University 316
Owens Community College: Toledo 316
Ozarks Technical Community College 316
Palomar College 317
Pellissippi State Community College 317
Pensacola State College 318
Pepperdine University 318
Pepperdine University's Graziadio School of Business
  and Management 319
Piedmont College 320
Pierce College (All) 320
Pima Community College 321
Pittsburg State University 322
Prince George's Community College 323
Quinnipiac University 323
Quinsigamond Community College 324
Radford University 324
Ramapo College - New Jersey 326
Raritan Valley Community College 326
Rensselaer Polytechnic Institute 328

Resurrection University	328
Rhode Island College	329
Roger Williams University	331
Roosevelt University	332
San Antonio College	333
San Diego State University	333
San Francisco State University	336
San Jacinto College - North	339
San Joaquin Delta College	340
San Jose State University	341
Santa Ana College	342
Santa Monica College	342
Seton Hall University	343
Seton Hill University	343
Shippensburg University of Pennsylvania	344
Sinclair Community College	344
Slippery Rock University	345
South Texas College (All Campuses)	345
Southeast Missouri State University	346
Southeastern Louisiana University	348
Southern Illinois University - Carbondale	349
Southern Methodist University	350
Spartanburg Methodist College	350
Spring Arbor University	351
St. Clair County Community College	351
St. Cloud State University	352
St. John Fisher College	353
St. Louis Community College at Forest Park	354
St. Louis Community College at Meramec	355
St. Petersburg College (All Campuses)	355
Stark State College	358
Stetson University	358
Stetson University College of Law	358
Stevenson University	359
Suffolk County Community College	359
SUNY Canton	359
SUNY Oswego	359
SUNY Plattsburgh	360
Susquehanna University	360

Tarrant County College (All) 360
Tennessee State University 362
Tennessee Tech University 363
Texas A&M University at College Station 364
Texas A&M University at Corpus Christi 364
Texas A&M University - San Antonio 367
Texas Christian University 368
Texas Southern University 369
Texas State University - San Marcos 370
Texas Tech University 371
The Catholic University of America 372
The College at Brockport (SUNY Brockport) 373
Thomas Jefferson University 375
Towson University 376
Trident Technical College 380
Troy University 381
Tulsa Community College 381
Tusculum College 382
University at Buffalo (SUNY Buffalo) 383
University of Akron 386
University of Alabama 387
University of Alaska Fairbanks 389
University of Arizona 391
University of Arkansas Fayetteville 393
University of Baltimore 393
University of California Berkeley 395
University of California Davis 395
University of California Los Angeles (UCLA) 396
University of California Merced 396
University of California San Diego 398
University of California Santa Barbara 399
University of Central Arkansas 400
University of Central Florida 402
University of Central Oklahoma 405
University of Cincinnati 408
University of Colorado Colorado Springs 409
University of Connecticut 410
University of Connecticut Avery Point 411
University of Denver 411

Contents xv

University of Findlay	412
University of Florida	414
University of Georgia	414
University of Hawaii at Hilo	415
University of Houston	416
University of Houston - Clear Lake	417
University of Illinois at Chicago	420
University of Illinois at Springfield	420
University of Illinois at Urbana-Champaign	421
University of Iowa	423
University of Kansas	425
University of La Verne	425
University of Louisiana at Lafayette	426
University of Louisville	426
University of Maine Orono	427
University of Maryland - Eastern Shore	428
University of Massachusetts	429
University of Massachusetts - Boston	431
University of Massachusetts - Dartmouth	433
University of Massachusetts - Lowell	434
University of Massachusetts School of Law	435
University of Memphis	437
University of Miami	439
University of Michigan	442
University of Minnesota Duluth	443
University of Missouri - Columbia	445
University of Montana - Missoula	445
University of Mount Union	445
University of Nebraska - Kearney	447
University of Nebraska - Omaha	447
University of Nevada - Reno	448
University of New Hampshire	450
University of New Haven	452
University of New Mexico	453
University of New Orleans	455
University of North Alabama	455
University of North Carolina	455
University of North Carolina at Charlotte	456
University of North Carolina Wilmington	457

University of North Dakota ... 458
University of North Florida ... 459
University of North Texas ... 464
University of Northern Colorado ... 465
University of Oklahoma - Health Sciences Center ... 466
University of Oregon ... 467
University of Pennsylvania ... 468
University of Pittsburgh ... 468
University of Puget Sound ... 469
University of Rhode Island ... 472
University of San Diego ... 472
University of South Carolina ... 473
University of South Carolina - Aiken ... 474
University of South Dakota ... 475
University of South Florida ... 475
University of Southern Maine - Gorham ... 477
University of St. Thomas ... 478
University of Tennessee - Chattanooga ... 481
University of Tennessee - Knoxville ... 484
University of Tennessee at Martin ... 485
University of Texas ... 486
University of Texas - Pan American ... 487
University of Texas at Brownsville ... 488
University of Texas at El Paso ... 490
University of Texas at San Antonio ... 491
University of Texas at Tyler ... 493
University of Texas Health Science Center ... 494
University of Texas Southwestern Medical Center ... 495
University of the Incarnate Word ... 496
University of the Pacific ... 496
University of Toledo ... 497
University of Toledo College of Law ... 499
University of Utah ... 500
University of Washington ... 502
University of West Georgia ... 503
University of Wisconsin - Eau Claire ... 505
University of Wisconsin - Madison ... 506
University of Wisconsin - Platteville ... 507

University of Wisconsin - Stout	508
University of Wisconsin - Parkside	510
Utah State University	510
Utah Valley University	511
Valencia College	514
Valparaiso University	516
Vanderbilt University	517
Vernon College	518
Virginia Commonwealth University	519
Viterbo University	522
Wagner College	522
Wake Technical Community College	523
Walsh College	524
Walters State Community College	524
Washburn University	525
Washington State University	526
Washington State University - Vancouver	528
Washington University in St. Louis	528
Waubonsee Community College	531
Wayland Baptist University	532
Wayne State University	532
Weber State University	536
Webster University	539
West Coast University Orange County	539
West Los Angeles College	540
West Valley College	541
West Virginia University	541
Western Connecticut State University	542
Western Illinois University	543
Western Kentucky University	544
Western Michigan University	545
Wichita State University	546
Wilkes University	546
William Paterson University	548
Williams Baptist College	551
Wilmington University	551
Winston-Salem State University	552
Wofford College	553

Woodbury University 553
Worcester State University 556
Wright State University 556
Xavier University 558
Youngstown State University 559
Index 561

Introduction

FacultyAwards.org is the first and only university awards program in the United States based on faculty peer evaluation. Faculty Awards was created to recognize outstanding faculty members (as viewed by their Faculty peers) at colleges and universities across the United States. Faculty members voted through the 2014–2015 academic year for their peers at their academic departments and schools within a number of categories.

Access to FacultyAwards.org to nominate and vote for Faculty was limited to university professors or faculty members at accredited U.S. institution of higher education.

Faculty members were nominated and voted for by other faculty members in their own academic departments and schools. We strove to maintain an accurate peer-review process. Voting was not open to students or the public at large. In addition, faculty members voted for educators only at their own college or university.

Winners for the 2014–2015 academic year, in all departments and colleges across U.S. institutions of higher education were announced in March 2015 and are permanently archived at FacultyAwards.org, as well as recognized in this 2015 print edition of the Faculty Awards Compendium.

For the academic year 2014–2015; thousands of votes were cast to nominate and vote for Faculty members, and no self-voting was allowed, to assure the integrity of the whole process. A total of 4106 winning Faculty members were determined after tallying the votes.

The awards are not intended to carry merit towards a typical faculty tenure or promotion process. However, as Faculty peer review of teaching, scholarship, research and service is a crucial component of the evaluation processes in U.S. institutions of higher education; the results provide a valuable resource to acknowledge the stellar perception of a particular set of distinguished Professors by their peers and, hopefully, will continue in the years to come to provide a further incentive for Faculty members to continue on their varying paths of academic and scholarly excellence so they would be recognized by their institutional peers.

FacultyAwards.org fully recognizes that peer evaluation is only one of many aspects that define a distinguished educator, scholar and researcher; which also include: student evaluations, scholarly production and publication records, innovative curricula development, distinguished teaching outcomes, outreach and service activities, external funding records, collegiality, research eminence and outcomes, and several other factors. The intention of this compendium is to highlight and reward *only* the peer-evaluation factor for Faculty members and none of the other factors involved in what typically would define an eminent and well-rounded Faculty member at a U.S. institution of higher education.

This volume of the Faculty Awards Compendium includes Faculty awardees within the Business, Education, Health Sciences and Nursing, Law, Medicine, Political & Policy Sciences, and Public Affairs Disciplines for the 2014–2015 academic year. We would like to thank all Faculty members who participated in the voting process and to wish all the Faculty awardees continued success in their academic endeavors. We look forward to resuming the voting process for the 2015–2016 academic year awards.

Biographies

Professor Tarek M. Sobh received the B.Sc. in Engineering degree with honors in Computer Science and Automatic Control from the Faculty of Engineering, Alexandria University, Egypt in 1988, and M.S. and Ph.D. degrees in Computer and Information Science from the School of Engineering, University of Pennsylvania in 1989 and 1991, respectively. He is currently the Senior Vice President for Graduate Studies and Research, Dean of the School of Engineering and Distinguished Professor of Engineering and Computer Science at the University of Bridgeport (UB), Connecticut; the Founding Director of the Interdisciplinary Robotics, Intelligent Sensing, and Control (RISC) laboratory; the Founder of the High-Tech Business Incubator at UB (CTech IncUBator), and a Professor of Computer Engineering, Computer Science, Electrical Engineering and Mechanical Engineering.

He was Vice President from 2008–2014, Vice Provost from 2006–2008, Interim Dean of the School of Business, Director of External Engineering Programs, Interim Chairman of Computer Science and Computer Engineering, and Chairman of the Department of Technology Management at the University of Bridgeport. He was an Associate Professor of Computer Science and Computer Engineering at the University of Bridgeport from 1995–1999, a Research Assistant Professor of Computer Science at the Department of Computer Science, University of Utah from 1992–1995, and a Research Fellow at the General Robotics and Active Sensory Perception (GRASP) Laboratory of the University of Pennsylvania from 1989–1991. He was the Founding Chairman of the Discrete Event and Hybrid Systems Technical Committee of the IEEE Robotics and Automation Society from 1992–1999, and the Founding Chairman of the Prototyping Technical Committee of the IEEE Robotics and Automation Society from 1999–2001. His background is in the fields of computer science and engineering, control theory, robotics, automation, manufacturing, AI, computer vision and signal processing.

Research Interests and Activities:

Dr. Sobh's current research interests include reverse engineering and industrial inspection, CAD/CAM and active sensing under uncertainty, robots and electromechanical systems prototyping, sensor-based distributed control schemes, unifying tolerances across sensing, design, and manufacturing, hybrid and discrete event control, modeling, and applications, and mobile robotic manipulation. He has published over 200 refereed journal and conference papers, and book chapters in these and other areas, in addition to 18 books. Professor Sobh is also interested in developing theoretical and experimental tools to aid performing adaptive goal-directed robotic sensing for modeling, observing and controlling interactive agents in unstructured environments.

Dr. Sobh serves on the editorial boards of 15 journals, and has served as Chair, Technical Program Chair and on the program committees of over 150 international conferences and workshops in the Robotics, Automation, Sensing, Computing, Systems, Control, Online Engineering and Engineering Education areas. Dr. Sobh has presented more than 100 keynote speeches, invited talks and lectures, colloquia and seminars at research meetings, University departments, research centers, and companies.

Some of the current research and development activities that he leads at the RISC laboratory include work on tolerance representation and determination for inspection and manufacturing, hybrid controllers for robotics and automation, service robots, prototyping and synthesis of controllers, simulators, and monitors for manipulators, algorithms for uncertainty computation from sense data, and web-based prototyping, control synthesis, and simulation of robots.

Dr. Sobh has supervised more than 50 award-winning graduate and undergraduate students working on different projects within robotics, prototyping, computer vision, control, and manufacturing; in addition to more than 300 undergraduate and graduate students working on their B.S. projects, Master's thesis or Ph.D. dissertations. Dr. Sobh is active in consulting and providing service to many industrial organizations and companies. He has consulted for several companies in the U.S., Switzerland, India, Malaysia, England, the United Arab Emirates, Kazakhstan and Egypt, to support projects in robotics, automation, manufacturing, sensing, numerical analysis, and control. He has also worked at Philips Laboratories in New York, and a number of companies in Egypt. Dr. Sobh has been awarded over 45 research grants to pursue his work in robotics, automation, manufacturing, and sensing. Dr. Sobh is a Fellow of the African Academy of Sciences and a member of the Connecticut Academy of Science and Engineering. He has received many awards and merits in recognition of his research and scholarly activities in engineering, education and computing and his services to the academic community.

Dr. Sobh is a Licensed Professional Electrical Engineer (P.E.) in the State of Utah, a Certified Manufacturing Engineer (CMfgE) by the Society of Manufacturing Engineers, a Certified Professional Manager (C.M.) by the Institute of Certified Professional Managers at James Madison University, a Certified Reliability Engineer (C.R.E.) by the American Society for Quality, a member of Tau Beta Pi (The Engineering Honor Society), Sigma Xi (The Scientific Research Society), Phi Beta Delta (The International Honor Society), Upsilon Pi Epsilon (The National Honor Society for the Computing Sciences), Phi Kappa Phi (The Academic Honor Society) and an honorary member of Delta Mu Delta (The National Honor Society for Business Administration).

Dr. Sobh is a member, senior member, founding or board member of several professional organizations including; ACM, IEEE, the International Society for Optical Engineering (SPIE), the National Society of Professional Engineers (NSPE), the New York Academy of Sciences, the American Society of Engineering Education (ASEE), the American Society of Quality (ASQ), the American Association for the Advancement of Science (AAAS), the Society of Manufacturing Engineers (SME), the International Association of Online Engineering (IAOE), the Discovery Museum, the Connecticut Pre-Engineering Program (CPEP), the Northeast Center for Computers and Information Systems Security (NECCISS), the International E-Learning Association (IELA), and the Society for Industrial Computing. Dr. Sobh is a graduate of Victoria College, Alexandria, Egypt, in 1983 and a life member of the Old Victorians Association. He is also a life Member of the Egyptian Engineering Syndicate, a licensed engineer of the Egyptian Engineers Association and the Alexandria Engineering Organization.

Professor Khaled Elleithy is the Associate Vice President for Graduate Studies and Research at the University of Bridgeport. He is a professor of Computer Science and Engineering. He has research interests include wireless sensor networks, mobile communications, network security, quantum computing, and formal approaches for design and verification. He has published more than three hundred fifty research papers in national/international journals and conferences in his areas of expertise. Dr. Elleithy is the editor or co-editor for 12 books published by Springer.

Dr. Elleithy received the B.Sc. degree in computer science and automatic control from Alexandria University in 1983, the MS Degree in computer networks from the same university in 1986, and the MS and Ph.D. degrees in computer science from The Center for Advanced Computer Studies at the University of Louisiana – Lafayette in 1988 and 1990, respectively.

Dr. Elleithy has more than 20 years of teaching experience. His teaching evaluations are distinguished in all the universities he joined. He is the recipient of the "Distinguished Professor of the Year", University of Bridgeport, academic year 2006–2007. He supervised hundreds of senior projects, MS theses and Ph.D. dissertations. He developed and introduced many new undergraduate/graduate courses. He also developed new teaching/research laboratories in his area of expertise. His students have won more than twenty prestigious national/international awards from IEEE, ACM, and ASEE.

Dr. Elleithy is a member of the technical program committees of many international conferences as recognition of his research qualifications. He served as a guest editor for several International Journals. He was the chairperson for the International Conference on Industrial Electronics, Technology & Automation. Furthermore, he is the co-Chair and co-founder of the Annual International Joint Conferences on Computer, Information, and Systems Sciences, and Engineering virtual conferences 2005–2014.

Dr. Elleithy is a member of several technical and honorary societies. He is a Senior Member of the IEEE computer society. He is a member of the Member of the Association of Computing Machinery (ACM) since 1990, member of ACM SIGARCH (Special Interest Group on Computer Architecture) since 1990, member of the honor society of Phi Kappa Phi University of South Western Louisiana Chapter since April 1989, member of circuits & systems society since 1988, member of the IEEE computer society since 1988, and a lifetime member of the Egyptian Engineering Syndicate since June 1983.

ABILENE CHRISTIAN UNIVERSITY

Don Pope

Most Helpful to Students, Management Sciences

"Don really tries to spend time helping each student individually."

ALFRED STATE COLLEGE

John Santora

Best Teacher, Education

Alfred State College, 1979–present Associate professor of culinary arts; responsible for instruction of first-and second-year students in food ingredients and products, hospitality cost control, furnishing and equipment, and facilities planning and energy conservation. Oversight of instruction and operation of the a la carte dining room during the lunch period and special events. Have also taught purchasing techniques, stewarding, and sanitation, preparing students to become certified food handlers.
 Southern Tier Health Network, 1994 and 1995 (summers) Consultant, advisor, and instructor to kitchen staff of five local hospitals
 Farm/Rural Training Program, 1980–85 Alfred/Wellsville For approximately 10 weeks each summer, taught people with limited or no exposure to the food industry, the basics of cooking, sanitation, and customer service.
 Fireside Restaurant, 1984 Port Allegany, PA Consultant for menu and recipe development
 Wellsville High School, 1982 (50 hours) Adult Education Course Instructor A 10-week course allowing adults to explore a variety of cooking techniques and food applications for the home
 Niagara County Community College (host site), 1976–79 CETA training program for unemployed and displaced workers (ages 16+); taught the essentials of food preparation, allowing for entry-level employment each program ran for 25 weeks.
 Niagara Hilton, 1978–79 Assistant Chef/Chef Responsible for all aspects of restaurant operation in the hotel, including the a la carte dining room, room service orders, and multiple banquets.

ALLAN HANCOCK COLLEGE

Yvon Frazier

Best Overall Faculty Member, Education
Best Teacher, Education
Most Helpful to Students, Education
Best Teacher, Education
Most Helpful to Students, Education
Best Overall Faculty Member, Education

Eliseo Munoz

Most Helpful to Students, Health Science

"Gives his knowledge of physical dynamics to his athletes, helps them with their recovery to return to the field."

Munoz begins his eighth year as the head trainer at Allan Hancock College and eleventh overall at the college. He earned a bachelor's degree in kinesiology with an emphasis on teaching from Cal Poly, San Luis Obispo, where he also served a graduate internship in athletic training. From 2001–03, he was a graduate assistant at San Jose State and the head athletic trainer at Scotts Valley High School. He earned a master's degree in kinesiology with an emphasis in athletic training from San Jose State University. During his time at San Jose State, Munoz also interned with the sports medicine team serving the San Francisco 49ers. Munoz is an instructor at Hancock, currently teaching athletic training, health education and Emergency Medical Services-CPR classes. Munoz played football at King City High School and at Hartnell College. Munoz and his wife Claudia reside in Santa Maria with their two-year-old son Valentino.

Thesa Roepke

Best Researcher/Scholar, Education
Best Researcher/Scholar, Education

ALLIANT INTERNATIONAL UNIVERSITY

Harriet Curtis-Boles

Best Overall Faculty Member, Psychology

"Highly regarded by colleagues and students especially for her contributions to multi-cultural education. She makes this perspective accessible, intelligent and imperative for working with clients in the field of psychology, especially clinical psychology."

Kristi Franklin

Best Overall Faculty Member, Psychology

Fred Heide

Best Researcher/Scholar, Psychology

"Dr. Heide is well published in the area of clinical Psychology. He is also Associate Editor of an on-line journal published by the American Psychological Association. His own research has combined Clinical Psychology with his own and his colleagues work in dramatic presentations. Perhaps most importantly in the context of the specific graduate school in which he teaches, he is tireless in his work with students on their dissertations in Psychology, especially using innovative research paradigms which stretch the limits of the field to include research focused on innovative topics, often combining work in the field of psychology with artistic endeavors."

Dr. Heide has been on the core faculty of the California School of Professional Psychology since 1983, and has won awards for both Master Teacher and Teacher of the Year. He's taught at the Medical College of Wisconsin's Door County Summer Institute since 1999. Dr. Heide is also co-founder, founding board president, board member, artistic advisor, playwright, and performer for American Folklore Theatre (AFT) in Fish Creek, WI, described by the Chicago Tribune (2012) as "one of the most exceptional professional troupes in the country" and winner of the 2012 Wisconsin Governor's Award for Arts, Culture and Heritage. His recent publications explore the application of theatre to psychological change. Among the 20 shows Dr. Heide has co-created are the sci-fi football musical "Packer Fans From Outer Space" and the metaphysical musical comedy "Belgians in Heaven" (both with Lee Becker and James Kaplan). These shows incorporate social learning principles to promote positive social behavior. Dr. Heide studied and co-created several shows with Paul Sills, founding director of Chicago's Second City Theatre, and teaches Mr. Sills' methods to augment mindfulness and expressive behavior for students at CSPP.

Randy Wyatt

Most Helpful to Students, Psychology

"Dr. Wyatt has been tireless in helping students address contemporary issues in clinical psychology. He does so in the classroom. But his most significant contribution continues to be leadership in developing and overseeing over 200 clinical agencies in which students obtain their practicum and internship training – which in CSPP is almost half of the education and training program. He is tireless, flawless and very creative in this work, often in situations in which complex professional and ethical issues arise."

Licenses
    Licensed Psychologist in California, 1991
    Professional/Honorary Memberships
    American Psychological Association Member
    California State Psychological Association Member
    Association of Family Therapists of Northern California Member
    Professional Practice Highlights
    Psychotherapy and Consultation Private Practice
    Board of Directors for Association of Family Therapists of Northern California (AFTNC)
    Council Member for Bay Area Practicum Collaborative (BAPIC)

ALLIANT INTERNATIONAL UNIVERSITY - LA

Judy Holloway

Best Teacher, Psychology

"Keeps students very interested, and they enjoy her classes while learning a lot. Many students have described her Ecopsychology class as a life-changing experience."

ANGELO STATE UNIVERSITY

Bruce Bechtol

Best Teacher, Law
Best Researcher/Scholar, Law

Walter Noelke

Best Teacher, Political Science

Deanna Watts

Best Overall Faculty Member, Political Science
Best Researcher/Scholar, Political Science
Most Helpful to Students, Political Science
Most Helpful to Students, Public Affairs, Political & Policy Sciences
Best Overall Faculty Member, Public Affairs, Political & Policy Sciences

"High caliber research on one of the most pressing political problems of our time, and one that has been sorely underexamined. Office is always open. Goes way beyond the call of duty."

ANNE ARUNDEL COMMUNITY COLLEGE

Lou Aymard

Best Overall Faculty Member, Medicine

Lou Aymard, Ph.D. is a retired child psychologist in Annapolis, Maryland. He has more than 40 years of clinical and teaching experience in the fields of parenting and child development. In 1977, he earned his doctorate in psychology from The Catholic University of America. Dr. Aymard is the founder of The Parenting Center at Anne Arundel Community College, which provides a broad range of educational and public service programs about parenting and family life. He received the 2002 International Play Therapy Award from the International

Society for Child and Play Therapy for significant career contributions to the field of child psychology. In 2004, the Warner Foundation of Baltimore, Maryland selected Aymard to receive the Humanitarian of the Year award for his work on behalf of children and families.

Tyrone Powers

Best Teacher, Criminal Justice
Most Helpful to Students, Criminal Justice

APPALACHIAN STATE UNIVERSITY

Ron Forster

Most Helpful to Students, Physical Education

ARKANSAS STATE UNIVERSITY

Natalie Johnson-Leslie

Most Helpful to Students, Education
Best Researcher/Scholar, Education
Best Teacher, Education
Most Helpful to Students, Education

Dr. Natalie A. Johnson-Leslie, is an associate professor of teacher education at Arkansas State University. She received her doctoral degrees from Iowa State University in the areas of Educational Leadership and Policy Studies as well as Curriculum and Instructional Technology. She teaches and/or supervises early childhood, mid-level and secondary education pre-service teachers. She is passionate about preparing pre-service teachers to be effective in the classroom. Her philosophy of education embraces "everyone can learn if given the support and tools needed for personal achievement."

Amy Pearce

Best Overall Faculty Member, Psychology
Best Researcher/Scholar, Psychology
Best Teacher, Psychology
Most Helpful to Students, Psychology
Best Researcher/Scholar, Medicine
Best Teacher, Medicine
Most Helpful to Students, Medicine
Best Overall Faculty Member, Medicine

In addition to teaching, Dr. Pearce conducts research in the Arkansas Biosciences Institute. With the help of student researchers, she examines the impact of oral nicotine on behavior and physiology in nonhumans. Recent studies show there is a strong, positive relationship between the amount of oral nicotine available in the environment and ingestion of oral nicotine by female rats. She is currently exploring changes in nicotine's metabolic pathway associated with chronic consumption of the drug.

ARKANSAS TECH UNIVERSITY

Christopher Housenick

Best Teacher, Political Science
Most Helpful to Students, Political Science

Mohamed Ibrahim

Best Researcher/Scholar, Education

ARMSTRONG ATLANTIC STATE UNIVERSITY

Barbara Hubbard

Most Helpful to Students, Education

Barbara S. Hubbard 120 East 45th Street Savannah, Georgia 31405 912-238-4613 barbara.hubbard@armstrong.edu

EDUCATION

Doctor of Education 2003 Educational Leadership Georgia Southern University, Statesboro, Georgia Dissertation: A Description of Gifted Elementary Service in Georgia Schools

Educational Specialist 2000 Educational Leadership Georgia Southern University

Leadership Training Consortium 1995 Master of Education in Leadership Georgia Southern University

Master of Education 1982 Early Childhood Education Armstrong State University

Bachelor of Arts 1975 Elementary Education The University of South Carolina

PROFESSIONAL LICENSURE

Georgia Educator Certificate 1975–2017 Early Childhood Education (P–5) Gifted Education (P–12) Educational Leadership (P–12)

ACADEMIC POSITIONS

Assistant Professor 2008 – Department of Early Childhood Education Armstrong Atlantic State University, Savannah, Georgia Supervise Interns I & II; class instructor' maintain accurate records; student advisor; collaborate with school administrators and teachers; revised lesson plan format; other duties as assigned. Courses Taught: Undergraduate EDUC 2130 Teaching and Learning – Educational Psychology ECUG 3750 Internship I ECUG 4300 Diagnosing and Prescribing for Reading Problems ECUG 4750 Internship II ECUG 4420 Special Topics in Education ECUG 4090 Classroom Management

Graduate: ECMT 6000 Teaching Reading and Diagnosis and Remediation ECMT 6010 Developmental Characteristics of Young Children ECMT 6020 Language Arts/Creative Activities ECMT 6040 Teaching Mathematics in Elementary School ECMT 6750 Internship II ECEG 7050 Advanced Methods in Elementary Mathematics ECEG 6090 Classroom Management ECEG 7070 Cross Cultural Communication MCEG 7070 Cross Cultural Communication ECEG 7060 Multi-Media Approaches to Children's Literature

Director of Field Experiences 2007 Department of Early Childhood Education The University of South Carolina, Beaufort Advised, placed and supervised interns; instructed classes; collaborated with school administrators and teachers; collaborated with school district administrators to meet security requirements and ensure safety of students; wrote section of NCATE re-certification application; revised student handbook. Courses Taught: Undergraduate BEDC 210 Observation and Analysis BEDE 469 Internship BEDE 476 Senior Seminar

P–12 TEACHING POSITIONS

Savannah-Chatham County Public School System (SCCPSS)

Director of Gifted Education Programs – SCCPSS District Office 2001–2007 • Responsible for daily operation and budget of gifted department • Placed and supervised qualified teachers in schools • Provided approved Professional Standards Commission professional development courses for teachers • Collaborated with school administrators • Ensured proper implementation of Georgia Gifted Standards and reporting of FTE (full time equivalent) • Oversaw testing and placement of students • Developed & implemented summer programs for students.

Acting Associate Superintendent of Curriculum & Instruction – SCCPSS District Office 2005–2006 • Supervised & evaluated Directors of English/LA, Math, Science, Social Studies, Visual & Performing Arts, Health & PE; facilitated curriculum guides for all subjects • Monitored testing data • Collaborated with parents, teachers, school administrators, superintendent, state officials • Planned & implemented programs & curriculum.

Assistant Principal - Butler Elementary School 1999–2001 • Served as liaison between faculty & administration • Created teacher schedules • Handled student discipline • Observed & evaluated teachers • Maintained inventory of books & supplies • Facilitated bus transportation for students. Gifted Facilitator – May Howard Elementary School 1987–1999 • Collaborated with classroom teachers to reinforce objectives • Taught higher level thinking skills • Facilitated competitions • Tested & placed qualified students following state guidelines for gifted education.

Elementary Classroom Teacher 1975–1987 • Planned & implemented lessons following state objectives to maximize student achievement & mastery of subjects. OTHER PROFESSIONAL EXPERIENCES Delegate to World Summit on Education – London 2007 Visited schools in London to observe use of technology in classrooms.

U.S. Delegate to China – Chinese Bridge for American Schools Program 2006 Visited Chinese schools in Beijing and Shangdong Province. Experienced Chinese culture and educational process. Integrating Technology in the Classroom (INTECH) 1998–2000 Taught INTECH courses (P–5)

Georgia Effective Teaching Strategies (GETS) 1998–1999 Taught courses for professional development (P–12)

Gifted Endorsement Courses 1997–2006 Wrote gifted endorsement course curriculum and taught courses (P–12)

PROFESSIONAL PRESENTATIONS

National College Board Forum, San Diego, CA 2007 Presented Chinese Bridge for American Schools Program

Georgia State Gifted Coordinators' Conference, Walton County, GA 2003 Presented State Gifted Education Service Models report

Georgia Association of Gifted Children (GAGC) Athens, GA 1998 Presented: "Engineering is Elementary" – An Exciting Partnership

GRANT and PROGRAMS

Armstrong College of Education Grant – Armstrong Future Teachers Program 2013–2014 Awarded funds for program to bridge the partnership between Armstrong State University and the local P–12 school systems to increase awareness of education as a career.

College Board Advanced Placement Summer Institutes 2006 – Created programs for teachers to receive AP certification locally SCCPSS – Savannah, Georgia

Savannah-Chatham County Public Schools Gifted Endorsement Program 2005 – Facilitated the Georgia Professional Standards Commission approval of Gifted Endorsement courses for teacher certification SCCPSS – Savannah, Georgia

College Board-Greenhouse Grants 2004–2007 Awarded funds for Pre-AP Summer Institute for rising AP students (held on AASU campus) Awarded funds for Junior Summer Academy for rising 6–8 grade students SCCPSS – Savannah, Georgia

21st Century Grant – a partnership with SCCPSS and AASU 2001 Awarded the following year SCCPSS – Savannah, Georgia

Federal Comprehensive School Reform Demonstration Program (CSRD) 2001 Awarded $57,750 per year for three years Butler Elementary School, Savannah, Georgia

Savannah-Chatham School Improvement Grant 2000 Awarded Magnet Status for Butler Elementary Micro-Society and funding Butler Elementary School, Savannah, Georgia

PROFESSIONAL IN-SERVICE TEACHER TRAINING

Making Standards Work – Presented by Douglas Reeves 2005 Teaching with the Brain in Mind – Presented by Eric Jensen 2005 Unwrapping the Standards – Presented by Larry Ainsworth 2005 Classroom Instruction That Works – Presented by Robert Marzano 2004 Differentiated Instructional Strategies – Presented by Carolyn Chapman 2003 The Parallel Curriculum – Presented by Carol Ann Tomlinson 2003 Using DATA to Improve Student Achievement – Presented by Deborah Wahlstrom 2002 Working on the Work – Presented by Phillip Schlechty 2001 Train Smart, Perfect Trainings Every Time – Presented by Rich Allen 2001 Georgia Institute for New Leaders 2000 Georgia Effective Teaching Strategies 1997 A Framework for Understanding Poverty – Presented by Ruby Payne 1996 Positive Classroom Discipline – Presented by Fred Jones 1994

PROFESSIONAL MEMBERSHIPS

Phi Delta Kappa Georgia Association for Gifted Children, Executive Board National Association for Gifted Children Association for Supervision and Curriculum Development National Institute of School Leaders – Chief Learning Officer Armstrong Atlantic State University Alumni Association Georgia Southern University Alumni Association College Board, Southern Regional Council Member Junior Achievement, Board Member Council on World Affairs Member St. John's Church – Chancel Society, Choir Member, Sunday School Teacher

PROFESSIONAL SERVICE Armstrong State University – Alumni Board Member 2013–15

Armstrong Atlantic State University – Faculty Senate 2012–15
Senate Constitution & By Laws Committee – Secretary 2012–13
Senate Academic Standards Committee – Senate Liaison 2012–14
Online Ad hoc committee 2012–13

University & College of Education Academic Appeals Committees 2012–14
College of Education, Dean's Leadership Committee 2012–14
Junior Achievement of Georgia – Liaison between schools and volunteers 2010–11
Co-chaired Leukemia and Lymphoma Society "Light the Night Project" for College 2009 Of Education, Armstrong Atlantic State University, Savannah, Georgia
Volunteered in Savannah-Chatham Schools and Blessed Sacrament for Career Day 2007–8 and test proctor
Early Childhood Education revision of Lesson Plan format - College of Education 2008 Armstrong Atlantic State University, Savannah, Georgia
Revised Intern Handbook, College of Education, University of South Carolina Beaufort, 2007 Beaufort, South Carolina
Chair, Standard III Committee for NCATE, College of Education, 2007 University of South Carolina Beaufort, Beaufort, South Carolina

Anne Katz

Best Teacher, Childhood and Exceptional Students

Patricia Parsons

Best Researcher/Scholar, Childhood and Exceptional Students

Regina Rahimi

Best Overall Faculty Member, Education
Best Researcher/Scholar, Education
Best Teacher, Education

"Regina is very progressive. She relates well to her students."

ATHENS STATE UNIVERSITY

John Berzett

Most Helpful to Students, Business

Emily Corzine

Best Teacher, Business

Linda Hemingway

Best Overall Faculty Member, Business

Stacie Hughes

Best Researcher/Scholar, Business

I love my job! I truly enjoy helping others achieve their goals through education and the field of accounting and I feel very fortunate to be able to do this at my alma mater, Athens State. When I'm not working, I enjoy painting, reading, cooking, and building things with my hands. I'm a huge football and hockey fan even though I have no skill in either sport. My favorite time is the time I spend with my husband, Scott, and my wonderful dogs, Archer and Buttercup.

Jim Kerner

Best Overall Faculty Member, Business
Best Researcher/Scholar, Business

Kim Roberts

Best Teacher, Business
Most Helpful to Students, Business

Kim Roberts is an Assistant Professor of Operations Management in the College of Business at Athens State University in Athens, Alabama. Kim holds a Bachelor of Science degree in Chemical Engineering from the University of Alabama, a Master of Business Administration degree from the University of North Alabama, and is a PhD Candidate in Instructional Leadership with a concentration in Instructional Technology at the University of Alabama. Kim has seventeen years of experience in the process industry sector. She has held positions of process engineer, shift supervisor, engineering lead, and operations manager for a specialty chemicals company. Kim now has eight years of experience as a full-time faculty member and adjunct instructor, teaching in both classroom and online learning settings. To keep current in the operations management field, Kim also serves as a consultant. She designs, develops, and delivers process technology training to process industries. Her research interests include process technology, instructional technology, and mobile learning.

AUBURN UNIVERSITY

James Barth

Best Researcher/Scholar, Finance

"Professor Barth is known throughout the world for his work on financial systems. He has published numerous books and articles that are widely cited. He has more citations than any other faculty member in finance at Auburn University."

James R. Barth is the Lowder Eminent Scholar in Finance at Auburn University, a Senior Fellow at the Milken Institute, and a Fellow at the Wharton Financial Institution Center. His research focuses on financial institutions and capital markets, both domestic and global, with special emphasis on regulatory issues. He served as leader of an international team advising the People's Bank of China on banking reform. Barth also participated in the U.S. Speaker and Specialist Program of the U.S. Department of State in China in 2007, India in 2008, Russia in 2009, and Egypt in 2010. Also in 2008, Barth spoke on "Competition in the Financial Sector: Challenges for Regulation" at the G-20 Workshop on Competition in the Financial Sector, Bali, Indonesia.

Barth was an appointee of Presidents Ronald Reagan and George H.W. Bush as chief economist of the Office of Thrift Supervision and previously the Federal Home Loan Bank Board. He has also held the positions of professor of economics at George Washington University, associate director of the economics program at the National Science Foundation, and Shaw Foundation Professor of Banking and Finance at Nanyang Technological University. He has been a visiting scholar at the U.S. Congressional Budget Office, Federal Reserve Bank of Atlanta, Office of the Comptroller of the Currency, and the World Bank.

Barth has testified before several U.S. Congressional Committees. He has authored more than 300 articles in professional journals and has co-authored and co-edited several books, including The Great Savings and Loan Debacle, American Enterprise Institute Press, The Reform of Federal Deposit Insurance, Harper Business, Rethinking Bank Regulation: Till Angels Govern, Cambridge University Press in 2006, Financial Restructuring and Reform in Post-WTO China, Kluwer Law International in 2007, The Rise and Fall of the U.S. Mortgage and Credit Markets: A Comprehensive Analysis of the Meltdown, John Wiley & Sons in 2009, and China's Emerging Financial Markets: Challenges and Opportunities, Springer in 2009.

His most recent books are Guardians of Finance: Making Regulators Work for Us, MIT Press in 2012, Fixing the Housing Market: Financial Innovations for the Future, Wharton School Publishing-Pearson in 2012, and Research Handbook on International Banking and Governance, MIT Press in 2012. He has been quoted in publications ranging from the New York Times and Wall Street Journal to Barron's and Newsweek, and has appeared on various broadcast programs and National Public Radio. Barth is the overseas associated editor of the Chinese Banker and included in Who's Who in Economics: A Biographical Dictionary of Major Economists, 1700 to 1995.

Tracy Richard

Best Teacher, Finance

"Ms. Richard is outstanding in her ability to convey finance to undergraduate students. She establishes a strong rapport with her students and is always willing to help them. There are always students in her office seeking help."

AUSTIN PEAY STATE UNIVERSITY

Stephanie Newport

Best Researcher/Scholar, Management

"She is a sincere and articulate person who believes in shared governance in a university."

BAKER COLLEGE OF MUSKEGON

Don Schrumpf

Most Helpful to Students, Education

BARRY UNIVERSITY (ALL)

Leonard Birdsong

Best Overall Faculty Member, Law

"Professor Birdsong is a good teacher, a good scholar, a good colleague, and a good student helper. He puts a lot of effort into both the substance and the theatre of the classroom, and his students appreciate it. But he also publishes regularly about topics that he truly finds intellectually interesting. He is friendly to his colleagues and acts as a supportive mentor, and he will champion people when he thinks they need help. He also frequently meets one-on-one with students even though he is a full, tenured, doctrinal professor who could rest on his laurels if he really wanted to."

Immediately after graduating from law school, Professor Leonard Birdsong was an attorney with the law firm of Baker & Hostetler. Later he served as a diplomat with the State Department with various postings in Nigeria, Germany, and the Bahamas. Professor Birdsong also worked as a federal prosecutor, first as an Assistant United States Attorney for the District of Columbia, and later as a Special Assistant United States Attorney for the U.S. Virgin Islands. After leaving government service, Professor Birdsong moved into private practice in Washington, D.C. where he specialized in trial work in areas ranging from criminal defense to political asylum.

While in private practice, Professor Birdsong also was involved with broadcasting. He has offered on-air TV legal analysis for Fox News, CNN, Court TV, BET TV News, and WUSA Channel 9 in Washington, D.C. Today, Professor Birdsong occasionally appears as a legal commentator for Fox News and MSNBC.

Helia Hull

Most Helpful to Students, Law

"Professor Hull is so focused on the students. She meets with them all the time, and she is involved in a number of student organizations. With respect to every concern, she always brings the question back around to how the students will be affected. She views them as whole persons and is concerned not just for their academics but for their stress."

Helia Garrido Hull is an Associate Professor of Law at Barry University Dwayne O. Andreas School of Law. She has also served as the Associate Dean for Student Affairs and the Associate Dean for Academic Affairs.

Professor Hull is a member of the Florida Bar and began her legal career practicing in the area of disability law, specializing in cases involving violations of the Americans with Disabilities Act. She also focused her practice on mental health law. Being consistently committed to public service, she was selected as one of ten Florida Bar Public Service Fellows while in law school. Currently, she serves as the Chair of the Board of Directors of the Advocacy Center for Persons with Disabilities, Inc. The Advocacy Center is charged under both federal law and the Florida Governor's Executive Order to advocate the legal, human and civil rights of individuals with disabilities in Florida.

Professor Hull teaches disability law, mental health law, legal research and writing, and appellate advocacy. Professor Hull's scholarship has focused on various issues affecting persons with disabilities. Her recent publications include: "Induced Autism: The Legal and Ethical Implications of Inoculating Vaccine Manufacturers from Liability," 34 Cap. U. L. Rev. 1 (2005) (Lead Article); "Equal Access to Post-Secondary Education: The Sisyphean Impact of Flagging Test Scores of Persons with Disabilities," 55 Clev. St. L. Rev. 15 (2007) (Lead Article); and "Electroconvulsive Therapy: Baby Boomers May Be in for the Shock of Their Lives," 47 U. Louisville L. Rev. 1 (2009) (Lead Article).

Taylor Simpson-Wood

Best Teacher, Law

"Professor Simpson-Wood is so loved by the students for her dynamics in the classroom. She has a theatre background and knows how to engage their attention, but she is also very logical and has such a good grasp of the material. They learn well from her."

Professor Simpson-Wood received her J.D. from Tulane Law School, graduating Magna Cum Laude and being elected as a member of the Order of the Coif. While earning her law degree, she was also an Associate Editor of the Tulane Maritime Law Journal. She then continued her legal education as the Maritime Law Center Fellow earning her LL.M. in Admiralty. Upon graduation, she received the Edward A. Dodd award for graduating first in her class.

Professor Simpson-Wood began her legal career as an associate in the maritime section of a major firm in Seattle. Her law practice experience also includes serving as an assistant vice president and associate general counsel for a national bank and as in-house counsel for a telecommunications company.

Professor Simpson-Wood previously taught at Appalachian School of Law, Regent University School of Law, and Tulane Law School and has published in the Tulane Maritime Law Journal and the Journal of Maritime Law & Commerce. She is extremely pleased to be joining the faculty at Barry University and will be teaching Civil Procedure and Admiralty.

Professor Simpson-Wood received her undergraduate degree in acting and had an extensive acting career on stage, screen and television before attending law school. She still acts professionally and hopes to continue performing in the Orlando area as time permits.

BARUCH COLLEGE

Ted Joyce

Best Teacher, Business

David Luna

Most Helpful to Students, Business

Nita Lutwak

Most Helpful to Students, Psychology

Dr. Lutwak is a licensed psychologist, and an Associate Professor of psychology at Baruch College. She received her PhD from Fordham University, and completed her postdoctoral analytic training in psychotherapy and psychoanalysis at NYU. She is the Director of the Masters in Mental Health Counseling program at Baruch College where she teaches and supervises graduate students. She is on the training faculty for the Institute for Psychoanalytic Studies where she teaches and trains candidates in character analysis and has a private practice in NYC.

Lin Peng

Best Researcher/Scholar, Finance

Elizabeth Reis

Best Teacher, Psychology

I received my PhD from Columbia University in Special Education. I also have two master's degrees in the areas of mental retardation and curriculum from Teachers College, Columbia University. My research interests relate to assisting classroom teachers to plan more effectively for meeting the learning needs of students with disabilities. Lately, I've been writing in the area of attention deficit hyperactivity disorder (ADHD). Additional lines of research in which I am interested include issues of gender.

Kristen Shockley

Best Researcher/Scholar, Psychology

I received my BS in psychology from the University of Georgia and my MS and PhD in Industrial/Organizational Psychology from the University of South Florida. I joined the faculty at Baruch College as an assistant professor in the Fall of 2010.

My main area of research focuses on understanding the intersection of employees' work and family lives. Specifically, I have conducted research aimed at understanding organizational initiatives to help employees managing competing life demands (i.e., flexible work arrangements); research that explores the relationship between work-family conflict and health outcomes, including eating behaviors and physiological indicators of health; research that addresses the theoretical foundations of work-family interactions; and research targeted at understanding how dual-earner couples balance work and family roles.

My secondary area of interest is in career development, with a specific focus on workplace and academic mentoring, people's idiosyncratic definitions of career success, and the consequences of career compromise.

I teach undergraduate Statistics, undergraduate I/O Psychology, undergraduate Introduction to Psychology, master's level Statistics, master's/doctoral level Psychology of Work and Family, and doctoral level Motivation.

Kristin Sommer

Best Overall Faculty Member, Psychology

Kristin Sommer is a Professor of Psychology at Baruch College, City University of New York. She also holds appointments on the doctoral faculties in Basic and Applied Social Psychology and Industrial/Organizational Psychology at the Graduate Center. Dr. Sommer's primary research interests lie with the effects of peer and coworker rejection on individual performance motivation and interpersonal behavior. She also conducts research on self-regulation, social influence, and motivated decision-making processes in small groups. Dr. Sommer teaches undergraduate and doctoral courses in research methods and social psychology, as well as a course on research design in work organizations as part of Baruch College's Executive Master's Program in Management of Human Resource and Global Leadership in Taipei, Taiwan and Singapore. She has received numerous internal awards as well as research and training grants from the National Institute of Mental Health and the National Science Foundation. She sits on the editorial boards of Social Influence and Basic and Applied Social Psychology.

BAY COMMUNITY COLLEGE

Brent Madalinski

Most Helpful to Students, Business

BENTLEY UNIVERSITY

Jean Bedard

Best Overall Faculty Member, Accounting
Best Researcher/Scholar, Accounting

Jean C. Bedard is the Timothy B. Harbert Professor of Accounting in the Department of Accountancy at Bentley University. Professor Bedard's research interests include individual auditor decision quality, risk assessment and adjustment in audit engagements, and the effects of computerization on the audit process. She has published in The Accounting Review, Journal of Accounting Research, Management Science, Auditing: A Journal of Practice and Theory, Behavioral Research in Accounting, and in a number of other scholarly and professional journals. Her primary teaching interests are financial accounting and auditing, and she has taught at the undergraduate, masters and doctoral levels. Professor Bedard has served as Vice President-Publications of the American Accounting Association, and has also been President, Secretary and Historian of the AAA's Auditing Section. Other service activities for the American Accounting Association include chairing the Deloitte Wildman Award Committee, and serving on the Nominations and Publications Committees. She has also served as Research Coordinator for the AAA's Accounting, Behavior and Organizations Section. Professor Bedard's work experience outside academia includes public accounting and management of public health services.

Mike Bravo

Best Overall Faculty Member, Management
Best Researcher/Scholar, Management
Best Teacher, Management
Most Helpful to Students, Management
Best Overall Faculty Member, Business

"Mike is very considerate of his students and gives them the best chance to succeed."

Primary teaching areas are Marketing/Operations Fundamentals, Integrated Business Project, and Managing Effective Teams. Serves as the coordinator for the Marketing/Operations Fundamentals course. Skills include operations management, process efficiency, quality management, policy and procedure, and tactical and strategic planning. Experience includes Manager of Corporate Policy and Procedure, Program Manager for Quality, Manager of Finance and Accounting, and Management Accountant. Current interests include learning about the benefits of sustainability and how it improves business performance. Some recent consulting projects include: (1) researching materials and developing a process flow for building an environmentally friendly manufacturing/warehouse facility; and (2) assessing the need for a Town Manager, screening applicants, conducting interviews, and recommending, to the Board of Selectmen, a candidate for the position of Town Manager of a municipality.

Alan Hoffman

Best Teacher, Management

John Landsman

Best Teacher, Business

Perry Lowe

Best Teacher, Business

Professor Lowe is a specialist in social media and has worked with more than a dozen companies in his Corporate Immersion classes in the past few years. These include Microsoft, LogMeIn, Groove, SocialVibe, KickApps, GreatPaths, FreeFi Networks, BuyDomains, SportsMates, Napster, Ruckus, SmartWorlds, NameMedia, CareerPath, Avenue 100, and others. Corporate Immersion project courses at the graduate and undergraduate level enable students and industry to jointly solve current business problems. Other recent companies include General Motors, Welch's, HP, Newsweek Magazine, The Federal Reserve Bank of Boston, The Boston Chamber of Commerce, Toshiba, Sony, Mazda, Dunkin Donuts, McLean Hospital, Fidelity Investments, Sperry Top-Sider, U Foods, Trans Lux, and numerous start ups. Perry Lowe's applied research has focused on the use of technology to accelerate learning and he is considered a national authority on Tablet PCs. Lowe has also worked closely with Microsoft Corporation on the development and testing of new software and hardware. He has also been a judge in the Microsoft U.S. and International 2008 Imagine Cup software development competition for college students. For Imagine Cup 2009, Lowe helped expand the scope of the competition to include primary market research and a Business Plan to increase the commercialization potential for all participants. His further work in the international arena includes the co-founding of The Consortium for the Study of Virtual Global Collaboration which enables students at the National University of Ireland in Galway, University of Ulster, Northern Ireland, University of Massachusetts in Amherst and Bentley to work collaboratively over the web in cross-cultural and cross-functional project teams. More recently, he was named the 2010 co-recipient of the EFMD Excellence in Practice Award at the EFMD Annual Conference in Wiesbaden, Germany for his Corporate Immersion projects and hosted the Bentley University Collaborative Consumption Summit featuring Rachel Botsman, and CEO's from Swap.com, RelayRides, and RentCycle. Prior to Bentley, he was a product manager at General Foods and Gillette, and also worked as director of marketing for Estee Lauder Cosmetics. His entrepreneurial interests include co-founding Cinema Centers Corporation and Theater Management Services as well as the Lowe Group of Companies, focusing on broadcasting, hospitality and real estate. Among his publication credits are co-editor, The Movie Business, Prentice Hall, 1983; and co-editor, The General Motors Marketing Internship Public Relations Handbook, Sgro Promo Associates, 1996. An expert in the development of long-term relationships between industry and education, Perry Lowe has been quoted in the Financial Times, The New York Times, Boston Globe, Investor's Business Daily, The Chronicle of Higher Education, Entrepreneur, Computerworld and by the Associated Press. His teaching has earned him Bentley's Teacher of the Year in 1995, Marketing Professor of the Year in 1997, and Who's Who Among America's Best Teachers Hall of Fame.

Karen Osterheld

Most Helpful to Students, Accounting

Teaching interests include the areas of financial accounting and corporate social responsibility. Formerly employed as an auditor at Ernst & Young.

Jim Salsbury

Most Helpful to Students, Business

Jim Salsbury is a senior lecturer and assistant chair of the Management Department at Bentley College in Waltham, MA. He is also Bentley's director of Assurance of Learning. Prior to joining Bentley College, Jim spent nearly 30 years as a health care executive, serving as chief executive officer and chief operating officer of community hospitals, vice president of quality management and improvement for a major health insurance plan, and project, operations and quality management consultant to service industry providers. His special interests are in quality management in the service sector.

Jay Thibodeau

Best Teacher, Accounting

Jay Thibodeau is a former auditor and a certified public accountant. He received his BS degree from the University of Connecticut in December 1987 and his Ph.D. from the University of Connecticut in August 1996. He joined the faculty at Bentley in September of 1996 and has worked there ever since. At Bentley, he serves as the coordinator for all audit and assurance curriculum matters. Professor Thibodeau's off-campus commitments include consulting with the Audit Learning and Development group at KPMG. Professor Thibodeau's scholarship is focused on audit judgment and decision making and audit education. He is a co-author of two textbooks and has written over forty book chapters and articles for academics and practitioners in journals such as Auditing: A Journal of Practice & Theory, Accounting Horizons and Issues in Accounting Education. Professor Thibodeau currently serves as the President of the Auditing Section of the American Accounting Association. Previously, he served on the Executive Committee for the Auditing Section from 2008–2010. He has received national recognition for his work four times. The first was for his thesis, winning the 1996 Outstanding Doctoral Dissertation Award presented by the ABO section of the AAA. Two other times it was for curriculum innovation, winning the 2001 Joint AICPA/AAA Collaboration Award and the 2003 Innovation in Assurance Education Award. Finally, he was recognized for outstanding service, receiving a Special Service Award from the Auditing Section of the AAA for his work in helping to create the "Access to Auditors" program.

Poh-Lin Yeoh

Best Researcher/Scholar, Business

"Poh-Lin is very concerned with the subject matter at hand and what the students must do to be successful."

Primary teaching interests are entrepreneurship, export marketing, and international marketing strategy. Current research interests include brand extensions in international marketing, exporting entrepreneurship and export performance, and information acquisition and firm performance. Has published articles in Strategic Management Journal, European Journal of Marketing, Industrial Marketing Management, Journal of Global Marketing, Journal of Public Policy and Marketing, and International Business Review. Previously taught at University of South Carolina, National University of Singapore and Michigan State University.

BINGHAMTON UNIVERSITY (SUNY)

Murnal Abate

Most Helpful to Students, Economics

Seden Akcinaroglu

Best Overall Faculty Member, Political Science
Best Researcher/Scholar, Political Science
Most Helpful to Students, Political Science

Seden Akcinaroglu specializes in international conflict and rivalries. Her main interest lies in unraveling the linkages between expectations, informational problems, learning and strategic action in the dynamics of civil wars or rivalries. Some of her past work focuses on expectations of rival aid and the duration of war, natural disasters and rivalry relations and the effectiveness of bluffing by external actors in civil wars. Her current project examines how expectations in the intensity of rivalry with one foe affect the actions of a strategic state with others. Akcinaroglu teaches courses on international conflict and international political economy. She joined the Binghamton faculty in 2008.

Manas Chatterji

Best Researcher/Scholar, Business
Best Teacher, Business
Most Helpful to Students, Business
Best Overall Faculty Member, Business

• Professor of Management, Binghamton University, State University of New York, 1976–Present. • Program Faculty, Department of Asian and Asian American Studies, Binghamton University 2003–Present. • Adjunct Professor of Economics, Binghamton University, State University of New York, 1986–Present. • Visiting Professor, National University of Distance Education, Madrid, Spain, Summer 2005 and Summer 2001. • Visiting Professor of Management, ESSEC Graduate School of Management, Cergy, (France), Summer 1994. • Adjunct

Professor of Human Services (Honorary), Cornell University, 1989–90. • Visiting Professor of Economics, Erasmus University, Rotterdam, The Netherlands, Summer 1988. • Visiting Professor of Economics, University of Karlsruhe, Germany, Summer 1984. • Guest Professor of Economics, University of Muenster, Germany, Summer 1980. • Visiting Professor of Regional Science, (part-time) University of Pennsylvania, Spring 1977. • Associate Professor of Management, School of Management, Binghamton University, 1968–1976. • Visiting Professor of Economics, Catholic University, Leuven, Belgium, July 1974–January 1975, June–August 1975. • Visiting Associate Professor of Regional Science (Honorary), 1973–74, and Visiting Associate Professor of Policy Planning and Regional Analysis, Cornell University, (part-time), Spring 1972. • Assistant Professor, Wharton School of Finance and Commerce, University of Pennsylvania, 1964–1968. • Assistant Professor of Industrial Management, University of Rhode Island, 1963–1964.

Kenny Christianson

Best Researcher/Scholar, Economics

Michael Little

Best Overall Faculty Member, Anthropology

Michael Little began research as a graduate student at the Pennsylvania State University on cold adaptation in the Peruvian Andes in 1962 and conducted studies of high-altitude adaptation in Quechua-speaking natives for about 15 years. In the 1970s, he began collaborating with the social anthropologist, Neville Dyson-Hudson, on ecological research of Turkana pastoral nomads from northwest Kenya. This 20-year project incorporated grazing-lands ecologists, cultural anthropologists, and biological anthropologists in multidisciplinary field investigations of the environment, biology, health, and behavior of these peoples. For the past decade or so, he has pursued historical and biographical investigations into the history of biological anthropology, largely through archival research. He has written or edited seven books and monographs and more than 150 papers, books chapters, and reviews. Throughout his career he has been committed to the promotion of collaborative international research and has served on, or been associated with, several committees linked to the National Academy of Sciences. These include: the International Biological Programme (IBP), UNESCO's Man and the Biosphere Program (MaB), the International Union of Biological Sciences (IUBS), and the International Union of Anthropological and Ethnological Sciences (IUAES). For these service activities, he was named a National Associate of the National Academies in 2001. He was elected as President of the American Association of Physical Anthropologists (AAPA, 1991–93) and the Human Biology Association (HBA, 1996–98), and most recently as Chair of Section H, Anthropology, of the American Association for the Advancement of Science (AAAS, 2010). He is the recipient of the Franz Boas Award from the HBA (2005) and the Charles R. Darwin Award from the AAPA (2007). In 2012, he was honored with a newly-named annual award – the Michael A. Little Distinguished New Scholar Award – which was initiated by the Human Biology Association.

Haim Ofek

Best Overall Faculty Member, Economics
Best Teacher, Economics

Christine Reiber

Most Helpful to Students, Anthropology

Chris Reiber is a biological anthropologist whose expertise revolves around evolutionary and epidemiological modeling of women's health issues, cardiovascular health and substance abuse. She has experience as an analyst in the pharmaceutical industry (pharmacokinetics and drug metabolism), and has done extensive university teaching at several schools within the University of Pittsburgh system, Maryville University and Carlow College. She has been an independent analytical consultant to federally-funded researchers in psychology and addiction medicine, and is the former director of Analytical Services for the Clinical Trials Group of UCLA Neuropsychiatric Institute's Integrated Substance Abuse Programs (ISAP), for which she and her team handled all statistical/analytical issues for all clinical trials, including complex single- and multi-site clinical trials containing biomedical, behavioral, demographic, economic, cost-benefit, safety, efficacy and pharmacokinetic data. Her research has included topics within women's health (Premenstrual Syndrome, menopause), cardiovascular health (perimenopausal changes in cardiovascular health in women, intra-abdominal adipose tissue as a cardiovascular risk factor), substance abuse and treatment (clinical trials of new pharmacological products for addiction, women and substance abuse), and evolutionary and epidemiological sciences.

BLINN COLLEGE

Debbie Horn

Best Teacher, Psychology

"Dr. Horn is helpful to students, is very interesting and keeps up to date on new information, research and trends."

BLOOMSBURG UNIVERSITY

Lawrence Kleiman

Best Teacher, Management

BLUEGRASS COMMUNITY & TECHNICAL COLLEGE

Mike Adkins

Most Helpful to Students, Economics

BOROUGH OF MANHATTAN COMMUNITY COLLEGE

Alyse Hachey

Best Teacher, Education

I have always been fascinated with young children, particularly their ability to see the world as a new and exciting adventure. I am consistently amazed at the creative ways that they come up with in order to deepen their understanding of our complex world. I love that to them, life is play.

I am in awe of the tremendous biological and cognitive development that happens in the first eight years of life. This has created for me a strong interest in cognition across the life span. In particular, as an educational psychologist, I have focused on the ends of the educational spectrum... the cognition of young children and college students, and the creation of curriculum to enhance learning for both of these groups based on what we know (and continue to find out) about the human mind.

Early in my career, I worked with pre-schoolers (admittedly my favorite age range) as a Lead Teacher and Center Administrator for Head Start in Detroit, Michigan. After moving to New York, I served as a research assistant for the pre-school T.V. show Blues Clues and I have conducted educational research for other children's television shows. In addition to being interested in young children and media literacy, I have also focused my attention on early childhood cognitive development, particularly related to intuitive mathematics and science.

As a professor at BMCC, I teach curriculum classes to prospective teachers. My goal is to help develop early childhood professionals that are advocates for teaching practices that respect young children as capable, curious and creative learners.

Percy Lambert

Best Teacher, Business

"Professor Lambert has been an institutional teacher for BMCC and the department. His commitment to teaching and education needs to be noted and appreciated."

Marie Padula

Most Helpful to Students, Business

Mahatapa Palit

Best Overall Faculty Member, Business

I started my career in Marketing Research and went on to get a doctoral degree in Business Management focused on Consumer Behavior. I worked for four years in a technology startup as their Marketing Director and joined BMCC in the Department of Business Management in 2003. I believe student engagement is the key to helping students succeed, and I have tried to cultivate different ways to do that through the use of digital technology, project-based learning and teamwork. I love to travel and believe that even as diverse as we are, there are more things that unite us than divide us.

Connett O Powell

Best Overall Faculty Member, Accounting

BOWIE STATE UNIVERSITY

Denyse Barkley

Best Teacher, Nursing

"Has worked tirelessly in the undergraduate and graduate programs since her employment in 2009. Dr. Barkley listens to the students and attempts to resolve issues timely and to the best of her ability. She is firm and ensures that all students, especially the graduate students, have a great understanding of nursing education and are prepared to use their advanced degree upon graduation."

BOWLING GREEN STATE UNIVERSITY

Steve Langendorfer

Best Teacher, Education

Education B.S.Ed., Cortland College - SUNY M.S., Purdue University Ph.D., University of Wisconsin-Madison

Research Interests: Evaluating coordination and control changes in motor skills such as throwing, hopping, and prone swimming and their relationships to physical activity. Effects of task and environment constraints on developmental change in motor skills

Courses Taught: Undergraduate: Introduction to Kinesiology; Motor Development; Water Safety Instruction; Facilitating Movement Change; Measurement and Evaluation; Senior Project; Swimming

Graduate: Measurement and Evaluation; Lifespan Motor Development; Developmental Kinesiology Seminar

BRADLEY UNIVERSITY

Heather McCord

Best Researcher/Scholar, Business

BRENAU UNIVERSITY

William Haney

Best Teacher, Business

"A professor that gets the students to want to learn and they see the value in what they learn in his classes."

William (Bill) Haney, Ph.D. Dr. Bill Haney has been at Brenau University since August, 2000. Bill is the Lead Professor for the Undergraduate and Gradute Programs in Organizational Leadership, and the former Dean of the College of Business and Mass Communication. He earned a M.S. degree in Counseling Psychology and Human Systems from Florida State University. He also holds an M.B.A. degree from Golden Gate University with an emphasis in Management, and a Ph.D. in Higher Education with an emphasis in Business Administration from Florida State University. Dr. Haney teaches in the areas of Organizational Behavior, Cultural Diversity, Human Resource Management, Team Building, Critical Thinking, Leadership, International Business, Managerial Communication, and Management Principles. He is a private consultant, and licensed clinical psychotherapist (Florida). For both business and pleasure, Bill has traveled and studied in France, England, Scotland, Ireland, Belgium, Holland, Luxemburg, Germany, Austria, Italy, the Vatican, Liechtenstein, Canada, Mexico, the Bahamas, Puerto Rico, and about half of the states in the USA.

BRIGHAM YOUNG UNIVERSITY - IDAHO

Kathleen Barnhill

Best Overall Faculty Member, Nursing
Most Helpful to Students, Nursing

Dean Cloward

Best Overall Faculty Member, Education

Susan Dicus

Best Teacher, Nursing

Jim Hopla

Best Teacher, Health Science
Best Teacher, Health Sciences and Nursing
Best Overall Faculty Member, Health Sciences and Nursing

Dana Johnson

Most Helpful to Students, Education

Lynn Perkes

Best Teacher, Health Sciences and Nursing

A Health Science degree with an emphasis in Health Promotion and Lifetime Wellness will prepare a student, both theoretically and experientially, for many career opportunities. Job titles or opportunities could include but not limited to: Community Health Educator, Health Promotion Specialist, Corporate Wellness, Health Counseling, Resort Wellness, Fitness Center, Educational Health promotion, Wilderness Health Promotion and Lifestyle Training Specialist. Students will receive training in a diversity of areas: kinesiology, ergonomics, lifestyle management, health and fitness appraisal and prescription, nutrition, and exercise physiology. Additional training can be attained in areas such as: sports medicine, gerontology, wilderness health promotion and environmental health.

Dennis Tolman

Most Helpful to Students, Health Science
Most Helpful to Students, Health Sciences and Nursing

Tyler Watson

Best Overall Faculty Member, Health Science
Best Researcher/Scholar, Health Science
Best Researcher/Scholar, Health Sciences and Nursing

BRONX COMMUNITY COLLEGE

Virgena Bernard

Most Helpful to Students, Nursing

"Volunteers time to make sure students are prepared for their clinical experiences."

Marcia Jones

Best Researcher/Scholar, Nursing
Most Helpful to Students, Nursing
Most Helpful to Students, Health Sciences and Nursing

*"She is outstanding, she loves her students, an excellent teacher. Unselfish
She is an all rounder, always ready to help"*

BROOKDALE COMMUNITY COLLEGE

Debbie Meyer

Most Helpful to Students, Economics

BROOKLYN COLLEGE

Beth Ferholt

Best Researcher/Scholar, Education

"Beth Ferholt has fostered an unprecedented relationship with Swedish preschools and NYC schools in an effort to support best practice in early education and care. Her research project on play inspired her to initiate reciprocal visiting of learned educators between Sweden and New York. She continues to bring questions of best practice to students, educators and leaders in our field."

Beth Ferholt taught undergraduate research courses in media and the design of social learning contexts in the Department of Communication at the University of California, San Diego. At Brooklyn College she teaches early childhood literacy and student teachers. Her research interests focus on emotional-cognitive development in adult-child play.

BROWARD COLLEGE (ALL CAMPUSES)

Natalie Butto

Most Helpful to Students, Business

Carlton Byrd

Best Overall Faculty Member, Health Science

1978–1996 Head Men's Basketball Coach 1986 Teacher of the Year 2002 Endowed Teaching Chair Recipient

Martin Green

Best Teacher, Wellness

Dr. Green has been teaching at Broward College since 2008. He is a Doctor of Chiropractic Medicine and holds the honor of being named in the Guide to America's Top Chiropractors in 2007, 2008 and 2009 by Consumers Research Council of America. He is certified as a

Chiropractic Sports Physician and is certified in Acupuncture/Meridean Therapy. Dr. Green is the Adjunct Professor Of The Year at South Campus of Broward College for 2014.

Gregory J Lindeblom

Best Overall Faculty Member, Economics
Best Researcher/Scholar, Economics
Best Teacher, Economics
Most Helpful to Students, Economics

Paul Moore

Best Teacher, Business

"Leadership"

Matthew Rocco

Most Helpful to Students, Business

"His exceptional leadership ability"

Shafi Ullah

Best Researcher/Scholar, Business
Best Teacher, Business
Most Helpful to Students, Business
Best Overall Faculty Member, Business

Dr. Shafi Ullah DBA, MBA, MA 1600 South West 100 Terrace Davie, FL 33324 954.638.9793 [Cell] 954.201.8937 [Office] 954.475.9748 [Home] sullah@broward.edu
    TEACHING EXPERIENCES: • Senior Professor, Department of Accounting and Business Administration, Broward College, Pembroke Pines, FL., from 1988–Present. Developed and taught both online and on-ground accounting, finance and management classes using Blackboard, Wiley-Plus software, Connect and My OM Lab. • Professor, American InterContinental University from 2002–2014. Teaching MGT 600, MGT 615, MGT 635, MGT 680, FIN 630, FIN 310, ACCT 205, ACCT 320, ACG 610, ACG 620. • Adjunct Professor (Online), Department of Management, UCLA Extension, Los Angeles, California, October 1998–2000. Taught online accounting classes. • Professor, Department of Business Administration, Fort Lauderdale College, Ft. Lauderdale, Florida from 1985–1988 • Assistant Professor, Department of Accounting, Dhaka University, Dhaka, Bangladesh, from 1973–1979 • Adjunct Professor, School of Business & Entrepreneurship, Nova Southeastern University, Fort Lauderdale, Florida 1988–2002. Taught at Masters in Accounting Programs. • Adjunct professor, School of Business, Barry University, Miami, FL 1989–1996.

PROFESSIONAL EXPERIENCES: • Chairman, Department of Accounting, Business Administration and Legal Assisting Programs, Broward College, South campus, January 1995 to December 1998. • CEO, Florida Co-Op Investment Corporation, FL 1994–1998.

EDUCATIONAL QUALIFICATIONS: • Online Teaching Certificate, UCLA Los Angeles, California. • Doctorate in Business Administration (DBA), Nova Southeastern University, Ft. Lauderdale, Florida. • Master of Business Administration, University Of Central Oklahoma, Edmond, Oklahoma. • Master of Accounting, University of Dhaka, Dhaka, Bangladesh. • Bachelor of Commerce with Honors in Accounting, University of Dhaka, Dhaka, Bangladesh.

ACADEMIC HONORS: • Recipient of the Holcombe Institute Classroom Research Awards. • Academic award from American Accounting Association for excellence in accounting education, research and practice. • Recipient of the prestigious Fulbright Scholarship to study in the University of Central Oklahoma. • Recipient of the Merit Scholarship in undergraduate and graduate programs in Bangladesh.

INVOLVEMENT IN COLLEGE ACTIVITIES: • Member of the College-wide IT Planning Committee. • Involved in Student Success/Mentor Programs and Workshop in South Campus. • Advisor to Phi Beta Lambda Group in South campus for more than 4 years. • College Ambassador for one year to represent BC at Fairs in South Florida. • Involved in registration, advisement, book selection, program development. • Member of the College-wide International Education Committee for two years. • Member of the College-wide Sick Leave Pool Committee for one year. • Chair of the College-wide Equity Committee for one year. • A mentor for Webpage Development in South campus for one year. • An active and a founding member of the "Right Click Academy" at BC.

PROFESSIONAL ACTIVITIES/EXPERIENCES: • A Re-Examination of Gray's Culture Framework Using a Modification of Nair's Accounting Practices Classification: Is There a Relationship with IAS Conformity? Presented at AAA Mid-Atlantic annual conference and published in the proceedings of AAA, 2003. • Reviewer panel in American Accounting Association [AAA]: "Enhancing Student Learning and Course Management Using Web-Based Publisher Developed Resources in Introductory Accounting?" and Review Processing: "Gender Issues in Accounting". • Presented a research paper on "An Empirical Analysis of Student Attitudes toward Financial Accounting Course [ACG 2001]: A Survey Analysis of Broward Community College Students of South Campus" presented in League Annual Conference, held in Dallas, Texas in October, 2005. • Attended and actively participated at the Thomson Business and Economics Publishing Seminar in a Financial Accounting Market Development Focus Group in Ft. Lauderdale in March 2005. • Conducted a Personal Finance Seminar on Money Management on March, 2005 and November, 2005 in South campus. • Involved in recommending the textbook for accounting program college-wide. • Reviewed Personal Finance Book from John Wiley and Sons and Accounting Book from McGraw-Hill publishing Company. • Have written and published a textbook on Estate Duty and Taxes in Bangladesh in 1979. • Have many publications in referred journals since 1975. [Available on request]. • Developed online courses for ACG 2001, ACG 2011, ACG 2071, GEB 1011 and FIN 1100, FIN 3405, MAN 4503 at BC.

OTHER ACTIVITIES AND HONORS: • An executive and a founding member of Darul Uloom Institute, a non-profit organization in Pembroke Pines. I was involved to raise funds for the organization as well as to oversee finances of the institute. • An active member of Feed the Homeless Programs at Sistrunk Blvd., Ft. Lauderdale in 1985–1987 serving breakfast to over 100 people on every Sunday. • Sponsored a high school dropout, a single parent working in fast food restaurant with GED exam fees. Passed the exam in one attempt and headed to BC for further studies. • Set up Book Club to help needy students with textbooks in accounting, business and finance for the semester. • President, Bangladesh American Association of Florida, 1997–2002.

PROFESSIONAL ASSOCIATIONS: • Member, American Accounting Association. • Former Board Member, National Social Science Association.

BUFFALO STATE COLLEGE

Pixita Hill

Best Teacher, Education

CABARRUS COLLEGE OF HEALTH SCIENCES

Valerie Rakes

Best Researcher/Scholar, Health Sciences and Nursing

Ms. Rakes received her Bachelor of Science in Nursing from Cabarrus College and a Master of Science in Nursing with a concentration in Leadership in Healthcare Education Systems from East Carolina University. She has over 20 years experience in the maternal-child setting with an emphasis in the neonatal intensive care unit. She also maintains her certification as a STABLE instructor for the Women's and Children's Division and is an active member of the Shared Governance Council in the neonatal intensive care unit at CMC-NorthEast.

Ms. Rakes' awards include the Elizabeth M. Mabry Distinguished Baccalaureate Merit Award and a first place poster presentation on Isolette Weaning of Premature Infants at the

2007 Cabarrus College Honor Society of Nursing Research Forum. She is currently a member of Sigma Theta Tau.

Other interests include reading, socializing with friends, and traveling.

CABRINI COLLEGE

James Hedtke

Best Teacher, Political Science

James Hedtke, Ph.D., came to Cabrini College to make a difference in students' lives.

"I am fully committed to Frances Cabrini's philosophy of providing students with an 'education of the heart.' This is a place where people strive to do extraordinary things," says Hedtke, who joined the history and political science department in 1973 and became a full-time faculty member in 1983. He says coming to campus every day is a joy.

Hedtke describes his teaching philosophy as simple and straightforward. "I always strive to be competent, fair, enthusiastic about the material, and genuinely concerned about the welfare of the students," says Hedtke.

Hedtke earned a bachelor's degree of science from Saint Joseph's University, a master's from Villanova University, and a Ph.D. from Temple University.

His research interests include the American presidency, the United States Civil War, and World War II, as well as research on the Freckleton Air Disaster, which took place in 1944 when a United States Army Air Force B-24 Liberator crashed into the center of the English village of Freckleton, Lancashire.

Hedtke's book, The Freckleton, England, Air Disaster, was published (McFarland Publishing) in December 2013.

Hedtke has authored two other books, Lame Duck Presidents: Myth or Reality, and Civil War Professional Soldiers, Citizen Soldiers and Native American Soldiers of Genesee County, New York: Ordinary Men of Valor, the latter of which he edited and researched with students.

He writes book reviews for Choice, a higher education magazine, and is a member of the Southeastern Pennsylvania Consortium for Higher Education (SEPCHE) Speaker's Bureau.

He also has made numerous presentations to area audiences on the presidency, terrorism, and the American Civil War, including one at the Pennsylvania Civil War Show commemorating the war's 150th anniversary.

A native of Batavia, N.Y., Hedtke has a passion for baseball, gardening, and traveling. He also maintains an intense curiosity about the American Civil War. He is married and lives with his wife, Judy, in Broomall, Pa. They have three daughters.

CALIFORNIA LUTHERAN UNIVERSITY

Ronald Hagler

Best Overall Faculty Member, Business
Most Helpful to Students, Business

Joe Huggins

Best Teacher, Business

Mr. Huggins developed and conducted the original Entrepreneur Academy small business classes for Ventura County. He has over twenty years of commercial banking and big 5 consulting experience. He has assisted clients in developing creative marketing strategies, strategic and business plans, loan proposals, and cash management analysis. He began his career at Bankers Trust, N.Y. where he was a Vice President in charge of the funds transfer, cash management, and internal consulting divisions. He was a senior manager with Price Waterhouse serving as national product manager for strategic planning consulting. At Deloitte Touche, Mr. Huggins served as Senior Manager for profitability and productivity consulting. He also worked for Citibank, Los Angeles, where he developed and managed the regional cash management center. Mr. Huggins has an MA in industrial psychology and an MBA in marketing from New York University. He specializes in the marketing of entrepreneurial service ventures. He is an adjunct professor/senior lecturer in the MBA program at California Lutheran University in marketing, strategic management, entrepreneurship and small business management. He has conducted numerous entrepreneurial training courses and consulted for hundreds of businesses throughout California.

Edward Julius

Best Teacher, Business
Best Teacher, Business

Professor Julius's specialty is financial accounting, which he teaches in the traditional undergraduate program. He is also CLU's Faculty Athletics Representative, and he has proudly served as the faculty mentor to the women's softball team since 2007. Prof. Julius has published numerous learning and teaching aids to accompany accounting textbooks, as well as six highly regarded crossword puzzle books. He has also published four books on rapid calculation, one of which appears in nine languages and was a Book-of-the-Month Club selection for nine years. His outside interests include vintage jazz, Broadway musicals, pop culture, wordplay, comedy, old movies, bowling, and the Boston Red Sox. A long-time member of Mensa, he is listed in Marquis Who's Who in the World, Who's Who in America, Who's Who in American Education, and Who's Who in Finance and Industry.

Charles Maxey

Best Researcher/Scholar, Business

Dr. Maxey teaches courses in human resources and organizational behavior. Author or co-author of numerous academic articles, he is also active as a business and litigation consultant and labor arbitrator.

His previous academic appointments were at Northwestern University, Loyola University of Chicago and the University of Southern California, where he also served as Senior Associate Dean of the Graduate School of Business.

Immediately prior to joining CLU's faculty, Dr. Maxey was visiting professor of management at the Sasin Graduate Institute of Business Administration in Bangkok, Thailand.

Renee Rock

Most Helpful to Students, Business
Most Helpful to Students, Business

Harry Starn

Best Overall Faculty Member, Business
Best Overall Faculty Member, Business

Harry Starn, Jr., MS, CFA, CFP®, is an executive faculty member in California Lutheran University's School of Management and serves as the Director of the Financial Planning Program and Director of Distance Learning. He has been teaching, creating and refining online courses for CLU's MBA-FP program since 2005.

In May of 2010, the United States Distance Learning Association (USDLA) presented him with the bronze award for Excellence in Distance Teaching. He serves on the CFP Board's Council on Education (2012–2015) and is a co-writer on the 11th Edition of "Practicing Financial Planning for Professionals."

Vlad Vaiman

Best Researcher/Scholar, Business
Best Researcher/Scholar, Business

"Exceptional scholar and prolific researcher"

CALIFORNIA STATE POLYTECHNIC UNIVERSITY

Thomas Spalding

Best Researcher/Scholar, Education

"He is innovative and thoughtful in Kinesiological and experimental research with high Integrity. Always involve students in his research."

CALIFORNIA STATE UNIVERSITY BAKERSFIELD (CSUB)

Mohsen Attaran

Best Overall Faculty Member, Management
Best Researcher/Scholar, Management
Best Teacher, Management
Most Helpful to Students, Management

Deborah Boschini

Best Overall Faculty Member, Nursing
Best Overall Faculty Member, Health Sciences and Nursing

Kathleen Gilchrist

Most Helpful to Students, Nursing

Michelle Kinder

Most Helpful to Students, Health Sciences and Nursing

Judy Pedro

Most Helpful to Students, Nursing
Best Teacher, Health Sciences and Nursing

John Stark

Best Researcher/Scholar, Business
Best Teacher, Business
Most Helpful to Students, Business
Best Overall Faculty Member, Business

Louis Wildman

Best Researcher/Scholar, Education

Debra Wilson

Best Teacher, Nursing

"Recently completed her PhD! Works with students all the time!"

CALIFORNIA STATE UNIVERSITY CHANNEL ISLANDS

Steve Fleisher

Best Overall Faculty Member, Psychology
Best Researcher/Scholar, Psychology
Most Helpful to Students, Psychology

In working construction, my first life, I learned a deep and abiding respect for determination and hard work, as well as for effective communication. After becoming proficient in my construction skills, I decided to pursue counseling studies to learn more about communication and relationships. The Bachelor's and Master's Programs in Counseling at Cal State Los Angeles provided an ideal fit, and there I came to love teaching and psychology.

Nancy Loman

Best Teacher, Psychology

CALIFORNIA STATE UNIVERSITY CHICO

Cathrine Himberg

Best Teacher, Education

Rebecca Lytle

Best Researcher/Scholar, Education

Program Coordinator. Adapted Physical Education Program, Department of Physical Education and Exercise Science, California State University, Chico. Academic advising. Program review and supervision for California Teacher Credentialing. Supervision of graduate students working in activity programs. Recruitment of students. Catalog revisions. Course scheduling. Purchasing and inventory of adapted physical education equipment. Consultant to faculty on issues related to modifications for students with disabilities in physical education settings. Development of APE recruitment brochure and APE website. (1993–96, 1997–to present)

Associate Professor. California State University, Chico. PHED 212 Introduction to Adapted Physical Activity. PHED 213 Programming for Individuals with Physical, Nuerological and Sensory Disabilities, PHED 215 Collaboration in Adapted Physical Education. PHED 216 Motor Assessment for Individuals with Disabilities. PHED 217 Field Experience. PHED 322 Adapted Physical Education Implementation. PHED 97 Adapted Physical Activity. PHED 98 Kids Integrated Development of Skills: Parental Leadership and Advocacy for Youth (service learning class). APE advising. (Fall 04 to present)

Maggie Payne

Best Overall Faculty Member, Education

Neil Schwartz

Best Researcher/Scholar, Psychology

CALIFORNIA STATE UNIVERSITY FRESNO (FRESNO STATE)

Danette Dutra

Most Helpful to Students, Nursing

Danette K. Dutra, Ed.D., M.S.N, F.N.P.-C Assistant Professor School of Nursing

Degrees Held: • Ed.D. University of San Francisco • M.S.N. F.N.P. California State University, Fresno • M.A. Exercise Physiology, California State University, Fresno • B.S.N. California State University, Fresno Research Interests: • Development of teaching pedagogy • Utilizing standardized patients in schools of nursing • Transition from student nurse to registered nurse • The implementation of active/learner centered methodologies • Holistic approach to care of the diabetic patient • Homeopathic Medicine and Complimentary Practice Approaches

Terea Giannetta

Best Overall Faculty Member, Nursing
Best Researcher/Scholar, Nursing
Best Teacher, Nursing
Most Helpful to Students, Health Sciences and Nursing
Best Teacher, Health Sciences and Nursing

Jenelle Gilbert

Best Teacher, Kinesiology

Wade Gilbert

Best Researcher/Scholar, Kinesiology

Dr. Wade Gilbert is an award-winning professor in the Department of Kinesiology at California State University, Fresno. Dr. Gilbert holds degrees in Physical Education, Human Kinetics, and Education from the University of Ottawa in Canada. Upon completion of his doctoral degree he was selected for a postdoctoral fellowship in the International Center for Talent Development at the University of California at Los Angeles (UCLA). Dr. Gilbert's areas of expertise include coaching science, talent development, sport and exercise psychology, physical education and youth sport. He has over 20 years of experience conducting applied research with partners around the world spanning the full range of sports and settings from local youth leagues through to World Cup Soccer. He is widely published and is frequently invited to serve as a featured speaker at national and international events. He is co-editor for the International Sports Coaching Journal and the Routledge Handbook of Sports Coaching, and is an associate editor for the Journal of Sport Psychology in Action. As a result of his many contributions, Dr. Gilbert has served as a scientific advisor to organizations ranging from school districts, collegiate teams, Olympic organizations and the United Nations. He currently serves as the chief scientific advisor to BeLikeCoach, a national non-profit dedicated to improving the quality of youth sport settings in the United States

Miguel Perez

Best Overall Faculty Member, Health Sciences and Nursing

Helda Pinzon-Perez

Best Researcher/Scholar, Health Sciences and Nursing

CALIFORNIA STATE UNIVERSITY FULLERTON

Stephen Aloia

Best Teacher, Education

Joanne Andre

Most Helpful to Students, Nursing

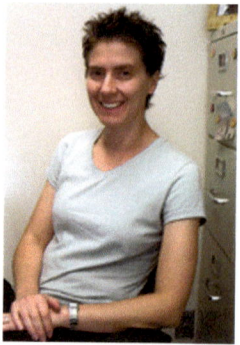

Kristin Beals

Best Teacher, Psychology

I am interested in studying the effects of stigma on psychological and physical well-being. Most of my research has focused on the experiences of lesbians, gay men, and bisexuals. However, I am interested in other stigmatized populations including HIV-positive individuals, overweight individuals, and ethnic minorities.

In addition, I sometimes wear the hat of a sport psychologist. I am currently conducting research on sport commitment with Titan athletes and exercise motivation among Fullerton students.

Melinda Blackman

Best Overall Faculty Member, Psychology
Best Teacher, Psychology

I received my PhD from UCLA in 1972. I taught at the University of Illinois, Urbana-Champaign from 1974–1986, where I was head of the Division of Quantitative and Industrial/Organizational Psychology, and promoted to Full Professor in 1982. I came to CSUF in 1986 on leave from University of Illinois, and made the position here permanent in 1988. I am founder and current co-director of the Decision Research Center. I was named Outstanding Professor for CSUF in 1991–92. I have published more than 125 scientific articles and three books. In the last decade, I have had five grants from the National Science Foundation and one from the American Psychological Association. I served as president of the Society for Mathematical Psychology, 2002–2003, president of the Society for Judgment and Decision Making, 2008–2009, and president of the Society for Computers in Psychology, 2009–2010.

Bert Buzan

Best Teacher, Political Science

Betty Chavis

Best Overall Faculty Member, Accounting
Best Researcher/Scholar, Accounting

Victoria Costa

Best Researcher/Scholar, Education

Donald Crane

Best Overall Faculty Member, Business

Dr. Crane is currently a tenured full professor of finance at California State University, Fullerton, where he is responsible for instruction in financial planning, portfolio management, retirement and estate planning, and investments. He is the Program Director for the Certified Financial Planner Board of Standards registered curriculum at the University.

He graduated from the University of Southern California with a doctorate in business administration with an emphasis in finance. He had previously been awarded an MBA and BA from USC and California State University, Fullerton, respectively. Dr. Crane is actively involved with the University's Continuing Learning Experience (CLE) Program for seniors, he chairs the President's Task Force on Affordable Housing and supports many student organizations on the campus. He is past chair and current member of the CSUF Foundation investment committee.

Dr. Crane has been a member of the Standards Writing Committee and the Examination Writing Committee for the CFP Board of Standards.

He is a Certified Financial Planner (CFP), a past member of the Institute of Certified Financial Planners, and was admitted to the Registry of Financial Planning Practitioners.

Dr. Crane is registered as an Investment Advisor with the California Department of Corporations. He also holds registrations as stockbroker, and is licensed as a real estate broker and life, disability and variable annuity insurance agent. He has been a consultant for numerous organizations, medical groups, the Small Business Administration and The Mortgage Bankers Association.

Dr. Crane has authored numerous articles on finance and financial planning and is a nationally recognized authority and speaker in the area of personal financial management.

Zvi Drezner

Best Researcher/Scholar, Business

Linda Fraser

Best Researcher/Scholar, Business

Linda Fraser specializes in the teaching of accounting communication, business writing, and critical thinking. She has received six awards from the University for her teaching and service. Currently, she serves on the WASC Writing Task Force and the MCBE Undergraduate Programs Committee. Dr. Fraser's research interests include assessment theory and methods, accounting communication, and the intertextuality of the visual arts and writing.

Dr. Fraser's experience includes employment by global, high-tech companies.

Dr. Fraser has presented papers at conferences of the Association for Business Communication, the California Joyce Conference, the Pacific Ancient and Modern Language Association, the American Literature Association, and the CSUF Assessment Conference.

Maryanne Garon

Best Researcher/Scholar, Health Sciences and Nursing

Barbara Glaeser

Most Helpful to Students, Education

Kevin Gotts

Best Teacher, Business
Most Helpful to Students, Business

Judith Hervey

Most Helpful to Students, Health Sciences and Nursing

I have been a Registered Nurse for 38 years. One of my greatest accomplishments is teaching nursing for CSUF. I have been teaching for the past 7 years. The profession has been so good to me. It was always an aspiration of mine to give back to our new generation of nurses.

Jay Hirsch

Most Helpful to Students, Accounting

Casey Kleindienst

Most Helpful to Students, Business

Irene Lange

Best Overall Faculty Member, Marketing

Richard Lippa

Best Researcher/Scholar, Psychology
Best Overall Faculty Member, Psychology

Laura Marcoulides

Best Overall Faculty Member, Business
Best Teacher, Business

Jack D. Mearns

Most Helpful to Students, Psychology

Marcella Mendez

Best Overall Faculty Member, Social Work

"Marcella is a remarkable woman in many ways, but as a faculty member she is about as good as it can get. She is organized and has created forms and systems to make our work more user-friendly. As a teacher, she is phenomenal at "connecting the dots", meaning that she can tie all the assignments to the skill or competency that they enhance. She is helpful to students in that she confronts them where they need to be confronted and at the same time she is nurturing and compassionate when that is called for. As the Director of Field Education

in this department, I rely heavily on Marcella for her feedback, her awareness of the student pulse, and her camaraderie as a colleague."

Marcella Mendez is CalSWEC Child Welfare Field Coordinator and lecturer in the Social Work Department. She began teaching Field Instruction part-time in 2009. In 2011, she became a full-time lecturer and CalSWEC Field Coordinator. Ms. Mendez teaches courses in Social Work Practice and Field Instruction. Ms. Mendez earned her BA from Cal State Northridge in Sociology, with the emphasis in Counseling and Interviewing, and her MSW from USC with a concentration in Families and Children and sub-concentration in School Setting Pupil Personnel Services. Ms. Mendez is a California state-licensed clinical social worker and has worked 20 years at several non-profit agencies within the Los Angeles area. Ms. Mendez has experience working with pregnant teens and teen mothers, children and adults with developmental disabilities and as a director of client care providing out-patient mental health services to children and families.

Alana Northrop

Best Overall Faculty Member, Political Science

This professor is now retired as can be seen on the webpage above.

Mitchell Okada

Best Teacher, Psychology

Rebecca Otten

Best Teacher, Nursing
Best Teacher, Health Sciences and Nursing

Jessie Peissig

Best Overall Faculty Member, Medicine

Melinda Pierson

Best Overall Faculty Member, Education

Elaine Rutkowski

Best Researcher/Scholar, Nursing

Dr. Elaine M. Rutkowski is an assistant professor in the School of Nursing at Cal State Fullerton. She currently teaches courses in community health, patient education and psychosocial issues. Dr. Rutkowski's research focuses on obesity risk knowledge, self-efficacy and activity levels in families of adolescents. She recently presented her research at the 10th Annual Nation/International Evidence-Based Practice Conference: Translating Research into Best Practice with Vulnerable Populations, February 19–20, 2009, Phoenix, AZ. Elaine has co-authored an article that has been accepted for publication, "Families on the Move: Development of a Weight Management Intervention for Latino Mothers and Children" in the Journal for Specialists in Pediatric Nursing. She has been involved in The Center for Promotion of Healthy Lifestyles and Childhood Obesity Prevention since it's inception in 2004.

Nancy Segal

Best Researcher/Scholar, Medicine

Vicki Sergent

Best Teacher, Medicine

Kimberly Tarantino

Best Teacher, Accounting

Jennifer Trevitt

Most Helpful to Students, Medicine

Stephanie Vaughn

Best Overall Faculty Member, Nursing
Best Researcher/Scholar, Nursing
Best Teacher, Nursing
Best Overall Faculty Member, Health Sciences and Nursing

Dr. Stephanie Vaughn is an assistant professor and the coordinator of the post-licensure nursing program in the Department of Nursing at Cal State Fullerton. Her areas of research include stroke prevention behaviors in Latin-American women, the management of stroke sequela in both men and women, and the development of culturally sensitive stroke/heart disease educational media. She teaches in both the undergraduate and graduate nursing programs; research methods, vulnerable populations, and professional nursing. Dr. Vaughn speaks locally, nationally, and internationally about stroke and stroke prevention. Her current research is a qualitative study that is exporing perceptions of caregiving needs and resources of family caregivers of Latino stroke survivors. She is a member of the American Heart Association/American Stroke Association, the Association of Rehabilitation Nurses and facilitates a monthly community Stroke Support Group in Orange County.

CALIFORNIA STATE UNIVERSITY LONG BEACH

Paul Boyd-Batstone

Best Overall Faculty Member, Education
Best Researcher/Scholar, Education

Maria Claver

Best Overall Faculty Member, Health Sciences and Nursing

Melissa Dyo

Most Helpful to Students, Health Sciences and Nursing

Constantine Glezakos

Most Helpful to Students, Economics

William Goeller

Best Overall Faculty Member, Gerontology

Betina Hsieh

Best Teacher, Secondary Education

"She is beloved by her students. She uses a variety of engaging best practice strategies in her instruction. She is a reflective educator ever seeking to be the best professor she possibly can. She extends herself willingly to assist her colleagues."

J Magaddino

Best Teacher, Economics

Tony McGuire

Best Researcher/Scholar, Nursing

Huong Nguyen

Best Overall Faculty Member, Secondary Education

Seiji Steimetz

Best Overall Faculty Member, Economics
Best Researcher/Scholar, Economics

CALIFORNIA STATE UNIVERSITY SAN BERNARDINO

Marilyn Stoner

Best Researcher/Scholar, Nursing

I am primarily interested in end-of-life, palliative and geriatric care. Researching the most effective methods to teach nursing students is also an interest. The use of patient simulators is a key part of my instructional professional development activities. The "Silver Hour" is a model I have developed to describe a unique time in the life of a patient and family. I define the Silver Hour as the 30 minutes before and after death.

CALIFORNIA STATE UNIVERSITY SAN MARCOS

Nancy Romig

Best Teacher, Nursing

CAMBRIDGE COLLEGE

Lyda Peters

Best Overall Faculty Member, Education

"Prof. Lyda Peters is a teachers teacher. Top advocate for students and faculty, alike. She is dedicated to the value of social justice in education and is a champion of academic excellence for herself and her colleagues. A fine person and educator. submitted by Dr Ethlyn Davis Fuller October 22, 2014"

Lyda Peters, B.A., M.Ed., part-time faculty member at Boston College, professor of Education at Cambridge College in Cambridge, MA, and Managing Partner of the Cambridge Leadership Consortium, LLC, a training and professional development corporation. She is the President of the Ruth M. Batson Educational Foundation, named in honor of one of Boston's most prominent African-American civil rights activists. While a public school teacher in the City of Boston, she became active in court-ordered school desegregation. She has worked for forty years in education as a practitioner, consultant, and trainer in organizational settings with expertise in social justice, social responsibility, and civil rights movement teacher training.

In 1986, Peters, as part of a team of consultants, began civil rights teacher training throughout the United States. After receiving a distance education certificate from the University of West Georgia in 1999, she developed a virtual class on America's civil rights movement, bringing her Cambridge-based students from the South, Midwest, Northeast, West Coast, and the Virgin Islands, for a richer teaching and learning experience. Since that first distance class, she has enriched the content, taught primarily through a learning management system, by applying technologies that enhance communication and reinforce learning. Her professional training and accomplishments include a Fellowship from Radcliffe Institute for Advanced Study - Harvard University, Cambridge, MA, a Carnegie Fellowship to pursue graduate studies at the University of Illinois, educational technology training at Mount Royal College, Calgary, Alberta, Canada, and additional technology training at Capella University and the University of Colorado at Denver.

CAPE COD COMMUNITY COLLEGE

Michael Bejtlich

Best Teacher, Business

CAROLINA LAW

Michael J. Gerhardt

Best Researcher/Scholar, Law
Best Teacher, Law
Best Overall Faculty Member, Law

"He has written nationally recognized award winning books that branch the disciplines of law, history and political science.
Michael Gerhardt is an amazing teacher. He has read every important presidential biography and teaches important historical context in constitutional law. He also uses many innovative teaching techniques to give students a lasting understanding of different conceptions of equality. He has students play the role of justices so that they genuinely learn the modalities each justice uses to make decisions."

Michael Gerhardt is Samuel Ashe Distinguished Professor of Constitutional Law. He specializes in constitutional conflicts and has written dozens of law review articles and five books, including The Power of Precedent (paperback, Oxford University Press, 2011). The Financial Times selected his most recent book, The Forgotten Presidents: Their Untold Constitutional Legacy (Oxford University Press, 2013), as one of its Best Books of 2013. Professor Gerhardt is also co-editor (with former Judge Abner Mikva and Hofstra Law School Dean Eric Lane) of the Fourth Edition of the casebook, The Legislative Process, published by Aspen Law and Business.

CARTHAGE COLLEGE

Richard Custin

Best Teacher, Business

Richard E. Custin exercises a practical approach to teaching based upon his experience as owner and manager of a law office and as a litigation attorney. As a member of the California Bar Association and The Association of Trial Lawyers of America he was appointed to a judge pro-tem assignment in Los Angeles, California. He served as a judicial law clerk in the Iowa District Court and was the recipient of the Hathaway Kemper Insurance

Law Fellowship. Custin holds active bar memberships in California and Illinois. He is further admitted to practice law before the District of Columbia Court of Appeals, The United States Court of Federal Claims, The United States Court of Appeals For The Ninth Circuit and United States District Courts for the Northern District of Illinois, Central District of California and Southern District of California.

CENTRAL CONNECTICUT STATE UNIVERSITY

Shelley Bochain

Best Researcher/Scholar, Nursing

Jill Espelin

Best Overall Faculty Member, Nursing

"Dr. Espelin is committed to her students and spends a good amount of time making interactive lectures and meaningful clinical experiences for these students. She receives excellent evaluations from her students. She is also committed to our nursing department and is willing to do anything to help out. She is well-respected in the nursing field and has a great amount of compassion for all who come in contact with her."

Leona Konieczny

Best Overall Faculty Member, Health Sciences and Nursing

Nancy Peer

Best Teacher, Nursing

Nathan Smith

Best Researcher/Scholar, Finance

Catherine Thomas

Most Helpful to Students, Nursing

CENTRAL NEW MEXICO COMMUNITY COLLEGE

Jesse Chenven

Best Overall Faculty Member, Education
Best Researcher/Scholar, Education
Best Teacher, Education
Most Helpful to Students, Education
Best Researcher/Scholar, Education
Most Helpful to Students, Education
Best Overall Faculty Member, Education

Kevin Dooley

Best Teacher, Education

Pat Dunworth

Best Teacher, Criminal Justice

Susan Ruth

Best Teacher, Anthropology

Melanie Upshaw

Best Teacher, Health Sciences and Nursing

Joanne Zakrzewski

Best Teacher, Education

CENTURY COLLEGE

Anna Gryczman

Best Teacher, Nursing
Best Teacher, Health Sciences and Nursing

CERRITOS COLLEGE

Jose Anaya

Best Overall Faculty Member, Business Administration

Adriana Flores

Best Teacher, Business Administration

CHABOT COLLEGE

Dmitriy Kalyagin

Best Teacher, Business
Most Helpful to Students, Business
Best Teacher, Business
Most Helpful to Students, Business
Best Overall Faculty Member, Business

Jan Novak

Best Overall Faculty Member, Business
Best Researcher/Scholar, Business
Best Researcher/Scholar, Business

CHADRON STATE COLLEGE

Patricia Blundell

Best Researcher/Scholar, Education

"Good research skills."

Lorie Hunn

Best Teacher, Education
Most Helpful to Students, Education
Most Helpful to Students, Education

"Excellent rapport with the students"

Linda Hunt-Brown

Best Overall Faculty Member, Education
Best Overall Faculty Member, Education

William Roweton

Best Researcher/Scholar, Education

Susan Schaeffer

Best Teacher, Education

CHAFFEY COLLEGE (ALL CAMPUSES)

Carol Dickerson

Best Teacher, Accounting
Most Helpful to Students, Accounting

Carol Dickerson earned her Ph.D. from Claremont Graduate University in 2009. Carol's dissertation investigated the ethical decision-making process of certified public accountants. Other research subjects explored during her graduate work included: contemporary issues in accounting education, philosophy of education, and teaching and learning in higher education.

She has a Master's degree in Education from UCLA, and a Bachelor's of Science from Cal Poly, Pomona. She has worked as a consultant for the Walt Disney Corporation and Nestle USA. Her corporate experience also includes: Wells Fargo, 20th Century Fox and Arthur Andersen.

Dr. Dickerson is a tenured professor at Chaffey College where she teaches accounting and business ethics classes and serves as the tri-chair of the President's Equity Council and a member of the Enrollment & Success Committee.

Patty Peoples

Best Teacher, Physical Ed
Most Helpful to Students, Physical Ed
Most Helpful to Students, Education
Best Teacher, Education

CHAMINADE UNIVERSITY

Peggy Friedman

Best Teacher, Business

Peggy Friedman is Associate Professor of Marketing at Chaminade University of Honolulu. She has an undergraduate degree in Sociology from Smith College, Northampton, MA. Her graduate work was completed at the University of Wisconsin-Madison where she received an MBA degree with a concentration in Marketing Research and a PhD degree in Business Administration, concentration, Marketing. Her dissertation research considered the role of the professional health care provider from a sales interaction perspective.

Peggy has worked as a market researcher in an advertising agency and as a business consultant in what is now called Accenture (formerly Arthur Andersen and Company), specializing primarily in health care consulting. Throughout her academic career, she has maintained ties to the "real world" through consulting experiences in a variety of contexts, including education, insurance, nonprofit organizations, to name a few.

Prior to moving to Hawaii from the mainland, Peggy taught at the University of Wisconsin-Whitewater as well as at the Management Institute, the continuing education division of the University of Wisconsin-Madison School of Business. In Hawaii, she first worked as Marketing Director at the Waianae Coast Comprehensive Health Center on the leeward coast of Oahu. However, she missed being in the classroom and so began teaching part time at Hawaii Pacific University, University of Hawaii and in the MBA program at Chaminade. Peggy joined the full-time Chaminade faculty in 2001 to revitalize the marketing program.

Peggy is the mother of three grown children, all of whom are currently living on the mainland. Her favorite hobbies are reading, origami and beading.

Richard Kido

Best Teacher, Business

An associate professor of accounting at Chaminade, Richard teaches a wide range of courses in accounting and research interests, including non-profit accounting, service learning, and accounting education pedagogies. Professor Kido will serve as regent for Chaminade University in an ex-officio capacity as part of his role as faculty senate president (2009.) Richard serves as the academic advisor to accounting majors. An inspiration to his students, he was instrumental in Chaminade University receiving the Presidential Award for its service-learning program. Prior to his entering academia in 2002, he was involved in several management positions in private industry, including chief financial officer positions in the savings and loan, real estate development and hospitality industries. Richard has also operated his own firm for several years specializing in accounting and financial consulting. A graduate of Saint Louis High School, he received his Bachelor of Arts degree from Michigan State University and his Masters in Business Administration degree from the University of Hawaii at Manoa.

Wayne Tanna

Best Overall Faculty Member, Business

CHAMPLAIN COLLEGE

Elaine Young

Best Teacher, Marketing

"Elaine is cutting-edge with her focus on social media marketing. With her publications, she consistently asks for faculty, student, and community feedback. Her students applaud her genuine commitment to both excellent teaching and cutting-edge research. She consistently updates her courses and designs new coursework that is relevant to the changes and trends in the marketing field. In addition, Elaine continues to be lead and to be involved with new initiatives in both the marketing division and the college campus. Specifically, Elaine has helped to organize the Sarah Ramsey Strong Scholarship Fund with its requirements, review of candidates, and award ceremony. In sum, Elaine exemplifies the true commitment to teaching, her students, and the marketing industry."

Since 2000, I have been a professor at Champlain. I currently teach courses in Digital Marketing, Digital Analytics and Analysis, Advanced Digital Marketing, Community Management and the Marketing Capstone. I also teach the Marketing course in Champlain's MBA Program.

I love teaching at Champlain because it allows me to connect with the students and truly mentor them. Class sizes allow me to know all my students by name (as long as my memory holds out) and get to know their strengths so I can teach more effectively. I enjoy watching students make connections—when the light bulb goes off it's great to see! I also understand that students today are juggling many different priorities, and it's important for them to be able to connect with someone in their chosen field of study. There is no better way to learn than by doing. It's hard, chaotic and messy. But, when you get through it, you can look back and realize that you have actually learned something. That's what makes teaching and learning at Champlain a meaningful experience.

My role as faculty advisor is one of the most challenging aspects of my job—and the one that I find most rewarding. Students seem to appreciate being able to come and talk about both challenges and successes and get advice from me about careers, internships and opportunities that come up outside of the classroom.

I have a Ph.D. in Organizational Management from Capella University, where my dissertation research examined technology use and adoption of college students, and I developed a teaching model for faculty to follow when teaching a technology application in a classroom environment. I also hold a M.S. in Internet Strategy Management from Marlboro College, a B.S. in Communication and Public Relations from SUNY Brockport and an A.S. in Communication from Genesee Community College. I have over ten years of experience in the Marketing and Public Affairs profession, specializing in non-profits.

CHANDLER-GILBERT COMMUNITY COLLEGE

Barbara Magenheim

Best Researcher/Scholar, Nursing

"Barb is always current on nursing theory/practice and closely follows NCLEX updates & revisions. She participates in scholarly book reviews and is a contributing author. Nationally, she travels to educate nursing students in other regions on NCLEX success. While traveling, she endeavors to learn about nursing and nursing education in each part of the country. She shares this insight with other faculty, staff, and students."

CHAPMAN UNIVERSITY

Margie Curwen

Best Researcher/Scholar, Education

Dr. Curwen earned her Ph.D. in Literacy, Language and Learning from the University of Southern California Rossier School of Education. She has been a classroom teacher and reading specialist for elementary grades. She currently teaches undergraduate and graduate courses in literacy, language arts, children's literature, and teaching methods. Research interests include qualitative approaches into the sociocultural resources accessed when individuals engage in reading and writing practices, out-of-school literacy contexts, and the teachers' role in instruction. Her research is focused on the differential academic achievement for students from diverse backgrounds, particularly multi-generational Latinos. She has published in the Journal of Adult and Adolescent Literacy, Urban Education, The Reading Teacher, Teaching Education, and Issues in Teacher Education.

Judy Montgomery

Best Researcher/Scholar, Education

Received B.S. and M.A. degrees in Communication Disorders from the University of Wisconsin-Milwaukee and California State University-Long Beach respectively; PhD. from Claremont University in Language and Cognition.

Extensive experience in public schools as a speech language pathologist, administrator; then in higher education as professor in special education, literacy, communication disorders. K–8 Principal and the Director of Special Education in Fountain Valley School District in California for 21 years. Wrote and directed an AAC Center (1975) with a five-year USDOE grant to establish the first school-based program to match students with severe disabilities with AAC devices to achieve academic goals. Began teaching at Chapman University in 1993, full professor in 2006.

Active in professional organizations; served as president of five national professional organizations, including ASHA (1995). ASHA Fellow and ASHA Honors (2010). Recipient of the Annie Glen Award for Outstanding Leadership in Speech Language Pathology (2008), Scudder Award for Outstanding Teaching and Service at Chapman University twice; Global Clinical Coordinator for Healthy Hearing, a healthcare initiative with Special Olympics to provide hearing screening for athletes with intellectual disabilities in 110 countries; consultant to the Speech-to-Speech Program with the California Public Utilities Commission and Hamilton Relay to provide telephone re-voicing services to persons with dysarthria and difficult to understand speech: International Association of Logopedics and Phoniatrics (IALP Committee chair

Authored or co-authored over 40 articles, four books, and eight clinical materials including The Bridge of Vocabulary, START-IN, Funnel Toward Phonics, and one standardized test-The Montgomery Assessment of Vocabulary Acquisition (MAVA). Editor-in-Chief of Communication Disorders Quarterly Journal.

Lilia Monzo

Best Overall Faculty Member, Education

Dr. Lilia D. Monzó is Assistant Professor of Education in the College of Educational Studies at Chapman University. She received the Ph.D. in Education from the University of Southern California in 2003 and followed that with a three-year Postdoctoral Fellowship at the University of California, Los Angeles. Her recent scholarship and research draw on revolutionary critical pedagogy to interrogate and confront the educational and sociopolitical contexts impacting Latino communities in the United States and América Latina. In her view, local realities and cultural formations, such as racism and patriarchy, must always be examined in relation to the totality of global capitalism. Dr. Monzó also draws on other theoretical frameworks to guide her work including Latina/Chicana feminist theory, decoloniality, and sociocultural theory. She uses critical ethnography and life history methods as a context within which to collaboratively improve existing social and material conditions for participants while simultaneously working to effect clarity, hope and vision for a societal transformation.

Dr. Monzó teaches courses in the Teacher Education Program that address critical and responsive approaches to working with students and families from diverse racial, ethnic, linguistic, and otherwise marginalized communities, including methods for teaching English learners and bilingual education. She also teaches qualitative research methodologies, including ethnography and life history methods in the Ph.D. program. She encourages students to challenge the ideologies that sustain existing capitalist social relations and to strategically adopt anti-racist, feminist, and critical pedagogy to create the spaces within which we can unite in solidarity and create a new social imaginary.

Kenneth Murphy

Best Overall Faculty Member, Business
Best Teacher, Business

Dr. Kenneth E. (Ken) Murphy teaches business analytics, operations management, statistics and information systems courses in the executive, graduate and undergraduate programs. The overriding theme in all of Dr. Murphy's courses is to enable current and future managers to improve organizational performance. The content of these courses emphasizes quantitative and technical analysis, effective decision making and the implementation of impactful change.

Dr. Murphy's current area of research is in analytics applied to college sports. Previously, he has published in Operations Research, Naval Research Logistics, Communications of the ACM and Information Systems Journal.

As Assistant Dean of Undergraduate Programs at the Argyros School, Dr. Murphy is responsible for undergraduate student academic and professional success while in school. He and his staff provide advising services and programming to support academic achievement and personal growth. They seek to empower students with the knowledge and practical experience necessary to meet their professional goals.

Throughout his career, Dr. Murphy has focused on curriculum innovation and modernization. He has introduced new coursework using technologies including SAP, @Risk, MS Excel and Visual Basic and has pioneered alternative methods of teaching and delivery. He also has experience in successfully leading outcome assessment development and accreditation activities with WASC and AACSB.

Jan Osborne

Best Teacher, Education
Most Helpful to Students, Education
Best Teacher, Education
Most Helpful to Students, Education
Best Overall Faculty Member, Education

Born and raised in Michigan and an alumna of The University of Michigan, Jan teaches in the Writing and Rhetoric Program in the English Department. Prior to coming to Chapman, Jan taught English in secondary public schools in Los Angeles and Orange counties, community colleges in Michigan and California, and at State Prison Southern Michigan. She is the faculty liaison for the Orange High School Young Writers' Collaborative and the Orange County Literary Society Partnership.

Her current research is focused on writing program assessment and chat discourse analysis.

Abel Winn

Best Researcher/Scholar, Economics
Best Overall Faculty Member, Business

Biography Assistant Professor of Managerial Economics, Dr. Winn's research interests include economic systems design and experimental economics. Prior to joining the faculty at Chapman University, he was director of experimental economics at the Market-Based Management Institute, in affiliation with Koch Industries and Wichita State University, where he conducted research on expectation formation, incentive compensation and hold-out problems. Dr. Winn helped to develop the auction system for the Commonwealth of Virginia's nitrous oxide emission credits, and has conducted research on buyer-only and two-sided multiple unit call markets.

CHATTANOOGA STATE COMMUNITY COLLEGE

Yolanda Green

Best Researcher/Scholar, Nursing

Lindsay Holland

Most Helpful to Students, Medicine

Shirley Kilgore

Best Overall Faculty Member, Nursing

Kathy Rose

Most Helpful to Students, Nursing

"Kathy works very hard with students to help them in life long learning and moving forward with education for a rewarding career in nursing."

Charlotte Webb

Best Teacher, Nursing

Betty Zmaj

Most Helpful to Students, Nursing

CHESTNUT HILL COLLEGE

Anne Pluta

Best Researcher/Scholar, Political Science
Most Helpful to Students, Political Science

CHICAGO STATE UNIVERSITY

Elizabeth Arnott-Hill

Best Overall Faculty Member, Psychology

Ivy Dunn

Best Teacher, Psychology

"people person, helpful, friendly"

CITY COLLEGE OF NEW YORK

William Crain

Best Researcher/Scholar, Medicine

CLAREMONT GRADUATE UNIVERSITY

Mark Abdollahian

Best Researcher/Scholar, Political Science

Dr. Mark Abdollahian delivers advanced analytics for data driven decision-making. His global experience spans national policy, corporate strategy, economic development, finance, public-private partnerships, M&A and business process reengineering. He creates, architects and implements enterprise class data and strategy analytics used by the US Government, the World Bank and the United Nations as well as private sector companies worldwide, including Arthur Andersen, Motorola, McKinsey, Raytheon, British Aerospace, Chevron and DeBeers. Dr. Abdollahian develops, manages and applies behavioral and predictive analytics to strategic and operational issues across government and business, focusing on bringing to market next generation innovations today.

Dr. Abdollahian is author of dozens of articles and two books on data driven strategy across business, politics and economics. He is a board member for several private and nonprofit enterprises and lectures to audiences worldwide. Dr. Abdollahian is a Full Clinical Professor at the Division of Politics and Economics, Claremont Graduate University, cofounder of Sentia Group, and currently chief executive officer of ACERTAS. In addition to Bachelors degrees in Political Science, History, and French from Case Western Reserve University, Dr. Abdollahian holds a Master's degree in Foreign and Defense Policy and a Ph.D. in Political Economy and Mathematical Modeling from Claremont Graduate University.

Professor Abdollahian's interests include strategic decision making, international political economy, sustainable development, economics, growth, human social cultural behavioral modeling, econometrics, predictive analytics, computational modeling and simulation, data analytics, and visualization.

CLAYTON STATE UNIVERSITY

Debra Cody

Best Teacher, Health Sciences and Nursing
Best Overall Faculty Member, Health Sciences and Nursing

Elicia Collins

Most Helpful to Students, Nursing

"There have been several reports by students regarding her caring attitude and helpfulness to students."

Han Dong

Best Teacher, Nursing

"Han is very dedicated to the students at Clayton State University School of Nursing. She spends countless hours preparing to ensure that everything is accurate."

Victoria Foster

Best Researcher/Scholar, Nursing

"Very knowledgeable about the research process; mentored several Nursing faculty to be co-authors. Willing to help others publish."

Deborah Gritzmacher

Most Helpful to Students, Health Sciences and Nursing

Betty Lane

Best Researcher/Scholar, Nursing
Best Researcher/Scholar, Health Sciences and Nursing

CLEMSON UNIVERSITY

Nancy Meehan

Best Researcher/Scholar, Nursing

Angela Pye

Best Overall Faculty Member, Nursing
Best Teacher, Nursing

Deborah Willoughby

Most Helpful to Students, Nursing

CLEVELAND STATE UNIVERSITY

Cheryl Delgado

Best Researcher/Scholar, Nursing

Dr. Delgado has been in nursing for more than 35 years and teaching at CSU for more than ten. She also holds a national certification as an adult Nurse Practitioner from the American Nurse's Association. Her interests include stress and coping, spirituality in illness, chronic illness, cultural competency and teaching with technology.

Research Interests: Stress and coping Chronic illness Under-served populations Cultural competency Teaching with technology

Teaching Areas: Research Nursing Theory Population Health

Professional Affiliations: Member of:

American Association of Nurse Practitioners Sigma Theta Tau (National Honor Society for Nursing), Nu Delta Chapter Midwest Nurses Research Society Ohio League for Nursing

Professional Experience: Previous employment as an RN and a Nurse Practitioner at St. Luke's Hospital (Cleveland), Marymount Hospital (Cleveland), and Mt Sinai Hospital (Cleveland).

University Service: Member of Committees in the School of Nursing: Graduate Curriculum committee, Graduate Admissions, Progressions and Standards committee, Member Undergraduate Curriculum committee, Undergraduate Student Affairs committee, Undergraduate Admissions, Standards and Progression committee (Chair), Director and faculty positions search committees,

In the COEHS member of: Faculty Affairs Committee, Doctoral Studies Committee, Sponsored Research Committee,

In the Graduate College member of: Faculty Council, Grade Dispute Committee, Faculty Review Committee

In the University: Faculty Senator, Admissions and Standards Committee, Academic Steering Committee, Subcommittee on Environmental Safety, Instructional Services Committee

Professional Service: Reviewer for Journal of Nursing Scholarship and the Western Journal of Nursing

Community Service: Volunteer providing primary care services to the homeless and underserved at the Free Clinic of Cleveland and Care Alliance for many years. Recipient of the Shining Star Award Care Alliance, 2007.

Anne Galletta

Most Helpful to Students, Education

Dr. Anne Galletta earned her Ph.D. in Social Psychology from the Graduate School and University Center of the City University of New York. She began her career as a language arts teacher at the middle school level, and later conducted research and provided technical assistance to schools and school districts at the Academy for Educational Development (AED) in New York City. Her research focus is educational change within newly created small public high schools and within desegregated schools, particularly in terms of policy development, classroom practice, and social context. Dr. Galletta collaborates with community-based organizations and public school systems in conducting research designed to engage educators and the community in school reform as well as contribute to the literature on educational change. She uses a qualitative approach in her research, with particular strengths in ethnography, participatory action research, and oral history.

Brian Harper

Best Overall Faculty Member, Education

Dr. Brian E. Harper is an Associate Professor in the Department of Curriculum and Foundations. A former classroom teacher in the Philadelphia, PA public school district, Dr. Harper completed his doctoral work at The Ohio State University in Educational Psychology. His research interests include African American racial identity development and motivational psychology, particularly as it applies to students in urban settings. His current work focuses on African American students and the factors that promote or inhibit academic self-regulation.

Grace Huang

Best Teacher, Education

Grace Huang is an Associate Professor in the Early Childhood Program, College of Education and Human Services, Cleveland State University. Dr. Huang teaches courses at the undergraduate and graduate levels including the Introduction to Early Childhood Education, Foundations of Early Childhood Education, and Curriculum Development in Early Childhood Education, Collaborations with Families and Professionals in Early Childhood Settings, Child Development. Her research interests focus on the parental involvement, early childhood family education, family child care, urban family education, Chinese families and education.

Joan Niederriter

Best Teacher, Nursing
Most Helpful to Students, Nursing

Albert F Smith

Best Overall Faculty Member, Psychology
Best Teacher, Psychology
Most Helpful to Students, Psychology
Best Researcher/Scholar, Medicine
Best Teacher, Medicine
Most Helpful to Students, Medicine
Best Overall Faculty Member, Medicine

Education: Ph.D., Psychology, Yale University M.S., Biostatistics, Columbia University

Kristine Still

Best Researcher/Scholar, Education

Dr. Kristine Lynn Still currently serves as Assistant Director of the Community Learning Center for Children and Youth, is an Assistant Professor of Early Literacy and is also the Co-Coordinator for the Literacy Education Program in the Department of Teacher Education at Cleveland State University. Prior to joining the faculty at Cleveland State in 2006, Dr. Still taught 1st grade for 10 years and also held the position of Visiting Assistant Professor at The University of Akron while working on her doctoral studies.

During her initial years at CSU, Dr. Still served as the Director of The Cleveland Schools Book Fund, a project funded by a $5 million dollar endowment and which created classroom libraries in all K–3 classrooms in CMSD.

Currently, Dr. Still is an active member of the Campus International School Design team where her efforts include serving as a resource for literacy related initiatives as well as developing research protocols for those interested in pursuing research studies at CIS.

Dr. Still's research interests focus on the integration of meaningful technology-based literacy events in primary grade classrooms, reading motivation, reading tutoring, and teacher professional development. Dr. Still is currently involved in three major research studies and one of which involves evaluating seven tutoring programs serving the St. Paul Public schools and which are all funded by The Greater Twin Cities United Way.

Dr. Still has established the first CSU Student Chapter of The International Reading Association and currently serves as the group's major faculty advisor.

Dr. Still remains active in the greater Cleveland community and beyond through her outreach and research initiatives with literacy related projects. Dr. Still is passionate about bringing the CLC into our local literacy service oriented network.

Joan Thoman

Best Overall Faculty Member, Nursing

Joan Thoman first received her diploma in nursing at St. Vincent Charity Hospital School of Nursing in Toledo, Ohio. Later, she earned both her Bachelor of Science in Nursing (BSN) and Master of Science in Nursing (MSN) from the University of Akron, College of Nursing. Her clinical experiences as a registered nurse includes hospital medical surgical units, charge nurse, public health, as a visiting home care nurse, and as a diabetes program coordinator. Joan has experience in staff development, informatics and management. Joan is credited with implementing and maintaining the Diabetes Self Management Skills and Training Program at St. Vincent Charity Hospital, Cleveland, Ohio which achieved recognition from the American Diabetes Association in August 2002. In 2003, Joan was selected by her nursing peers to receive the award of Outstanding Excellence in Clinical Nursing. She started a multidisciplinary Diabetes Resource Team to address both in-patient and out-patient needs. In collaboration with the hospital's community outreach department, she started a community health worker program for the diabetes center. In 2004, Joan was key in assisting the hospital become an affiliate of the Joslin Diabetes Center which has a philosophy of an integrated team approach for the care of people with diabetes. Joan assisted in implementing a medical clinic portion to augment the education program based on the Joslin diabetes standards of care. In October 2004, St. Vincent Charity Hospital recognized Joan as "Caregiver of the Month" for her work with the Joslin Diabetes Center. In June, 2005, Joan was selected to receive the Distinguished Women in Health Care award for clinical practice and was featured in Cleveland Magazine. She has been a registered nurse for over 25 years.

COLLEGE OF CHARLESTON

Mary Blake Jones

Most Helpful to Students, Education

Mary Blake Jones Professor of Education College of Charleston
Mary Blake Jones has devoted her career to the education of children. She obtained a Bachelor of Arts degree from St. Joseph College in West Hartford CT.
Mary spent 8 years as an elementary school teacher, teaching first and second grades. She has spent the last 26 years "teaching teachers" as a professor in the Education Department of the College of Charleston, in South Carolina. Language Arts (English) is her area.
Dr. Blake Jones continuously seeks out new ways to instruct and to improve our education system. She builds a strong bond with students and colleagues with her interpersonal skills and has a commitment to the quality education of our future teachers.
Mary has been written about in the Post and Courier Newspaper. She has also been published in numerous journals, including The Reading Teacher, Journal of Adolescent and Adult Literacy, The New England Reading Association Journal and The Australian Journal for Remedial Education. Mary is co-author of the textbook Integrating the Language Arts.

COLLEGE OF LAKE COUNTY

Imelda N Forsberg

Best Teacher, Health Science
Best Researcher/Scholar, Health Sciences and Nursing
Best Teacher, Health Sciences and Nursing
Most Helpful to Students, Health Sciences and Nursing
Best Overall Faculty Member, Health Sciences and Nursing

Margaret Kyriakos

Best Overall Faculty Member, Health Science
Best Researcher/Scholar, Health Science
Most Helpful to Students, Health Science

COLLEGE OF NEW ROCHELLE

Joseph Biscoglio

Best Overall Faculty Member, Counseling

Marie Hurrell

Best Researcher/Scholar, Psychology
Best Teacher, Psychology
Most Helpful to Students, Psychology

Rose Marie Hurrell is a New York State Licensed Psychologist, has held several professional consultancies, and has served as a U.S. Navy-ASEE Faculty Research Associate. Dr. Hurrell has presented at professional conferences, including the Annual meetings of the American Psychological Association, given talks at academic and community events, and published on a variety of topics, including factors associated with regular exercise, intelligence testing in mildly retarded children, and attitudes toward women in the military. She currently teaches courses in health psychology, developmental psychology, and creativity.

Bob Wolf

Best Overall Faculty Member, Psychology

Professor Robert Irwin Wolf is a New York State licensed creative art therapist and psychoanalyst with over 35 years of experience in teaching, private practice and clinical supervision. He is currently the President of the Institute for Expressive Analysis and for many years, has been on the faculty of The College Of New Rochelle, Pratt Institute and The Training Institute for the National Psychological Association for Psychoanalysis. He has also been a past President of Institute for Expressive Analysis and the New York Art Therapy Association and Clinical Director of the Henry Street Settlement School. He has published numerous articles and book chapters on art therapy, countertransference, expressive therapy and photo therapy and his work as a fine art sculptor and photographer have been exhibited internationally.

COLLEGE OF SOUTHERN MARYLAND

Bruce Fried

Best Teacher, Accounting
Best Overall Faculty Member, Business

COLLEGE OF SOUTHERN NEVADA

Starvos Anthony

Best Overall Faculty Member, Political Science

Councilman Stavros S. Anthony was elected to represent Ward 4 on June 2, 2009. Councilman Anthony, a Las Vegas resident for 29 years, has made community service and public safety his focus as a Las Vegas Metropolitan Police Department officer, a member of the Nevada System of Higher Education Board of Regents and now as a City Councilman.

Councilman Anthony began his career with Metro Police in 1980 and most recently served the community as a captain overseeing the Financial and Property Crimes Bureau of the department. As a captain, Anthony spent time in charge of many divisions within Metro including Professional Standards, Personnel, Vice/Narcotics, Northeast Area Command and the Transportation Safety Bureau.

Along with his experience in the realm of public safety, Councilman Anthony also brings an education background to the City Council. He was elected in 2002 to a four-year term as a regent with the Nevada System of Higher Education, and was re-elected to a six year term on the board in 2006. As chairman of the Board of Regents, Anthony led the way in developing a master plan, system goals and a value statement.

Councilman Anthony graduated with a Bachelor of Science in Criminal Justice from Wayne State University in Detroit, Michigan in 1980. In 1987 he graduated with a Master of Arts in Political Science from the University of Nevada Las Vegas (UNLV), and in 1999 he received his Ph.D. in Sociology from UNLV. He has also attended the University of Louisville Southern Police Institute Administrative Officers Course and the FBI National Academy in Quantico, Va.

Councilman Anthony serves on the board of directors for Goodwill of Southern Nevada and is a member of the National League of Cities 2011 Public Safety and Crime Prevention Policy and Advocacy Committee. He is a past member of the International Association of Chiefs of Police, the Police Executive Research Forum, the Academy of Criminal Justice Sciences and the Association of Governing Boards of Universities.

Councilman Anthony is also a past member and president of the Board of Directors for the St. John Greek Orthodox Church. Anthony received an Exemplary Service Award from Metro Police in 2006, and received an Award of Excellence in 2002 from the Community College of Southern Nevada where he was an adjunct faculty member.

Married for 30 years, the Anthonys have two daughters who are both graduates of the University of Nevada Reno. The Councilman is an avid golfer and runner.

Alan Balboni

Best Teacher, Political Science
Best Teacher, Public Affairs, Political & Policy Sciences

Earnest Bracey

Most Helpful to Students, Public Affairs, Political & Policy Sciences

Eric Davis

Most Helpful to Students, Political Science
Best Researcher/Scholar, Public Affairs, Political & Policy Sciences

Lisa Finnegan

Best Teacher, Health Sciences and Nursing

Ron Gonzalez

Most Helpful to Students, Nursing

"Not only teaches students in the lecture, clinical, and lab setting, but acts as a mentor, advisor, and role model for the students. Also helps all of the students in the nursing program with their course schedules so that they don't interfere much with their personal schedule - thereby helping promote their success in nursing school."

Royse Smith

Best Overall Faculty Member, Public Affairs, Political & Policy Sciences

Steve Yuen

Best Researcher/Scholar, Political Science

COLLEGE OF STATEN ISLAND
(CUNY - STATEN ISLAND)

Karen Arca-Contreras

Most Helpful to Students, Nursing

"Always working with students to assist them with academic success."

MS Medical-Surgical Nursing/Nursing Administration Advanced Certificate in Nursing Education DNP in Progress NYCNECT Transformer Scholar Areas of interest/research Flipping the classroom Nursing student success Simulation Type 2 Diabetes

Arthur Binford

Best Researcher/Scholar, Medicine

Biography/Academic Interests: Leigh Binford came to the College of Staten Island from Mexico's Autonomous University of Puebla where he worked following twelve years at the University of Connecticut. He has carried out fieldwork in highland and lowland Mexico and highland El Salvador, focusing on rural social economies, peasantries, international migration, human rights, civil war and post-war reconstruction. His current project (2010–2012), titled "From Wartime to Peacetime: Post-insurgent Individuality in northern Morazan, El Salvador," addresses the consequences of almost two decades of post-civil war neoliberal policies for former rebels of the Farabundo Marti Front for National Liberation and their supporters. The work is supported by a grant from the National Science Foundation.

Katie Cumiskey

Best Teacher, Medicine
Most Helpful to Students, Medicine
Best Overall Faculty Member, Medicine

Kathleen M. Cumiskey is an Associate Professor in the Psychology Department and Women, Gender and Sexuality Studies Program at CSI. She is director of the Social Media Lab, and she is currently engaged in research that looks at how the use of mobile phones leads to risk-taking as well as the ways in which mobile phones impact women's perceptions of public safety. She has presented at conferences in London, Budapest, Dresden and Sydney and has organized conferences that have attracted scholars on mobile technology from all over the world.

Dr. Cumiskey is an advocate for girls' in confinement in New York State. She serves on the community advisory board of a residential facility that is run by the Office of Children and Family Services. She has worked in collaboration with other organizations to stop the expansion of youth detention facilities in New York City and to change state-run facility operations manuals to ensure the safety of youth who identity as lesbian, gay, bisexual, or transgender or who are gender non-conforming.

Dr. Cumiskey is committed to involving students in "community action projects" and has developed service learning courses at CSI. She serves as the faculty advisor to CSI's "Gay-Straight Alliance". She also serves on the board of the Staten Island LGBT Community Center.

Danna Curcio

Best Overall Faculty Member, Nursing
Best Researcher/Scholar, Nursing
Best Teacher, Nursing
Most Helpful to Students, Nursing
Best Researcher/Scholar, Health Sciences and Nursing
Best Teacher, Health Sciences and Nursing
Most Helpful to Students, Health Sciences and Nursing
Best Overall Faculty Member, Health Sciences and Nursing

Barbara DiCicco-Bloom

Best Researcher/Scholar, Nursing

"Intense and prolific researcher"

A clinical nurse for 17 years, Dr. DiCicco-Bloom's professional background is in primary care, home care, and hospice. She was the first nurse Director and Principle Investigator of the National Health Services Research Fellowship in Primary Care from 2007–2009 in the Department of Family Medicine at the University of Medicine and Dentistry/Robert Wood Johnson Medical School in New Brunswick. Dr. DiCicco-Bloom's research employs qualitative and mixed methodologies and focuses on relationships between health care professionals. She is concerned with locating the intractional and cultural elements that facilitate or interfere with safe and effective patient care. She recently completed two studies. The first identified issues that interfere with information sharing among oncologists, primary care clinicians (nurse practitioners and family medicine physicians), and their patients, and the second explored the relationships of nurse practitioners and physicians in family medicine practices. At present, she is conducting an ethnographic study of a hospice setting located in a large medical center. Her focus is on the relationships among the members of the hospice team, and between the hospice team and other members of the larger hospital system.

Arlene Farren

Best Overall Faculty Member, Nursing

"Relentlessly works to assist the students to reach their academic goals."

Daniel McCloskey

Best Researcher/Scholar, Psychology
Best Researcher/Scholar, Medicine

Dr. McCloskey's research is focused on understanding the ways that neurons participate in networks to produce complex behaviors such as learning, memory, and mood. The brain uses patterns of rhythm and synchrony to accomplish these tasks, but the amount of synchrony required is a delicate balance. Too little, and the brain will not develop properly. Too much, and seizures can occur. The McCloskey Laboratory at the College of Staten Island uses a combination of anatomy, physiology, and behavior to study the mammalian brain to understand how networks achieve this balance, and what interventions can help to restore it.

Susan Mee

Best Teacher, Nursing

"Excellent educator always trying new teaching strategies to reach the students."

Jonathan Peters

Best Teacher, Business

Jonathan R. Peters is a professor of finance in the Business Department at The College of Staten Island of The City University of New York and a Research Fellow at The University Transportation Research Center at The City College of New York. He received his Ph.D. in Economics from the City University of New York. He has previously published in The Journal of Applied Finance, Transportation Quarterly and most recently in Public Works Management & Policy. He currently conducts research in the areas of regional planning, road and mass transit financing, corporate and public sector performance metrics, capital costs and performance management.

Vasilios Petratos

Best Overall Faculty Member, Economics

Thomas Tellefsen

Best Researcher/Scholar, Business
Most Helpful to Students, Business
Best Overall Faculty Member, Business

At CSI, Professor Tellefsen teaches courses in marketing and management at the undergraduate and graduate levels. He has also served as an advisor for students conducting independent studies and a mentor to CUNY baccalaureate candidates. Professor Tellefsen is a member of the doctoral faculty of the Graduate School of the City University of New York. In that role, he has served as a dissertation committee member for Ph.D. candidates in the doctoral program in business.

Nelly Tournaki

Best Overall Faculty Member, Education
Best Researcher/Scholar, Education
Best Teacher, Education
Most Helpful to Students, Education
Best Researcher/Scholar, Education
Best Teacher, Education
Most Helpful to Students, Education
Best Overall Faculty Member, Education

Dr. Tournaki has taught at the college level for over 20 years. She is dedicated to inclusive education. To that effect, she opened and directed one of the first inclusive preschools in Athens, Greece. She currently sits in the Board of Directors of the Lavelle Preparatory Charter School, the first Charter school designed to teach students with special needs in inclusive classrooms.

Her research has two areas of focus. The first area is mathematics. More specifically, she evaluates strategies and tools (e.g., manipulatives, technology) to improve mathematics achievement for students with and without disabilities. The second area is teacher efficacy and effectiveness. In this area, she has examined differences in efficacy between special and general education teachers; the effects of different approaches to graduate level teacher preparation programs on teacher efficacy and effectiveness; the effects of professional development on teacher effectiveness. Her research has been consistently supported by grants from PSC-CUNY.

Given her strong background in psycho-metrics she is the internal evaluator for the STEAM and NSF-STEP grant at the College of Staten Island.

Simone Wegge

Best Researcher/Scholar, Economics
Best Teacher, Economics
Most Helpful to Students, Economics

Professor Wegge teaches a range of courses in the PEP Department, including microeconomics, macroeconomics, statistics, econometrics, industrial organization, and economic history. She also teaches regularly in the Ph.D. program at the CUNY Graduate Center, where she is member of the doctoral faculty. Professor Wegge is an active member of the Cliometric Society, the Economic History Association, and the Social Science History Association. Since 2005, she has represented the PEP Department at the College Council and the Faculty Senate bodies of CSI. For the spring semester of 2009, she is serving as Deputy Chair of the PEP Department.

Professor Wegge's research focuses primarily on improving our understanding of historical migration processes. A second and new area of research consists of the comparative analysis of child labor incidence across industrializing countries in nineteenth-century Europe. Both topics concern nineteenth-century Europe, in particular Germany and draw on various literature in economics, economic history, history, and other social sciences. Her research on migration seeks to identify the factors that influenced migrants and explain who emigrated, why they may have left, and what sorts of communities they left from.

COLLEGE OF THE REDWOODS

Ryan Emenaker

Best Overall Faculty Member, Political Science
Best Teacher, Political Science

Professor Emenaker has taught Political Science at College of the Redwoods since 2006. He received a BA in Political Science, an MA in Social Science from Humboldt State University, and another MA in Government from The Johns Hopkins University. His research interests focus on judicial politics especially in relation to federalism and separation of powers conflicts. He is a regular commentator in the press on issues involving state and local politics, making guest appearances on NBC affiliates as a political analyst and on local radio to discuss political controversies. He has also written numerous articles on the history and role of the Supreme Court. His most recent article, on why the Supreme Court should uphold the Voting Rights Act, appeared on SCOTUSblog, the premier news and research site on the Supreme Court. You can check out some of his recent scholarship at http://ssrn.com/author=1789397

Chris Gaines

Best Teacher, Business

COLLIN COLLEGE

Susan Evans

Best Teacher, Physical Education

"Susan is enthusiastic about health and she's very helpful to her students. She is also a great mentor for women in the health and fitness field."

Michael McConachie

Best Teacher, Political Science
Most Helpful to Students, Political Science
Most Helpful to Students, Public Affairs, Political & Policy Sciences

Debra John

Best Overall Faculty Member, Political Science
Best Researcher/Scholar, Political Science
Best Overall Faculty Member, Public Affairs, Political & Policy Sciences

Tyler Young

Best Researcher/Scholar, Public Affairs, Political & Policy Sciences
Best Teacher, Public Affairs, Political & Policy Sciences

COLORADO STATE UNIVERSITY

Marie Legare

Best Teacher, Enviornmental Health

"Dr. Legare is an excellent and committed teacher. She is respected and well-liked by her students."

Michele Betsill

Best Researcher/Scholar, Political Science

"She continues to do research and get grants to further her knowledge on Global Climate Change."

Michele Betsill, associate professor in the Department of Political Science, takes an all encompassing approach to studying the ebb and flow of climate change politics. She looks at all of the components – national governments, international businesses, non-governmental organizations, the market, individual choices – on a "big picture scale" and assesses how the pieces fit together. Or how they don't.

"My research allows me to have a better sense of the large landscape of the politics of climate change from the local to the global level. I watch the debate about who gets what, when and how," she said.

"By this point most people recognize that climate change is a problem but what everyone is asking now is what we are going to do about it. Climate change has huge implications because it fundamentally changes the earth's natural systems. It is directly related to how we produce and use energy to meet our daily needs. Tough ethical issues arise and there are no clear or easy solutions," said Betsill.

Beyond the big picture aspect, she is particularly interested in mitigation policies that are aimed at limiting greenhouse gas emissions, and has hope that the United States will become a bigger player in this area in the future.

Her research focuses on the politics of global climate change, from the local to the global level. She is particularly interested in the various policy approaches to controlling greenhouse gas emissions. She is the co-author, with Harriet Bulkely from Durham University, of "Cities and Climate Change: Urban Sustainability and Global Environmental Governance," (Routledge 2003), which analyzed climate change decision making in six cities in the United States, United Kingdom and Australia, with a focus on the energy, transport and planning sectors.

Betsill is involved with the university's School of Global Environmental Sustainability and helps lead the Environmental Governance Research Working Group within the school.

She has been a long-time observer of international climate change negotiations and is currently working on a project that traces the evolution of emissions trading markets as a policy instrument for mitigating greenhouse gas emissions. In this project, she and Matthew Hoffman, University of Toronto, are examining how the idea of emissions trading developed and spread through more than 30 policy venues, how trading rules reflect the specific context of each venue, and the broader implications for the global politics of climate change.

She received her doctorate in political science from the University of Colorado-Boulder. Prior to coming to CSU, she was a post-doctoral fellow with the Global Environmental Assessment project at Harvard's Kennedy School of Government. She spent her sabbatical, 2006–07 academic year, as a visiting scientist with the Institute for the Study of Society and the Environment at the National Center for Atmospheric Research.

Doug Hoffman

Most Helpful to Students, Marketing

Dr. K. Douglas Hoffman is Professor of Marketing, Partners for Excellence Fellow, and University Distinguished Teaching Scholar at Colorado State University. Doug's teaching experience at the undergraduate and graduate levels spans nearly 30 years while holding tenure track positions at Colorado State University, the University of North Carolina at Wilmington, and Mississippi State University. In addition, Doug has taught as a visiting professor at the Helsinki School of Business and Economics (Helsinki, Finland), the Institute of Industrial Policy Studies (Seoul, South Korea), Thammasat University (Bangkok, Thailand), and Cornell-Nanyang Technological University (Singapore).

Professor Hoffman is an accomplished scholar in the services marketing area specializing in the field of service recovery research. In addition, Doug has written numerous journal and conference proceedings articles on teaching scholarship that have appeared in a variety of outlets such as the Journal of Marketing Education, Marketing Education Review, Journal of Services Marketing, and Innovative Higher Education. His teaching scholarship has also expanded into the co-authorship of three textbooks including Services Marketing: Concepts, Strategies & Cases (4e) published by Cengage. Professor Hoffman served as editor of Marketing Education Review from 2010–2012. Doug has been the recipient of 16 teaching awards at the college, university and national discipline levels including the prestigious Board of Governors Excellence in Undergraduate Teaching Award. Doug was named a University Distinguished Teaching Scholar—a lifetime appointment in 2007.

Dr. Hoffman is the founder of Colorado State University's Master Teacher Initiative (MTI). The primary focus of the MTI is to enhance the quality of teaching and the visibility of the teaching mission across Colorado State University. The Master Teacher Initiative has been adopted by all nine colleges on campus and has been singled out by The Association to Advance Collegiate Schools of Business (AACSB), the lead international accreditation body, as a benchmark program for other universities to follow.

Doug's current research and consulting activities are primarily in the areas of sales/service interface, customer service/satisfaction, service failure and recovery, and services marketing education.

Jennifer McLean

Best Teacher, Microbiology

"She is an innovative and enthusiastic teacher."

Jennifer McLean is an assistant professor in the Department of Microbiology, Immunology and Pathology at Colorado State University where she also received her Ph.D. Her current research efforts are focused on identifying immune correlates of protection against tuberculosis in terms of vaccine development, as well as UDL strategies in the classroom. Courses she has taught include general microbiology, immunology, immunology laboratory, survey of microbiology, microbial biology laboratory, and freshman microbiology seminar. Teaching awards include the Provost's N. Preston Davis Award for Instructional Innovation, Innovative Instructional Methodology Award in Undergraduate Education, and the Dr. Blanche M. Hughes Distinguished Faculty Staff Award.

Susan Opp

Most Helpful to Students, Political Science

"Many of my students have told me how helpful Susan has been to them with advising and ideas for future career pursuits."

Ian Orme

Best Researcher/Scholar, Health Sciences and Nursing

COLUMBUS STATE COMMUNITY COLLEGE

Lisa Cerrato

Most Helpful to Students, Health Sciences and Nursing

Peggy Mayo

Best Teacher, Multi-Competency Health

COMMUNITY COLLEGE OF BALTIMORE COUNTY

Judy Blum

Best Overall Faculty Member, Medicine

CONTRA COSTA COLLEGE - SAN PABLO

Miguel Johnson

Most Helpful to Students, Health

Miguel Johnson was born in Oakland, CA. Miguel attended high school at Richmond High, where he was on varsity for three years, captain of the team for a year and leading scorer for two years, scoring over 1,000 points. He also attended Contra Costa College where he earned his A.A. degree, Eastern Washington University where he earned his Bachelor of Science Degree and John F. Kennedy where he earned his M.A. degree.

Miguel coached at Solano Middle School for two years, Greater Vallejo Recreation District for two years, and Jesse Bethel High School for two years.

His college coaching experience includes six years at Contra Costa College, two years at Merritt College, and two years at Solano College

Jeannette McClendon

Most Helpful to Students, Education

Aminta Mickles

Best Overall Faculty Member, Health Sciences and Nursing

Richard Ramos

Best Overall Faculty Member, Administration Of Justice
Best Overall Faculty Member, Law

Nader Sharkes

Best Overall Faculty Member, Culinary Arts

"The department is successful because of chef Sharkes"

COSUMNES RIVER COLLEGE

Matthew McHugh

Most Helpful to Students, Health Education

"Students respect him and look forward to taking classes from him."

CULINARY INSTITUTE OF AMERICA

Richard Coppedge

Most Helpful to Students, Culinary Arts

"Rich always supports students and helps them as much as he can."

CUNY QUEENS COLLEGE

John Bowman

Most Helpful to Students, Political Science

John Bowman has been on the Queens College faculty since 1984. His principal area of research is comparative political economy, with a focus on advanced industrial societies. In 1996 he was a guest researcher at the Institute for Social Research in Oslo, Norway. He returned to Norway in 2014 as a professor in the University of Oslo's Summer School in Comparative Social Analysis. His current research concerns the politics of labor markets and social policy in Europe and the US.

He is the author of Capitalisms Compared: Work, Welfare, and Business (Sage/CQ Press, 2014) and Capitalist Collective Action: Competition, Cooperation, and Conflict in the Coal Industry (Cambridge University Press 1989, paperback edition 2006); as well as articles in several journals, including Politics & Society, Comparative Political Studies, and (with Alyson Cole) Signs: Journal of Women in Culture and Society.

At Queens, Professor Bowman regularly teaches Introduction to Political Science (PSCI 101); Current Political Controversies: How Democratic are US Elections (PSCI 102); The Politics of the Welfare State (PSCI 242), Comparative Political Economy (PSCI 232) and Business and Politics (PSCI 210). At the CUNY Graduate Center, where he has been a faculty member since 1990, he teaches Comparative Political Economy.

Professor Bowman is a member of the Editorial Board of Politics & Society and has also served on the Editorial Committee of the journal Comparative Politics.

Diane Coogan-Pushner

Best Researcher/Scholar, Business

Diane Coogan-Pushner is Distinguished Lecturer at Queens College, City University of New York (CUNY) and Director of the Graduate Program in Risk Management. At various times, Ms. Coogan has been the Associate Dean of Social Sciences for the College. Ms. Coogan was a Managing Director at Swiss Re and as team leader for structured reinsurance deals, closed over one billion dollars of transactions helping clients transfer risk and manage their balance sheets. Ms. Coogan was a portfolio manager for the longest running US equity hedge fund and created and launched a new long-short product for the financial services sector. She was also a portfolio manager for a private equity fund, also dedicated to financial services, which included fundraising for and recapitalization of a portfolio company. At PricewaterhouseCoopers, Ms. Coogan created and led the Value Based Management practice for financial services clients. She developed valuation and risk models in support of transactions and other value based management engagements. Prior to joining PwC, Ms. Coogan led forecasting, planning and corporate strategy for AT&T's $18 billion consumer markets division reporting to the

CFO, with responsibilities for earnings attribution, investigation of earnings anomalies, and cash flow shortfalls. Ms. Coogan has served as a member of Standard & Poor's Insurance Advisory Council, served as a director of a privately held insurance company, and provides asset management and risk management consulting services to insurance companies. Ms. Coogan began her career at the World Bank on a project to recapitalize the banking sector in Uganda. She has published research on topics of risk and investment performance and is the co-author of The Handbook of Credit Risk Management. Ms. Coogan earned her PhD in economics from Boston University in 1992 and wrote her dissertation on insurance profit cycles. Ms. Coogan is a CFA charter holder.

Julie George

Best Overall Faculty Member, Political Science
Best Researcher/Scholar, Public Affairs, Political & Policy Sciences

Professor George specializes in comparative politics, focusing on ethnic politics, democratization, and state building. Her current research focuses on how states undergoing significant transformation and reform address ethnic minorities. Professor George has conducted research in the former Soviet Union, primarily in the Russian Federation and in Georgia, where she was funded by the Fulbright Association.

Professor George is the author of The Politics of Ethnic Separatism in Russia and Georgia (Palgrave Macmillan, 2009), as well as articles in Europe-Asia Studies, Post-Soviet Affairs, European Security, and Central Asian Survey. She has written chapters for inclusion in The Politics of Transition in Central Asia and the Caucasus: Enduring Legacies and Emerging Challenges (Routledge, 2009) and Conflict in the Caucasus: Implications for International Legal Order (Palgrave Macmillan, forthcoming 2010).

Keena Lipsitz

Best Teacher, Public Affairs, Political & Policy Sciences

Keena Lipsitz is an Associate Professor of Political Science at Queens College, City University of New York. She is the author of Competitive Elections and the American Voter (University of Pennsylvania, 2011) and a co-author of Campaigns and Elections: Rules, Reality, Strategy and Choice (W.W. Norton, 2012) and Democracy at Risk: How Political Choices Undermine Participation and What We Can Do About It (Brookings Institution, 2005). She has also published numerous articles in the areas of political communication, political behavior, and democratic theory.

Marvin Milich

Most Helpful to Students, Accounting

Patricia Rachal

Best Researcher/Scholar, Political Science

Patricia Rachal is the principal investigator of the New York Deaf-Blind Collaborative, a five-year, $2.875 million grant from the U.S. Department of Education, Office of Special Programs (OSEP).

Alex Reichl

Best Teacher, Political Science

Alex Reichl is the proud recipient of the President's Award for Excellence in Teaching at Queens College. His teaching, as well as his research, is in the area of US politics with a particular focus on the politics of New York and other cities. His classes combine big-picture theories with real-world examples in order to address fundamental issues of democracy and power in the US.

Prof. Reichl is the author of Reconstructing Times Square: Politics and Culture in Urban Development (University Press of Kansas, 1999). He has published articles on urban development and housing in Urban Affairs Review, the Journal of Urban Affairs, and the Journal of Affordable Housing and Community Development Law, and his book reviews have appeared in the International Journal of Urban and Regional Research, Urban Affairs Review, Polity, and Architecture. He is a contributor to Understanding the City: Contemporary Perspectives in Urban Studies (Blackwell, 2002) and Cities in American Political History (CQ Press, 2011). His current research projects include: an analysis of the changing nature of parks and public spaces in post-industrial New York City; an assessment of the changing socio-demographic conditions of New York City neighborhoods; and a study of urban planning in the effort to rebuild New Orleans (a beloved city where he lived, taught, and conducted research for five years).

Prof. Reichl brings years of experience working in politics and government at the city and state levels. He has worked as a political activist in Chicago, an analyst for the Illinois House of Representatives, and a researcher and program assistant for the New York City Department of Housing Preservation and Development.

Like many in the Queens College community Prof. Reichl is a second generation American – a true Bohemian (capital B).

CUYAHOGA COMMUNITY COLLEGE

Michael Hall

Best Teacher, Physical Education
Most Helpful to Students, Education

Gail Nelson

Best Researcher/Scholar, Early Childhood Education

Eric Primuth

Best Teacher, Accounting

Danielle Stanaczyk

Best Teacher, Early Childhood Education

Mrs. Danielle Stanaczyk has been in the education field since 1998. She received her BA from Baldwin Wallace College and her Masters in Curriculum and Development from Cleveland State University. In addition to teaching preschool, Mrs. Stanaczyk has taught first and second grade and spent 9 years at the college level, instructing pre-service teachers in early childhood education at Cuyahoga Community College, Lorain County Community College, and the Ohio State University. She is passionate about education and enjoys sharing this passion with the families, students and children that she encounters!

Mrs. Stanaczyk has chaired and presented at CAEYC conferences. She enjoys hiking, swimming at the beach, and reading. She loves cheering for her boys during soccer, basketball and baseball games, and worshiping together with her family at Bethany.

Jeffrey Zola

Best Overall Faculty Member, Psychology
Best Teacher, Psychology
Most Helpful to Students, Psychology

DALTON STATE COLLEGE

Orenda Gregory

Best Teacher, Education

DAYTONA STATE COLLEGE

Harold Orndorff

Most Helpful to Students, Public Affairs, Political & Policy Sciences

"Understands students and is always available to them especially when the students are in need of additional explanation."

DE ANZA COLLEGE

Roger Mack

Best Teacher, Economics

DELTA COLLEGE

Renee Hoppe

Best Overall Faculty Member, Physical Education
Best Researcher/Scholar, Physical Education
Best Teacher, Physical Education

Carla Murphy

Best Researcher/Scholar, Education
Best Teacher, Education
Most Helpful to Students, Education
Best Overall Faculty Member, Education

John Neal

Most Helpful to Students, Physical Education

Betty Rickey

Best Researcher/Scholar, Health Science

Ski Vanderlaan

Best Teacher, Accounting

DELTA STATE UNIVERSITY

Cameron Mongomery

Best Researcher/Scholar, Business

DEPAUL UNIVERSITY

Scott Hibbard

Best Overall Faculty Member, Public Affairs, Political & Policy Sciences

Scott Hibbard teaches courses on American foreign policy, Middle East politics, and international relations. He has been at DePaul since 2005 and spent the 2009–2010 academic year teaching at the American University of Cairo as part of a Fulbright Award from the U.S. Department of State. Professor Hibbard worked in the U.S. government for twelve years–as a program officer at the United States Institute of Peace and as a legislative aide in the United States Congress. He is the author of Religious Politics and Secular States: Egypt, India, and the United States and co-author (with David Little) of Islamic Activism and U.S. Foreign Policy.

Catherine May

Most Helpful to Students, Public Affairs, Political & Policy Sciences

Phillip Stalley

Best Teacher, Political Science

Phillip Stalley has been an assistant professor at DePaul University since 2007. He teaches courses on a variety of subjects including Chinese politics, environmental politics, and international relations. Prior to joining DePaul, Phillip was a visiting research fellow at Princeton University in the Princeton-Harvard China and the World program. Professor Stalley also served as a visiting scholar in the environmental economics department at Fudan University in Shanghai. He is a member of the Public Intellectuals Program of the National Committee on U.S.-China Relations. Phillip's research focuses primarily on Chinese environmental politics. He is the author of Foreign Firms, Investment, and Environmental Regulation in the People's Republic of China (Stanford University Press, 2010) and his work can be found in academic journals such as The China Quarterly and Journal of Contemporary China. His current research project focuses on China's environmental diplomacy and its approach to international environmental institutions.

DEVRY UNIVERSITY

Neisa Jenkins

Best Teacher, Health Sciences and Nursing

"Neisa is always helpful to students and a great resources to students. She goes above and beyond the call of duty to assist students."

John Kyser

Best Researcher/Scholar, Business

Before coming to DeVry University, I worked at AT&T/Lucent Bell Labs just outside Chicago for 21 years. For the last six years, I hired - and admired - interns from DeVry University to assist my team in providing engineering and operations support to over 70 broadband, wireless, and digital switching test and development lab environments. I then took early retirement from Bell Labs and DeVry University hired me. So our association goes back longer than the six years I have been teaching at DeVry University Chicago.

I now teach subjects and skills that used to be my work tools. My career began at Argonne National Lab. For six years, I was part of the technical staff developing and programming a large 3D computer simulation of nuclear reactor failure scenarios. At Bell Labs, I started on the telecommunications field support staff and worked up to various technical management positions. That's where I acquired my programming and network operating systems expertise - and my belief that trouble-shooting is what you get paid for in industry.

Michael Morrison

Best Teacher, Business
Most Helpful to Students, Business
Best Teacher, Business
Most Helpful to Students, Business

Robert Salitore

Best Overall Faculty Member, Business
Best Overall Faculty Member, Business
Best Researcher/Scholar, Business

Michael Vasilou

Best Teacher, Business

WHERE I COME FROM As an undergraduate, I majored in history at Loyola and wanted to teach, but I thought it was impractical to become a professor. With a group of my friends, I went to law school for the material rewards. I worked my way through law school as a waiter; it was a rough experience. I'd leave after a day of classes, read on the subway and in the back of the restaurant. Then I was a lawyer for a while and didn't like it.

I went to business school at night, and got an MBA from the University of Chicago. With that combination of business and law, I got work I liked, writing and managing 401k plans for small business. One day, 17 years ago, I answered an ad for someone to teach a 7:00 am law class at DeVry University. I'd always wanted to teach, and I figured I could fit it in before starting my business day in the Loop.

Later the chairman offered me a 9:00 am class if I'd teach accounting, and I'd do anything for two hours more sleep. For a while I juggled teaching and my pension practice, but it began to feel like working my way through college - torn and tiring. When I received an offer to chair the accounting department 12 years ago, I became a full-time professor.

"A YOUNG PERSON WITHOUT CONNECTIONS IN THE WORLD" I had a student who was waitressing her way through college. She kept falling asleep in class. One day she came into the office and I smelled that restaurant smell on her and I knew why. She said the work was killing her and she didn't have time to look for a job in her field. I called everyone I knew to try to find her an accounting job in the city and finally got her in somewhere.

That's what got our internship program going. There is nothing that students like her can do outside of school that looks good enough on their resume to interest employers, even though they are working their tails off and would be outstanding employees wherever they go. That's what happened to me when I got out of law school. As a young person without connections in the world, you are adrift. Once you break in, life is different.

GETTING SYSTEMATIC ABOUT INTERNSHIPS We needed an internship system; we couldn't afford to be hit or miss like that. My colleague Professor Stubb and I called all our connections. We got very lucky with the Institute of Real Estate Managers, whose members manage large office buildings in the city. Many had similar backgrounds to our students and they were very sympathetic. They let us pitch DeVry University's internship program at their annual meeting. In one year, we've placed about 25 interns through them. Of the first five interns, two got full-time job offers from it, and those who didn't still made very influential contacts. Other companies hear about the program from their peers and call us, so I think the program will expand astronomically in the next few years.

TRANSFORMATIVE TEACHING My job is to transform students. DeVry University trains people to succeed in the corporate environment. They come to us as rookies and they leave as business people. You can see it. A class of freshmen and a class of seniors on campus look like completely different people. We call on former students who are managers at Big 4 accounting firms; when they were here, they were just kids who worked in their dad's dry-cleaner at night. Some of them could barely speak English when they enrolled here.

WHAT DRIVES STUDENTS The main thing that DeVry University students share is the outsider's drive to get inside, to make it. Many students on the Chicago campus come from immigrant families. Their parents often run small businesses where everyone works, making constant sacrifices and living on the edge. They don't want to live that way. They want to have a corporate career that provides benefits and a middle-class status. A small percentage of our students want to be entrepreneurs. The skills we teach will help them succeed at that too.

FINDING A NICHE We prepare students for their first position. For example, the tax departments of accounting firms are the more mechanical, less prestigious parts of the firm. Students from prestige schools tend to avoid these jobs. We saw that as an opening, where our students could learn how to operate in that environment without having to compete immediately with students from more privileged backgrounds that taught them the social skills for client contact.

We started working this niche years ago at Price Waterhouse, DeVry University's auditor at the time. Professor Stubb and Professor Monbrod were instrumental in getting them to hire our top students. Once our students got in, they were able to compete on technical ability. Many successful graduates start in tax departments and rise to general managerial positions.

DIVERSITY HIRING Another angle we work is diversity hires. The most diverse student body in America is right here on the Chicago campus, ready for the work force. Corporations are looking for minorities and particularly for bilingual and international employees because business is global. Are you doing business in China? My students are fluent in Mandarin and Cantonese, and they know the culture. They have an advantage over students from more prestigious schools who have only lived in America, only speak English, and don't know Bulgaria from Romania.

"LIVING ROOM" LEARNING Maybe because of my history training, I teach using original source documents, which in the business world are financial statements. I don't plod through journal entries. Financial statements provide real world information about how things work. That gets students interested and we just talk our way through business problems.

Once a new kid joined the accounting class and after a while he said, "What are you doing here? Nobody raises a hand, but nobody interrupts anybody. You are just all talking." And someone answered, "That's just how it is here. It's like being in the living room with your family and your dad is explaining something to you." It was just what I'd always dreamed teaching would be. That was one of the highlights of my teaching career.

STREET CREED DeVry University started here in Chicago, where some people still think of us as a trade school instead of a business school. "Pioneers in accounting education" is our new reputation among academics, and the business world is beginning to realize that. DeVry University Chicago has been here for over 80 years and we have a lot of good contacts in industry. Our business programs are 15 years old and we are building our contacts and networks.

WHY DEVRY? DeVry University provides excellent courses, but coursework is not our edge. Acculturation is. Our students are not taught by teaching assistants, they are taught by professors. Almost all of our accounting faculty are first-generation Americans; most are the first generation in their family to go to college. We see in our own lives that college can be a transformative experience, and we want to make that happen for our students. We have good graduate employment ratios for all students. We put in an extra effort to help top students get employed in high profile positions, where they can blaze a trail for others. - See more at: http://www.devry.edu/academics/university-faculty/michael-vasilou.html#sthash.F4wfdhIA.dpuf

DIABLO VALLEY COLLEGE

Leonard Chaplin

Best Teacher, Physical Education

DOMINICAN UNIVERSITY OF CALIFORNIA

Ellen Christiansen

Best Researcher/Scholar, Nursing

DRAKE UNIVERSITY

Mary Edrington

Best Teacher, Marketing

Matthew Mitchell

Best Teacher, Business

Matthew is an author, consultant, and assistant professor of international business and strategy at Drake. He also serves as the executive editor of the journal International Business: Research, Teaching, and Practice and is the chair of the largest national chapter of the Academy of International Business.

Matthew earned his Ph.D. in International Business from the University of South Carolina—the world's leading program in international business. His research investigates the relationship between firm strategy politics and religion in different global markets. He has written many conference papers, book chapters, and journal articles on these topics. For example, his most recent research projects have investigated: 1) the returns of conventional vs. Sharia-compliant stock indices, 2) the interpretation of intellectual property protection efforts by religious communities in East Asia, and 3) the evolution of the life insurance industries in India and China among others.

Matthew has traveled, lived, and worked in more than 75 countries and has been invited as a consultant and guest lecturer at companies, NGOs, and universities around the world.

Troy Strader

Best Researcher/Scholar, Business

DREXEL UNIVERSITY

Dali Ma

Best Researcher/Scholar, Business

Fariborz Partovi

Best Overall Faculty Member, Business

Srini Srinivasan

Best Researcher/Scholar, Marketing
Best Teacher, Marketing

Srini Srinivasan has developed unique expertise in the areas of Retailing and Promotions. He has conducted extensive research in Marketing Research and Strategy, Pricing and Promotions, Loyalty and Satisfaction. His research results have appeared in the Journal of Advertising, Journal of Retailing, Journal of Marketing Research, the Journal of Professional and Personal Selling, and the Journal of Product and Brand Management. He has held positions with several firms. Dr. Srinivasan is Associate Professor of Marketing, Drexel University, Philadelphia.

Rajneesh Suri

Best Overall Faculty Member, Marketing
Most Helpful to Students, Marketing

A professor in Drexel's LeBow College of Business, Suri's area of research includes pricing, promotions and branding. He has published numerous papers on consumer behavior, including developing effective guidelines and slogans, the effects of scarcity on the evaluation of prices, the impact of gender differences on the evaluation of promotional emails, and research on online and print coupons.

Joan Weiner

Best Teacher, Business

DUQUESNE UNIVERSITY

Michael Essig

Best Researcher/Scholar, Health Sciences and Nursing
Best Overall Faculty Member, Health Sciences and Nursing

Vincent Giannetti

Best Researcher/Scholar, Pharmacy
Best Teacher, Pharmacy

"The relationship of his teaching to population health, patient care and health outcomes."

Vincent Giannetti is a Professor of Social and Behavioral Sciences in Pharmacy at the School of Pharmacy where he teaches courses in Social and Behavioral Pharmacy, Health Care Ethics, Substance Abuse, Behavioral Health, and Psychiatry. In addition, Dr. Giannetti has taught Organizational Behavior and Health Care Systems in the School of Business Administration and Stress and Health in the Masters of Liberal Studies Program.

He has offered numerous continuing education programs at Duquesne University and the University of Pittsburgh in addition to other organizations.

Dr. Giannetti has published articles concerning substance abuse, mental health, health care ethics and behavioral health. He is the co-editor of two books regarding brief psychotherapy in the Applied Clinical Psychology Series published by Plenum Press. He also contributed chapters to these texts regarding brief approaches to substance abuse counseling and mental health policy and published three additional chapters in textbooks regarding health care ethics, medication problems among the elderly and contemporary workplace issues. His current research is focused upon adherence, quality of life and health outcomes. Dr. Giannetti is a member of The Editorial Board of Social Work in Public Health and The Journal of Evidence Based Social Work published by Hayworth Press. He is a licensed psychologist in Pennsylvania and holds graduate degrees in the fields of psychology, public health, and social work.

Dr. Giannetti has maintained a practice as a psychologist providing assessment and counseling services as well as consultation to both human service agencies and corporations.

Joan Kiel

Best Teacher, Health Sciences and Nursing
Most Helpful to Students, Health Sciences and Nursing

Prior to Duquesne University, Dr. Kiel was an administrator at the Lutheran Medical Center in Brooklyn, New York and the Pittsburgh Mercy Health System. She continues to be active in the community as an analyst in managed care and practice management and has developed and implemented a unique HIPAA training program for physicians and other health care professionals. At Duquesne University, she teaches in both the undergraduate and graduate programs. Dr. Kiel was a 1997 nominee for the Robert Foster Cherry Award for national teaching excellence, and a 1998 recipient of the Duquesne University Creative Teaching Award. She has also been the keynote speaker for Faculty Development Presentations at various schools including: The University of North Carolina, Thomas Jefferson University, Howard University, and East Tennessee State University. Her teaching motto is to "tie theory to practice" and all of her classes include experiential exercises. Dr. Kiel is the author/editor of the book, "Information Technology for the Practicing Physician," and "Healthcare Information Management Systems" - A Practical Guide, and served as section editor for MD Computing. Her next book "Healthcare Information Management Systems" is due out in 2004. Dr. Kiel has numerous publications and national presentations on health management systems' topics. In 2004, Dr. Kiel was awarded the Dean's Award for Excellence in Scholarship and the Duquesne University Apple Polishing Award for Excellence in Employee Performance. Her research interests include utilizing information technology to effectuate organizational objectives, intellectual property, and adaptation to technology. She holds a doctorate degree from New York University.

Christine O'Neill

Best Overall Faculty Member, Pharmacy
Most Helpful to Students, Pharmacy

Charles Rubin

Best Overall Faculty Member, Political Science
Best Teacher, Political Science

Rubin's The Green Crusade: Rethinking the Roots of Environmentalism (1994), is a critical look at key figures of the environmental movement like Rachel Carson, Barry Commoner and Paul Ehrlich. In 2000, he published an edited collection of essays titled Conservation Reconsidered: Nature, Virtue and American Liberal Democracy, containing fresh looks at key figures in the conservation movement and those who influenced them. Since then he has published on a variety of topics at the intersection of science, public policy and political philosophy, e.g., the problem of global climate change, the difficulty of applying the precautionary principle to measures dealing with Earth/asteroid collisions and conceptual flaws in the scientific search for extraterrestrial intelligence. More recently, he has published studies of literary figures ranging from Henry Adams and Flannery O'Connor to Neal Stephenson and Karl Cepak. His forthcoming book, The Progress of Inhumanity, is a critical look at advocates of redesigning human beings, a topic he also blogs about at Futurisms (http://futurisms.thenewatlantis.com/).

Sarah Wallace

Best Overall Faculty Member, Health Science
Best Researcher/Scholar, Health Science
Best Teacher, Health Science
Most Helpful to Students, Health Science
Best Researcher/Scholar, Health Sciences and Nursing
Best Teacher, Health Sciences and Nursing
Most Helpful to Students, Health Sciences and Nursing
Best Overall Faculty Member, Health Sciences and Nursing

Sarah E. Wallace, Ph.D., is an assistant professor in the Department of Speech-Language Pathology at Duquesne University. She teaches courses on aphasia, cognitive-communication disorders, and augmentative and alternative communication (AAC). Her current research interests include semantic interventions to improve word retrieval of people with aphasia as well as development of appropriate AAC strategies for people with aphasia and traumatic brain injury.

DURHAM TECHNICAL COMMUNITY COLLEGE

Micara Lewis

Best Teacher, Education
Best Overall Faculty Member, Education

EAST CAROLINA UNIVERSITY

Laura King

Best Overall Faculty Member, Special Education
Best Researcher/Scholar, Education

Research Interest: Assistive Technology Universal Design for Learning Disability in Education Distance Education in Higher Education

EAST STROUDSBURG UNIVERSITY

Nurum Begum

Most Helpful to Students, Education

EASTERN ILLINOIS UNIVERSITY

Andreea Chiritescu

Best Teacher, Economics

Linda Ghent

Best Overall Faculty Member, Economics
Best Researcher/Scholar, Economics
Best Teacher, Economics
Most Helpful to Students, Economics

Mukti Upadhyay

Best Overall Faculty Member, Economics
Best Researcher/Scholar, Economics
Most Helpful to Students, Economics
Best Researcher/Scholar, Business

Melody Wollan

Most Helpful to Students, Business

EASTERN KENTUCKY UNIVERSITY

Joe Beck

Best Teacher, Environmental Science

On Saturday, August 9th, 2014, beloved EHS Professor, Joe Beck, passed way after suffering a massive heart attack, a result of complications related to end stage renal disease.

Words cannot replace the stories, feelings, or support provided by this man. Professor Beck served Eastern Kentucky University and many others up unto his last day. His contributions to the university, alumni, friends, family, the EHS profession, and many others will be missed and will be greatly appreciated well into the future.

Professor Beck will be fondly remembered for his love of teaching and for his devotion to students. He joined the EHS faculty in 1995 and played a key role in the rapid growth and lofty national reputation of the program. A gifted storyteller in the classroom, he was always available outside of class to help his students in any way possible. They never forgot Professor Beck's advocacy and acts of kindness, typically staying in touch with him long after graduation.

Robert Brubaker

Most Helpful to Students, Psychology

Myra Beth Bundy

Best Teacher, Psychology
Best Overall Faculty Member, Psychology
Most Helpful to Students, Psychology

"Good at making learning accessible/understandable. Commitment."

Dr. Bundy is a Professor of Psychology at Eastern Kentucky University and a licensed psychologist. Her graduate and post-graduate training specialized in developmental disabilities. She interned at the University of North Carolina TEACCH program, a pioneering program in autism intervention. Dr. Bundy coordinates the EKU Developmental Disabilities Specialty clinic, which provides opportunities for EKU graduate students to work with children, adolescents, and adults with Autism Spectrum Disorders and other developmental disabilities. She collaborates with EKU faculty from Occupational Therapy, Communication Disorders, and Special Education to direct the EKU Autism Spectrum Disorder Certificate Program. She collaborates with EKU students to write and conducts research in the area of autism. She enjoys spending time with individuals with autism spectrum disorders and their families. Dr. Bundy is also interested in research and clinical practice related to adoption.

Connie Callahan

Best Researcher/Scholar, Counseling

John Ferguson

Most Helpful to Students, Education

Jonathan Gore

Best Researcher/Scholar, Psychology

"Involves students in research as well as doing his own regularly."

Dr. Jonathan Gore received his Ph.D. in Psychology at Iowa State University, and is an active member of the Society of Personality and Social Psychology and the Southeastern Society of Social Psychologists. He is also the faculty supervisor of the EKU Chapter of the National Society of Collegiate Scholars.

Currently, he is an associate professor in the Psychology Department with a research focus on goal motivation, self-concept, and culture. Since joining the faculty at EKU, Dr. Gore has published over 20 articles and book chapters on these topics as well as topics developed over the years with his students. His research assistants are actively pursuing their own interests, and they have presented at such venues as the Undergraduate Presentation Showcase at EKU, the Kentucky Psychological Association, and the Posters at the Capitol in Frankfort.

Dr. Gore's research interests also correspond with his teaching interests. He teaches social psychology, cultural psychology, and research methods to both undergraduate and graduate students. He also teaches his "Boot Camp" course every summer, which gives students the experience of writing and submitting their own research paper for publication.

Neal Gray

Best Overall Faculty Member, Counseling
Best Teacher, Counseling
Most Helpful to Students, Counseling

Neal D. Gray earned his Ph.D. in Counselor Education from the University of New Orleans. He currently is an Associate Professor and Chair of the School of Counseling at Lenoir-Rhyne University. Before joining the counseling department at LR he spent six years at Eastern Kentucky University. He has counseling experience in both schools and agencies. He is licensed as a Professional Counselor (LPC) in North Carolina and a Certified School Counselor in Louisiana. His research interests include professional identity of counselors, post-master's degree supervision, and school counseling.

Samuel Hinton

Courses Schedule Vita Community Services Links
  Samuel Hinton
  PERSONAL
  Married to Mary Hinton for twenty-three years. Three adult children. Residence: 326 Timothy Way, Richmond, KY 40475. Telephone: Home (859) 626-0099. Work (859) 622-1127. Fax (859) 622-2004 Email: Samuel.Hinton@eku.edu; also hintn@yahoo.com.
  EDUCATION
  D.Min (Doctorate in Ministry) Andersonville Theological Seminary, Camilla, Georgia. 2001.
  Ed.D (Doctorate in Education) University of Virginia, 1981. Charlottesville, Virginia. Major: Foundations of Education. Specialization -Comparative and International Education. Minor International Relations.
  M.Ed. (Master of Education) University of Virginia, 1978. Charlottesville, Virginia. Major: Sociological and Cultural Foundations of Education.
  M.Ed. (Master of Education) Kent State University, Kent, Ohio 1989. Major: Higher Education Administration. Internship in office of international student admissions.
  B.A. (Bachelor of Arts) University of Durham, England, 1967. Attended Fourah-Bay College Freetown, Sierra Leone. Major: English and Economics. Minor: International Relations.

CURRENT POSITION

Professor of Educational Studies, Department of Curriculum and Instruction, Eastern Kentucky University, Richmond, KY. Full Graduate Faculty membership.

TEACHING: Teaching and have taught courses in comparative education, multicultural education, critical thinking in education, philosophy of education, research methods in education, and advanced research methodology at the graduate level. Supervised graduate student research proposals. Directed dissertations of education specialist students. Also teaching and have taught undergraduate courses in introduction to education, school and society, and human development and learning. Proficiency in educational technology (Teaching courses on the Internet and on television). Academic advising.

ADMINISTRATION: Coordinated educational foundations unit for four years working with school leaders, teachers and students in the public school system. Coordinated student services admissions, freshman orientation, and enrolment management program for three years in a major university. Coordinated educational studies program including field experiences for pre-service teachers. Completed a one-year administrative leadership internship – College of Arts and Humanities, Eastern Kentucky University.

SUMMARY OF QUALIFICATIONS

Cumulative work experience

Teaching - Taught at the elementary, secondary, and university levels in Sierra Leone and the United States. Taught adult basic-education and English as a second language in the United States. Extensive teaching experience at the public school and college levels - classroom teaching at the undergraduate, graduate, and post-graduate levels. Experience in curriculum design, distance education (teaching by television and teaching on the Web). Full-time tenured graduate faculty. Counseling - Counseling youth and families in crisis in residential social services agencies. College enrollment management and programming and college recruitment & retention programs in major mid-western university in the United States. Higher education administration – Developed and implemented programs in student services - new student orientation, enrollment management and admissions. Academic administration/coordinated graduate degree program in Virginia, and coordinated educational foundations undergraduate program in Kentucky. Internship in international admissions-Office, Kent State University. Fellowship in college administration - Office of the Dean, College of Arts and Humanities, Eastern Kentucky University. Fulbright - Hays Fellowships to Brazil 1993 and Czech and Slovak Republics 2003. United Nations Development Program Educational Consultant – Milton Margai College of Education, Sierra Leone. Leadership in professional university professor's organization at Eastern Kentucky University. Project Director – Citizenship Education Internship (PiE).

WORK EXPERIENCE

Professor, Educational Studies Department of Curriculum and Instruction. College of Education, Eastern Kentucky University. Present.

Professor, Educational Studies Department of Administration, Counseling and Educational Studies. College of Education, Eastern Kentucky University. 1995–2003.

Associate Professor, Department of Administration, Counseling, and Educational Studies. Full graduate Faculty membership 1993–1995.

Assistant Professor, Department of Administration, Counseling, and Educational Studies, Eastern Kentucky University 1990–93.

Program Officer, New Student Orientation,/Coordinator, Early Intervention – Office of Admissions. Kent State University, Kent, 1987–1989.

Senior Staff Associate. Foundations of Education University of Virginia, Falls Church, Virginia 1986–1987.

Teacher, English As A Second Language. Arlington Public Schools, Arlington Virginia. 1980–1983

Youth and Family Counselor, Southern Area Youth Services, Friendly, Maryland. 1977–1980.

Teacher - Sierra Leone Grammar School/Instructor Freetown Technical Institute. Employer - Ministry of Education, Sierra Leone, West Africa. 1967–1977.

PROFESSIONAL ORGANIZATIONS

Comparative and International Education Society
American Educational Research Association
American Association of University Professors
American Association of Colleges of Teacher Education (Institutional Representative)
Mid-South Educational Research Association
Phi Delta Kappa.

HONORS AND AWARDS

Fulbright - Hays Award (Brazil), 1993.
Fulbright - Hays Award (Czec and Slvak Republics, 2003.
Kappa Delta Pi - Academic Honor Society 1989.
Cultural Diversity Award Kent State University Division of Student Affairs 1989. (For furthering cultural diversity in new student orientation programming).
Phi Delta Kappa - Academic Honor Society 1990.
Phi Delta Kappa Leadership Training 1991.
Minority Caucus Award Louisville, Kentucky Gas & Electric Company 1993.

LEADERSHIP IN PROFESSIONAL ORGANIZATIONS

President - American Association of University Professors Eastern Kentucky University Branch 1993–97.

Member: Statewide Multicultural Education Task Force - Kentucky Department of Education 1995.

American Association of Colleges of Teacher Education – AACTE. Committee on Global and International Teacher Education (Member). 2000–2003

BOARD MEMBERSHIPS

Madison County Kentucky Literacy Council (Project Read) 1990–2000.
Madison County Adult Education 1995–1997.
Leonenet Street Children Project Inc. A Project providing foster-care scholarships for child-victims of war in Sierra Leone. Founder, and President 1996–2003.

REVIEWED ACADEMIC MANUSCRIPTS FOR THE FOLLOWING

Conference proposals for the Mid-South Educational Research Association, 1990, 1991, 1992, 1994.

Action in Teacher Education 1993–1994, 1996.

Research in the Schools. Journal of the Mid-South Educational Research Association. 1993–95. Editorial Board Member.

Action In Teacher Education, 1994,95, 1995/96, 1997. Editorial Board Member.

NCEOA JOURNAL: Journal of the National Council of Educational Opportunity Associations 1993–95. Editorial Board member.

Reviewed Duane E. Campbell (1996). Choosing Democracy: A Guide to Multicultural Education. Merrill, Prentice Hall, Englewood Cliff. (Text).

Segal, William, E (1997). Introduction to Education, Columbus, Ohio, Merrill. (Text).

Gollnick, Donna, M and Chinn, Philip, C (1997). Multicultural Education in a Pluralistic Society (Fifth Edition). Columbus, Ohio, Merrill. (Text)

Quintero, Elizabeth P, and Rummel, Mary, Kay (1997). American Voices: Webs of Diversity. Columbus, Ohio, Merrill. (Text).

Reader: Fulbright-Hays applications for 1995/1996/1998. United States Department of Education, Washington, D.C. December 1995.

REFERENCES

Available upon request.

PUBLICATIONS

Hinton, Samuel. (1982). Education, Social Responsibility and the University Student In Africa: Case Study on Sierra Leone. Journal of Abstracts in International Education. 11. (2). 54–56.

Hinton, Samuel, (1982). Intellectual Commentaries on the Probable Impact of Western Education on Africans. Journal Of Abstract In International Education. 11. (1). 25–37.

Hinton, Samuel, (1988). Graveyard Shift. Poem in American Poetry Anthology, 8 (4) 13.

Hinton, Samuel, (1988). Book Review. Michael The well. Duties, Pleasures and Conflicts: Essays In Struggle. In Explorations in Sight and Sound. National Association of Ethnic Studies. 87–88.

Hinton, Samuel, (1989). Book Review. John Craig-Holt. Ethnic 1. In Explorations In Sight and Sound. National Association of Ethnic Studies, 20.

Hinton, Samuel, (1990). Outcomes Assessment In Student Affairs. Division Dialogue: Student Affairs Newsletter. Kent State University, Kent, Ohio.

Hinton, Samuel, (1991). Education For The People: Old Concepts and New Realities In Sub-Saharan Africa. Journal Of Abstracts In International Education. 18, (1).

Hinton, Samuel, (1991). Ethnicity In The United States: Dehomogenizing Blackness In American Society. Ethnic Forum: Journal Of Ethnic Studies And Ethnic Bibliography. 11, (1) 31–38.

Hinton, Samuel, (1991). Educating For Diversity: A Module For Freshmen Orientation. (Summary). Proceedings Of the 21st Annual Conference of the National Association For Ethnic Studies, In Explorations In Ethnic Studies 14, (2) 77–78.

Hinton, Samuel, (1991). Thinking About College: A First Step in Early Intervention. Resources In Education, Washington D.C. The George Washington University. (ERIC) No. ED. 341341.

Hinton, Samuel, (1992). Capital City - Poem. EKU International Magazine, p.18. Creative.

Hinton, Samuel, and Stockburger, Muriel, (1992). Personality Trait and Professional Choice: A First Step In Early Intervention. Resources In Education, Washington D.C. The George Washington University. (ERIC Clearinghouse on Higher Education), No. ED341672.

Hinton, Samuel, (1992). Political Participation and Perceived Role. A Self-Report of Students At The University Of Sierra Leone. Abstracts of Proceedings. Comparative and International Education Society. 36th Annual Meeting. Annapolis, Maryland.

Hinton, Samuel, (1992). AMERICA 2000: A Review of the Literature. Abstracts of Proceedings. Comparative and International Education Society 36th Annual Meeting, Annapolis, Maryland.

Hinton, Samuel, (1993). Supplementary schooling in Japan, and Sierra Leone. Abstracts of Proceedings. Comparative and International Education Society 37th Annual Meeting, Kingston, Jamaica.

Hinton, Samuel and Lanier, Linda, (1993). At-Risk Urban and Rural High School Students Ratings of A College Campus Visitation Program" The High School Journal, Vol. 76, No.3. pp.221–225.

Hinton, Samuel, (1993). "The Learning Style Preferences of Students in Graduate School". Resources in Education, Document No. ED354807, July, Washington DC. The George Washington University ERIC Clearinghouse on Higher Education.

Hinton, Samuel and Stockburger, Muriel, (1993, Fall). An Assessment of the Relationship Between Personality Trait and Professional Choice Among Preservice Teachers, Using the Myers-Briggs Type Indicator. Eastern Kentucky University Educational Review. Vol. 17, No.1, pp.26–34.

Addington, Brenda, and Hinton, Samuel, (1993). Developmentally Appropriate Practices: A Survey of Primary School Teachers. Resources in Education. ERIC Document No. ED368494.

Hinton, Samuel, (1993). Thinking About College: A First step in Early Intervention. NCEOA Journal: Journal of the National Council of Educational Opportunity Associations. Vol.8, (I), pp.14–18.

Hinton, Samuel, (1993). Multicultural Challenges in Higher Education: Some Suggestions for U.S. Educators. Ethnic Forum: Journal of Ethnic Studies and Ethnic Bibliography. Vol.13, No 1, pp.28–33.

Hinton, Samuel, (1994). Education in Brazil: Some Issues on Reform in Report: Fulbright-Hays Seminar on History and Culture of Brazil, 1993. Brasilia, Brazil, Fulbright Commission, Brasilia.

Hinton, Samuel, (1996). Reforming Teacher Education in Brazil: Issues and Prospects. Eastern Kentucky University Educational Review. Vol.19, No.2, pp.21–34.

Hinton, Samuel (1996). The Need For Consciousness-Raising and Community Education in Sub-Saharan Africa. Proceedings of the Annual Meeting of the Comparative and International Education Society. Williamsburg, Virginia, March 6–10.

Hinton, Samuel (1996). From Gullah to Krio: Language Convergence in St. Helena and Sierra Leone. Proceedings of the Annual Meeting of the Comparative and International Education Society. Williamsburg, Virginia, March 6–10.

Hinton, Samuel, and Oleka, Sam (1996). College Students' Assessment of Teaching By Television. Proceedings of the 25th Annual Meeting of the Mid-South Educational Research Association. Tuscaloosa, Alabama, November 6–8.

Hinton, Samuel (1997). Instructional Technology and School Reform Proceedings of the Annual Meeting of the Comparative and International Education Society. Mexico City, Mexico, March 10–15.

Hinton, Samuel (1997). Resource Groups For Sub-Saharan Africa. Social Education, pp.423–425, November/December.

Hinton, Samuel, and Oleka, Sam (1997). College Students' Assessment of Teaching By Television. ERIC Resources In Education.

Hinton, Samuel (1997). The Leonenet Street Children Project: Using Technology in Social Development. Proceedings of the Annual Meeting of the Comparative and International Education Society. Mexico City, Mexico, March 10–15.

Al-Khatab, Anisa, Gabbard, Carol, Hinton, Samuel, and Pullins, Eddie (Editors), (1997). EDF 103 Early Field experiences Handbook. Eastern Kentucky University, Richmond, Kentucky.

Acker, Dean, Downing, Jan, Hinton, Samuel (1997). EDF 203 Field Experiences Handbook. Eastern Kentucky University, Richmond, Kentucky.

Hinton, Samuel (1997). Resource Groups For Sub-Saharan Africa. Social Education, pp.423–425, November/December.

Hinton, Samuel (1998). Adventure is in store when students are invited to plan the itinerary. Book Review of Student- Oriented Curriculum: Asking The Right Questions (1995). Alexander, Wallace, M, McAvoy, Kathy and Carr, Dennis. National Middle School Association, Columbus, Ohio. In Journal Of Staff Development, National Staff Development Council.

Hinton, Samuel (1998). Consciousness-Raising and Community Education In Africa: Musings on Sierra Leone. Proceedings of the annual meeting of The Comparative and International Education Society. Buffalo, New York, March 18–20.

Hinton, Samuel (1998). Learning Cultures and Distance Education: A Comparative Assessment. Proceedings of the annual meeting of The Comparative and International Education Society. Buffalo, New York, March 18–20.

Hinton, Samuel, and Downing, Jan (1998).Team-Teaching A College Core Foundations Course: Comparison of Instructors' and students' Assessments. Proceedings of the Twenty Seventh Annual Meeting of the Mid-South Educational Research Association. New Orleans, Louisiana, November 5.

Hinton, Samuel (1999). School reform and multiculturalism in the classroom. The Kentucky Example. Proceedings of the Annual Meeting of The Comparative and International Education Society. Toronto, Canada, April 14–18.

Hinton, Samuel (1999). Perspectives on western education by the African writer. Proceedings of the Annual Meeting of The Comparative and International Education Society. Toronto, Canada, April 14–18.

Hinton, Samuel (1999). School Reform, Multiculturalism and Equal Educational Opportunity: Classroom Possibilities. Education and Society: International Journal in Education and Sociology. Vol. 17, No. 2, 1999. pp.-18

Hinton, Samuel (2000). Information Age Technology and School Reform: The Kentucky Paradigm. Proceedings of the International Conference on Designing Education For The Learning Society. SLO. Netherlands Institute of Technology Development. Enschede, The Netherlands. November 5–8.

Hinton, Samuel (2001). Information And Communication Technology (ICT), School Reform and Teacher Standards: A Triadic Connection. Abstracts of the 45 Annual Conference of the Comparative and International Education Society. Washington, D.C. March 14–17.

Hinton, Samuel (2002). "Poverty in Sierra Leone: Vignettes from the Leonenet Street Children Project". Book Chapter in 'Suffer the Little Children:' National and International Dimensions of Child Poverty and Policy. Elsevier Inc., Oxford, England. In Progress.

Hinton, Samuel (2002). University Student Protests and Political Change In Sierra Leone. The Edwin Mellen Press, Lewiston, New York.

Hinton, Samuel (2003).Test Critique. "Management Development Questionnaire" in The Fifteenth Mental Measurements Yearbook, Buros Institute Of Mental Measurements, Lincoln, Nebraska.Lincoln.

Hinton, Samuel (2003).

Hinton, Samuel (2003).The Road To Kenema: And Other Poems. Africa Future Publishers, Schriesheim, Germany.

SPECIALIST IN EDUCATION SUPERVISED COMPLETED THESES

Don Cravens - "Principals' Attitudes Toward Block -Scheduling" (1998).

Patricia Cox - "Relationships between Primary Hands-on Science Instruction and Kentucky Instructional Results Information System (KIRIS) Science Scores" (1999).

Jimmy Pack - "Technology Integration by Teachers In the Region Eight Service Center Area of Kentucky" (1999).

Elizabeth Norris-Hieronymous - "The Impact of Junior Mentor Courses on Eleventh Grade Student Test Performance on the Kentucky Instructional Results Information Test" (1999).

Robin, Richmond Mason - "High School Drop Out and Retention Rates of Kentucky School Districts" (2000).

Wilds, Sonya – "Effect of Extended School Services on The Reading Achievement of Primary Students."
2001.

Willis, Carla – "Using Multiple Intelligence To Teach Primary Grade Mathematics". 2001.

Bowling, Dessie – "The Effectiveness of Seventh Grade Students Using Word Processors to Complete Written Compositions. 2003.

Work in Progress

Hinton, Samuel. THE CULTURAL TRADITIONS OF THE GULLAHS OF ST. HELENA ISLAND, SOUTH CAROLINA. I applied for a National Endowment For The Humanities grant in 1995 to research this topic. Not funded. Would like to pursue this again.

Hinton, Samuel. TRANSFORMATIONAL TEACHER LEADERSHIP IN SIERRA LEONE. A United Nations Development Project grant enabled me to start this project. I spent eight weeks at Milton Margai Teacher's College from May–June 1997 and 1998 working with school teachers. A coup aborted the continuation of this project.

WORKSHOPS PRESENTED

Hinton, Samuel, (1992). Using Basal Readers. Literacy Council Of Madison County, Training Of Tutors Workshop. Richmond, Kentucky.

Hinton, Samuel, (1992). Introduction To IBM Computers For Teachers. Educational Excellence Consortium - Eastern Kentucky University, and Pulaski County Schools, Kentucky.

Hinton, Samuel, (1992). Aesthetics In Education: The Oodi Weavers Of Botswana. University of Virginia, Falls Church, Virginia Regional Center.

Hinton, Samuel, (1992). Joining In The American Conversation: A Multicultural Education Workshop. Camp Anytown, Kentuckiana. National Conference Of Christians and Jews, Louisville, KY.

Hinton, Samuel, (1992). Diversity in the Workplace: Empowerment For increased Productivity. Louisville Gas and Electric Company and Minority Caucus Committee. Louisville, Kentucky.

Hinton, Samuel, (1997). Shaping Campus Climate and Culture Through Academics and Student Services. Kentucky Association of Blacks in Higher Education. Richmond Kentucky.

Hinton, Samuel (1997). Educational Planning and Economic Development. Workshop Series presented at Milton Margai College of Education, University Of Sierra Leone.

Hinton, Samuel (1998). Educational Planning and Economic Development. Workshop Series presented at Milton Margai College of Education, University Of Sierra Leone.

Hinton, Samuel (2001). Adult Literacy In Eastern Kentucky: The Work Of Project Read. Episcopal Diocese of Lexington Bishop's Conference. Eastern Kentucky University, October, 2001.

GRANT ACTIVITY

Hinton, Samuel and Stockburger, Muriel, (1992). Pilot Study using the Myers Briggs Type Inventory-Personality Trait and Professional Choice Among Preservice Teachers in Eastern Kentucky. Funded by the Office For Educational Research and Evaluation, Eastern Kentucky University.

Henson, Kenneth, and Hinton, Samuel, (1993). Minority Recruitment and Retention Scholarship Proposal, Kentucky Department of Education.

Hinton, Samuel (1993). Fulbright - Hays Study Seminar Grant. History and Culture of Brazil.

—(1995). National Endowment For The Humanities Summer University Teacher Stipend - "Cultural Renaissance Among The Gullahs of St. Helena Island, South Carolina. Not funded.

—(2000) Provost's Professional Development Grant, Eastern Kentucky University for attending Summer Online computer workshop, Eastern Kentucky University.

—(2001) Provost's Professional Development Grant, Eastern Kentucky University for travel and research - British Library, London, England.

Hinton, Samuel (2003). Fulbright - Hays Study Seminars. Czech and Slovak Republics.

Hinton, Samuel and Brickley, Bethany(2003). American Councils for International Education. Partners in Education Grant to host seven Ukrainian Educators for five weeks in a Citizen Education Seminar Series.

PRESENTATIONS

Hinton, Samuel, (1989). "Minority Students Attitudes Towards Pre- Orientation In a Predominantly White Campus." National Orientation Directors Association National Conference. Traverse City, Michigan.

—(1989, March). "Cultural Diversity: A Module for the Orientation Course." National Orientation Directors Workshop. Otterbein College, Westerville, Ohio.

Hinton, Samuel (1990, March). "Pathways In Sub-Saharan African Education: Reviewing the Past, Acknowledging the Present, and Proposing the Future." Comparative and International Education Society Annual Conference. Anaheim, California.

Hinton, Samuel(1990, October) "Orientation, Early Awareness, and the Precollegiate Focus: A Cooperative School and University Model." National Orientation Directors Association Conference. Hartford, Connecticut.

Hinton, Samuel(1991, March) "Educating for Diversity. A Module for Orientation Instructors." Annual meeting of the National Association for Ethnic Studies Cal Polytechnic and State University, Pomona, California.

Hinton, Samuel, (1991, October 4–5). "Learning: A Self Report by Preservice Teachers in Eastern Kentucky. 21st Annual Conference of the Mid West Association of Teachers of Educational Psychology. Richmond, Kentucky.

Hinton, Samuel and Chapman, Ann (1991, October). "Multiculturalism: What School Counselors Ought to Know", with Ann Chapman. Conference-Kentucky Association for Counseling and Development. Louisville, Kentucky.

Hinton, Samuel, (1991, October) "Thinking About College: A First Step in Early Intervention." Mid-South Educational Research Association Annual Conference. Lexington, Kentucky.

Hinton, Samuel, and Stockburger, Muriel, (1991, October). "Personality Trait and Professional Choice: A Case Study of Preservice Teachers in Kentucky." Mid-South Educational Research Association. Lexington, Kentucky.

Hinton, Samuel, (1992, March). "Political Participation and Perceived Role: A Self Report by Students at the University of Sierra Leone." Comparative and International Education Society Annual Conference. Annapolis, Maryland.

Hinton, Samuel, (1992, March). "America 2000: A Comparative Literature Review." Comparative and International Education Society Annual Conference. Annapolis, Maryland.

Hinton, Samuel(1992, April). "The Kentucky Education Reform Act: The Challenge of Multiculturalism." College of Arts and Humanities Annual Maywood's Symposium. Eastern Kentucky University, Maywoods, Kentucky.

—(1992, October). "What Is Learning Style Theory Doing In My Classroom?" 22nd Annual Meeting of the Mid-West Association of Teachers of Education Psychology. Highland Heights, Kentucky.

—(1992, November). "The Learning Style Preferences of Graduate Students." Mid-South Educational Research Association Annual Conference. Knoxville, Tennessee.

—(1993, March). "Supplementary Schooling in Japan and Sierra Leone". Annual Meeting of the Comparative and International Education Society, Kingston, Jamaica.

Addington, Brenda, and Hinton, Samuel, (1993, October). "Developmentally Appropriate Practices in the Primary Program: A Survey of Primary School Teachers". Kentucky Association of Teacher Educators. Lexington, Kentucky.

Addington, Brenda and Hinton, Samuel, (1993, November). "Developmentally Appropriate Practices in the Primary Program: A Survey of Primary School Teachers". Paper presented at the annual conference of the Mid-South Educational Research Association. New Orleans, Louisiana.

Pitman, Janean and Hinton, Samuel, (1993, October). "Children Attitudes Towards School Reform: A Focus On Kentucky. Paper presented at the annual conference of the Kentucky Association of Teacher Educators. Lexington, Kentucky.

Pitman, Janean, and Hinton, Samuel, (1993, November). "Children Attitudes Towards School Reform: A Focus On Kentucky. Paper presented at the annual conference of the Mid-South Educational Research Association. New Orleans, Louisiana.

Hinton, Samuel, (1994, March). "Education In Brazil: Some Issues on Reform". Comparative and International Education Conference. San Diego, California.

Hinton, Samuel (1995). Kentucky Education Reform: Helping the Monocultural Teacher In a Diversified Classroom. Paper presented at the Conference - Diversity Spoken Here. Morehead State University, March 23.

Hinton, Samuel, and Harris. Margaret (1995). Globalizing Social Studies in the American Classroom: Making Africa Relevant. Paper presented at the annual meeting of the Comparative and International Education Society. Boston, Massachusetts, March 28.

Hinton, Samuel (1996). The Need For Consciousness-Raising and Community Education in Sub-Saharan Africa. Paper presented at the annual meeting of the Comparative and International Education Society. Williamsburg, Virginia, March 6–10.

Hinton, Samuel (1996). From Gullah to Krio: Language Convergence in St. Helena and Sierra Leone. Paper presented at the annual meeting of the Comparative and International Education Society. Williamsburg, Virginia, March 6–10.

Hinton, Samuel (1997). Leonenet Street Children Project Paper presented at the Annual Meeting of the Comparative and International Education Society. Mexico City, Mexico, March 10–15.

Hinton, Samuel (1997). Technology and School Reform. Presented at the Annual Meeting of the Comparative and International Education Society. Mexico City, Mexico, March 10–15.

Hinton, Samuel (1998). Learning Cultures and Distance Education: A Comparative Assessment. Presented at the annual meeting of The Comparative and International Education Society. Buffalo, New York, March 4.

Hinton, Samuel, and Downing, Jan (1998).Team-Teaching A College Core Foundations Course: Comparison of Instructors' and students' Assessments. Presented at the Twenty Seventh Annual Meeting of the Mid-South Educational Research Association. New Orleans, Louisiana, November 5.

Hinton, Samuel (1999). School Reform: Its Relevance To Multiculturalism And Equal Educational Opportunity. Annual Conference of the Comparative and International Education Society, Toronto, Canada, April 14–18.

Hinton, Samuel (1999). Perspectives on Western Formal Education by the African Writer. Annual Conference of the Comparative and International Education Society, Toronto, Canada, April 14–18.

Hinton, Samuel (2000). Information Age Technology and School Reform: The Kentucky Paradigm. International Conference on Designing Education For The Learning Society. SLO. Netherlands Institute of Technology Development. Enschede, The Netherlands. November 5–8.

Hinton, Samuel (2001). Adult Literacy In Eastern Kentucky: The Work Of Project Read. Episcopal Diocese of Lexington Bishop's Conference. Eastern Kentucky University, October, 2001.

Hinton, Samuel (2002). The marginalization of street children in Sierra Leone. Annual Conference of the Comparative and International Education Society, Toronto, Canada, April 14–18. Orlando, Florida, March 6–9, 2002.

PROFESSIONAL DEVELOPMENT WORKSHOPS AND CLASSES ATTENDED

Alternatives To The Lecture. Leader - Prof. Jack Hillwig, Department of Mass Communications, Eastern Kentucky University. Fall, 1991.

Writing For Publication - Dr. Kenneth Henson. Dean, College of Education. Eastern Kentucky University. Fall, 1991. Grantwriting. Leaders - Evans Tracy, Director of University Grants Office, and Dr. John Rowlett, Vice President, Eastern Kentucky University. Fall, 1991.

Writing For Publication. American Association of Colleges of Teacher Education. San Antonio, Texas. March, 1992.

Writing For Publication. 3 hour graduate course. Dr. Kenneth Henson. Dean, College of Education. Eastern Kentucky University. Fall, 1992.

Multicultural Education: Trainers Workshop. Effective Schools of Pascagoula Mississippi, Fall, 1992.

Teaching Consultation Process, Eastern Kentucky University. Faculty Consultant - Dr. Hal Blythe, Department of English, EKU. Spring, 1994.

Multimedia Authoring. Class taught by Mr. Kevin Wallace, EKU Telecommunications Networking Analyst. Fall, 1995.

Kentucky Teacher Intern Program: Teacher - Educator training. January. 1995. Certified as an Intern Teacher-Observer.

Teaching By Television (Distance Education). Division of Media Resources Television Section, Eastern Kentucky University: Workshop by Dr. Larry C. Bobbert and Dr. Fred Kolloff. May 16,1995.

Teaching By Television(Compressed Video Orientation). Dr. Fred Kolloff, Division of Media Resources, Eastern Kentucky University, January 12, 1996.

Inclusion: An In-Depth Look At One Model. Phi Delta Kappa Professional Development Seminar, presented by Pulaski County Schools, Kentucky. EKU Model Laboratory School Auditorium. March 12,1996.

Power Point in Teaching By Television. Division of Media Resources, Eastern Kentucky University. April 17, 1996.

Teaching on the Web using Web Course In A Box. Eastern Kentucky University Summer Technology Institute. 1997.

Teaching on the Web using Web Course In A Box. Eastern Kentucky University Summer Technology Institute. 1998

Teaching on the Web using BLACKBOARD Eastern Kentucky University Summer Technology Institute. 1999.

Teaching on the Web using BLACKBOARD Eastern Ken

Stephen Kappeler

Most Helpful to Students, Criminal Justice

Stephen Kappeler is a lecturer/instructor at the Corbin and regional campuses. Mr. Kappeler has been with the Department of Criminal Justice and Police Studies since 2005. In addition to instructing courses, he also serves as academic advisor to all our criminal justice and police studies majors at all our regional campuses. Mr. Kappeler received his B.S. in Criminal Justice from Central Missouri State University and his M.A. in Criminal Justice Administration from Radford University.

Victor Kappeler

Best Teacher, Criminal Justice

Laurie Larkin

Most Helpful to Students, Health Sciences and Nursing

Dr. Laurie Larkin, assistant professor of public health, was named the KAHPERD 2014 College/University Health Education Teacher of the Year. A member of the EKU faculty since 2010, she holds a bachelor's degree from the University of Wisconsin-River Hills, a master's degree from the University of Wisconsin-La Crosse, and a doctoral degree from Purdue University.

Betsy Matthews

Best Overall Faculty Member, Criminal Justice

Dr. Betsy Matthews' primary areas of focus are community corrections and correctional rehabilitation. She has published several articles and book chapters on both of these issues.

Dr. Matthews joined the EKU faculty in 1999 and received her Ph.D. in criminal justice from the University of Cincinnati in 2003. Dr. Matthews has a blend of practical and academic experience. She began her career as a child care worker in a residential treatment facility for behaviorally disordered adolescents before moving into an adult probation officer position in Greene County, Ohio. After earning her master's degree, Dr. Matthews accepted a position with the American Probation and Parole Association, serving as a research associate on federally funded grant projects.

Robert Mitchell

Best Researcher/Scholar, Psychology

Rose Perrine

Best Overall Faculty Member, Psychology

Gary Potter

Best Overall Faculty Member, Criminal Justice
Best Researcher/Scholar, Criminal Justice

Troy Rawlins

Most Helpful to Students, Loss Prevention & Safety

"As a new faculty member Dr. Rawlins is showing his true character toward his students. He engages his students, recruits new partnerships, and recruits new students through college and career readiness partnerships."

Dr. Troy A. Rawlins Email: Troy.Rawlins@eku.edu
    Biographical Information Assistant Professor, Department of Safety, Security & Emergency Management Eastern Kentucky University
    Formal Education: B.S. Kentucky State University, Frankfort, KY, M.P.A. Kentucky State University, Frankfort, KY, Ed.D. Spalding University, Louisville, KY
    Dr. Rawlins began his career in safety and health within the Kentucky Occupational Safety and Health Program (KYOSHP) as an Industrial Hygiene (IH) Consultant Senior in Frankfort, KY. During his tenure (2001–2005) he consulted on over 500 voluntary consultative health surveys with construction and general industry private and public sector employers within the Commonwealth of Kentucky. He also worked as a member of KYOSHP consultant on

Voluntary Protection Partnership sites (VPP) and Safety and Health Achievement Recognition Partnerships (SHARPS) towards achieve recognition of the best of the best companies in safety and health.

Judah Schept

Best Researcher/Scholar, Criminal Justice

Judah Schept is an Assistant Professor of Justice Studies at Eastern Kentucky University. He holds a Ph.D. in Criminal Justice from Indiana University and a BA in Sociology from Vassar College. Judah is under contract with New York University Press for his book, Left and Locking Up: Capital Departures, (Neo) Liberal Politics, and Carceral Expansion. In addition, he has articles in print and forthcoming in journals such as Radical Criminology, Theoretical Criminology, Social Justice, and the Journal of Criminal Justice and Popular Culture, and a book chapter in an edited volume on popular culture and criminology. Judah is an invited blogger for the Reclaiming Justice Network based out of the Centre for Crime and Justice Studies in the United Kingdom.

Judah is a scholar-activist whose community organizing informs his research and teaching. Judah practices a critical interdisciplinary scholarship drawing from diverse academic literature including American studies, critical and cultural criminology, cultural anthropology, cultural and political geography, the sociology of punishment, and postcolonial studies. Judah is particularly interested in the political economies, geographical histories, and cultural politics of various sites of the prison industrial complex. His current research projects include examinations of jail growth and resistance to it in a small and progressive city, prison expansion in Appalachia, and the rise of penal tourism in Kentucky.

Irina Soderstrom

Best Teacher, Criminal Justice

Dr. Irina R. Soderstrom is an associate professor in the Department of Criminal Justice. She received her B.A. in sociology/pre-law at the University of Illinois, Urbana-Champaign in 1987. She received her M.S. in administration of justice in 1990 and her Ph.D. in educational psychology/statistics and measurement in 1997 from Southern Illinois University at Carbondale.

Her primary teaching interests include statistics, research methods and research seminar courses at both the undergraduate and graduate levels. Her primary research focus is in program evaluation and she has conducted considerable evaluative research on parole programs, boot camps, correctional industries, teen courts and school safety.

Linda Wray

Best Overall Faculty Member, Nursing

EASTERN MICHIGAN UNIVERSITY

Martha Baiyee

Best Teacher, Education

Dr. Baiyee is a professor of education. She completed her BS., MS., and Ph. D. at Tuskegee University, Ball State University, and Virginia Polytechnic Institute and State University, respectively. She has been teaching at EMU since 1989.

Joe Bishop

Best Teacher, Education

Laurie C. Blondy

Best Teacher, Nursing
Most Helpful to Students, Nursing
Best Teacher, Health Sciences and Nursing
Most Helpful to Students, Health Sciences and Nursing

"Dr. Blondy is an amazing and innovative educator. She understands curriculum, teaching and learning theories and processes. She always has very clear presentations and a variety of learning activities and strategies to support student learning. She goes the extra mile for her students, meeting with them and supporting them in class, as an advisor and a mentor. Dr. Blondy truly cares about her students and values their input in her quest for continuous quality improvement.

Dr. Blondy is a sincere student advocate. She understands and respects the rules and policies of the University and educates students about them. She meets with students in office hours and is available answering emails and phone calls even on nights and weekends. She always ensures that each of her advisees is on the right track and provided with the most current and accurate information possible. She assists them with scholarships, fellowships and mentoring. Whether as a course instructor, an advisor or a program coordinator, Dr. Blondy gives her all to make sure students have a positive experience here at EMU."

Sherry Bumpus

Best Researcher/Scholar, Nursing
Best Teacher, Nursing
Best Researcher/Scholar, Health Sciences and Nursing

"As a researcher Dr. Bumpus has a comprehensive program of research that is multidisciplinary, multi institutional and employs both quantitative and qualitative study. She works to improve care transitions for cardiac patients and patients with fibromuscular dysplasia. In addition to her own program of research, she volunteers over the summer to mentor undergraduate and graduate students in a research fellowship program at the Michigan Cardiovascular Outcomes Research and Reporting Program. Each year she mentors an EMU student in the fellowship and regularly supports students for the Undergraduate Symposium, as well as helps students develop research, investigate and present findings at local and national conferences such as The Midwest Nursing Research Association, and The American Heart Association Quality of Care and Outcomes Research. She is well liked by students and has a passion not only for improving patient care, but for sharing research experiences with students."

Robert Carpenter

Best Researcher/Scholar, Education

I serve as the director of the Educational Studies doctoral program and a faculty member in Educational Psychology. I believe scholarship is a valuable avenue to affect the social change necessary for a more just, equitable, and humane existence. My varied interests are rooted in my experience as a classroom teacher, community activist, and a person living in our complex socio-political milieu. I work with colleagues from a variety of institutions, such as the National Institute of Education in Singapore, Loyola University-Baltimore, and the University of Michigan and have been honored with the Outstanding Faculty Mentor Award (2009) and the Dean's Award for Creative Scholarship in 2008.

Caroline Gould

Best Overall Faculty Member, Education
Best Teacher, Education
Best Overall Faculty Member, Education
Best Teacher, Education

Sandra Hines

Best Overall Faculty Member, Nursing

Marilyn Horace-Moore

Most Helpful to Students, Criminal Justice

Marilyn Horace-Moore (MA, Eastern Michigan University 1989) is a Lecturer B, Internship Coordinator and Undergraduate Criminology/Criminal Justice Advisor in the Department of Sociology, Anthropology, and Criminology. She has been teaching Criminology and Sociology courses at Eastern Michigan University since 1994. Prior to becoming a lecturer and advisor in the department, Ms. Horace-Moore was a Police Officer with the City of Ypsilanti Police Department, retiring as a Police Lieutenant in 1994. A few of her research issues include, race, discrimination issues and career advising. She work with students at Fast Track, athletic recruiting and believes in equal opportunity for all. I will celebrate 37 years of marriage, have two children, 11 grandchildren and enjoy fishing, traveling and meeting new people.

Sylvia Jones

Best Overall Faculty Member, Education
Most Helpful to Students, Education

Pat Pokay

Best Researcher/Scholar, Education

Phil Smith

Most Helpful to Students, Education

Mary Vielhaber

Best Teacher, Management

Dr. Vielhaber served two years as an Acting Associate Graduate Dean, and three years on a special assignment in the President's Office at Eastern Michigan University. She has served as the co-director (with Dr. Fraya Wagner-Marsh) of our Master's program in Human Resources and Organizational Development. Dr. Vielhaber co-authored with Dr. Rick Camp a book, Strategic Interviewing: How to Hire and Keep Good People, published by Jossey-Bass. For nearly twenty years, Dr. Vielhaber has been on the faculty of The Executive Education Program at the University of Michigan. She has also provided executive coaching and communication consulting for a variety of clients.

Stephen Wellinski

Best Overall Faculty Member, Education

"While teaching and assisting all the students who count on him, he has done an amazing job helping the department and the school tread through the difficult politics of the EAA — all because he really cares about education and teachers."

Dr. Stephen Wellinski is an Associate Professor in the Department of Teacher Education at Eastern Michigan University, where he teaches undergraduate and graduate literacy courses. His teaching often focuses on secondary content area literacy. As a high school teacher in Indiana, Steve had the opportunity to work on a renowned research project funded by the National Reading Research Center that focused on supporting struggling adolescent readers. This opportunity led him back to his Alma Mater, Purdue University, to work on his doctorate in Literacy and Language. Dr. Wellinski's research focus is looking at the adolescent literacy practices inside and outside of school. Layered within this focus, is the idea of helping secondary teachers on ways to best support adolescent learners.

Donna (Kay) Woodiel

Best Overall Faculty Member, Health Science

Dr. Woodiel has been at Eastern Michigan University for 15 years and is serving as the Graduate Coordinator for Health Education. She has received 2 teaching awards including the Alumni Teaching Award and the Holman Learning Center Teaching Award. She has been recognized as a Woman of Excellence by the EMU Women's Center and as a Role Model and Mentor by the LGBT Resource Center. She has received the Provost's Distinguished Faculty Award for Service. She received the Michigan Campus Compact Faculty/Staff Community Service-Learning Award and was a finalist for the Governor's Service Award-Volunteer category in 2009. She has received 3 Gold Medallions from the Division of Academic and Student Affairs. She chairs the Women's Commission, is a member of the MLK Planning Committee and serves as the Co-Chair for the University's involvement in the AHA county heart walk.

Tsu-Yin Wu

Best Overall Faculty Member, Health Sciences and Nursing
Best Researcher/Scholar, Health Sciences and Nursing

"Tsu-Yin is an amazing researcher. Her focus area of health equity makes a substantial contribution to nursing and all of health sciences. Her work has secured numerous external awards and has been recognized at the state level. Tsu-Yin shares her expertise with other faculty and with students. She is passionate and collegial.

Tsu-Yin is an amazing researcher in the field of health equities. She is passionate about her work and always finds time to support other faculty and students. Her work is extremely relevant to all of our science and is recognized by the state."

EASTERN NEW MEXICO UNIVERSITY

Bill Gaedke

Best Teacher, Education

EASTERN UNIVERSITY

Jack Bower

Best Overall Faculty Member, Business

Jack Bower is an Associate Professor of Accounting at Eastern University, where he has taught since 1984. He served for 18 years as the elected auditor of Upper Merion Township and was a candidate for PA State Auditor in 1992. He also has served on the board of directors of several nonprofit organizations, including 10 years as the Treasurer of Norristown Ministries, which operates a day-shelter for the homeless in Norristown. Professor Bower has work experience in Thailand and has also worked as an international auditor in the Caribbean and Kenya. Dr. Bower's latest book is Accounting Through the Eyes of Faith, 3rd edition. Dr. Bower is also the recipient of the 2004 Lindback Award for Excellence in Teaching.

EASTFIELD COLLEGE

Dora Falls

Best Overall Faculty Member, Psychology
Best Researcher/Scholar, Psychology
Best Teacher, Psychology
Most Helpful to Students, Psychology

R Hendrickson

Best Researcher/Scholar, Public Affairs, Political & Policy Sciences
Best Teacher, Public Affairs, Political & Policy Sciences
Most Helpful to Students, Public Affairs, Political & Policy Sciences
Best Overall Faculty Member, Public Affairs, Political & Policy Sciences

Rhonda Miller

Best Teacher, Public Affairs, Political & Policy Sciences

EDGEWOOD COLLEGE

Moses Altsech

Best Researcher/Scholar, Business

Moses Altsech is the Chair of the Marketing discipline in the School of Business. He teaches Strategic Marketing, Marketing Research, Consumer Behavior, Advertising & Promotion, and Healthcare Marketing. Dr. Altsech holds a Bachelor's degree in Marketing and International Business from the University of Cincinnati and a Ph.D. in Marketing from the Pennsylvania State University. He is a dual citizen of the United States and the European Union, and has lived in the U.S. since 1987. Moses is fluent in three languages, and has traveled to more than 40 countries and territories in 5 continents, acquiring a wealth of international experience and a global perspective on business and social issues. Professor Altsech is an expert on Customer Satisfaction and Customer Service, Employee Retention, Health Care Marketing, and Strategic Planning and also an avid student of history, literature, politics, social-demographic trends and international affairs. Over the years, Dr. Altsech's research has been presented at state, national and international conferences and published in several prestigious marketing publications. He has appeared as a marketing expert on numerous television and radio programs, has been featured in several magazine and newspaper articles, and is the recipient of numerous professional and academic awards and distinctions.

Mark Barnard

Most Helpful to Students, Business

Dr. Barnard has taught at Edgewood College since 2003. He has a BA in Intercultural Studies (Biola University), MA in Asian Studies (University of Hawaii), an MSc and a PhD in Business (National University of Singapore). Mark teaches courses in Human Resource Management, Organizational Behavior and Business Strategy. He also teaches and leads the International Study Tour for the RAAD and MBA programs. At the College, Mark currently serves on Planning and Budget, RAAD Council and the General Education Task Force. Mark's research interests include strategy, employment relationships, psychological contracts and high performance work systems. He also is a reviewer for the annual Academy of Management conference for which he received an Outstanding Reviewer Award in 2007. He also serves as a judge for the annual Wisconsin Governor's Business Plan Contest. Mark lived in China for one year where he taught English to undergraduate students, graduate students and faculty. He also lived in Singapore for eight years where he earned his PhD and taught at the National University of Singapore. He has also spent considerable time in Indonesia and has traveled extensively throughout Asia.

Elaine Beaubien

Best Teacher, Business
Best Overall Faculty Member, Business

Professor Beaubien is a tenured faculty member teaching management, and marketing to graduate and undergraduate students since 1980. After receiving her undergraduate degree in business administration and economics from the UW-Platteville, she joined the JC Penney Company as a department manager and fashion merchandiser. Leaving retailing, she earned an MBA from the University of Wisconsin where a Teaching Assistantship launched a career in higher education. Prior to Edgewood, she taught economics at the Detroit College of Business, marketing at Mercy College in Detroit and management at the University of Wisconsin-Whitewater. Elaine, who is a respected educator, an experienced trainer, an accomplished speaker, a published writer and a successful entrepreneur has been the recipient of many awards including: the Underkofler Excellence in Undergraduate Teaching Award; the "Madison 100 recognizing her contributions to the City; the Estervig-Beaubien Excellence in Teaching and Mentoring Award created in her honor by undergraduate business students; and the Athena Award Nomination that honors a person who exemplifies excellence in his or her profession. In addition to teaching, training, speaking and consulting, Elaine writes and had published business advice for several publications. She is also very active in civic and community activities, lending her time and expertise to several boards and committees

Denis Collins

Best Overall Faculty Member, Business
Best Researcher/Scholar, Business

Denis Collins (born January 12, 1956)[1] is an American business ethicist and tenured professor of business at Edgewood College in Madison, Wisconsin.[2] He is an expert in the areas of business ethics, leadership, and service-learning, and is considered a leader among business ethics teachers.

Lynea Lavoy

Most Helpful to Students, Business
Best Teacher, Business

LLynea holds an Ed.D. in Educational Leadership from Edgewood College, an MS in Organizational Leadership from Regis University, an MBA in Organizational Strategy from Roosevelt University, and a BS in English and Communications from UW at La Crosse. She is the Manager of Training & Development at TDS Telecom in Madison, WI. Her doctoral research focused on the use of virtual social learning in the college classroom in order to enhance student professional skills.

EDINBORO UNIVERSITY OF PENNSYLVANIA

Mary Nientimp

Most Helpful to Students, Education
Most Helpful to Students, Education

Bethany Scullin

Best Researcher/Scholar, Education

Doctor of Philosophy (Ph.D.), Curriculum and Instruction (Literacy and Urban Education) 2011–2014 Curriculum and Instruction with Literacy and Urban Education Specializations
    Dissertation Title: "Being True": How African American Adolescent Male Students Participate in a Culturally Relevant Literature-Based Reading Curriculum
    Committee Members: Dr. William Bintz, Dr. Lori Wilfong, and Dr. Vilma Seeberg Master of Education (M.Ed.), Educational Leadership and Administration 2006–2008 Bachelor's degree 1995–2000 Dual Major: Special Education (K–12) and Elementary Education (K–6)

Nick Stupianski

Best Teacher, Education

Whitney Wesley

Best Overall Faculty Member, Education
Best Teacher, Education
Best Overall Faculty Member, Education

EDISON STATE COLLEGE

Terri Heck

Best Teacher, Psychology
Most Helpful to Students, Psychology

Haili Marotti

Best Overall Faculty Member, Psychology
Best Researcher/Scholar, Psychology

EL CAMINO COLLEGE

Tanja Carter

Best Teacher, Economics

Kofi Yankey

Best Teacher, Business

EL CAMINO COLLEGE COMPTON CENTER

Saundra Bosfield

Best Teacher, Health Sciences and Nursing

Wanda Morris

Best Researcher/Scholar, Nursing

Ms. Morris recently assumed the position of Interim Director of Nursing at El Camino College, In this position, she provides administration, management, organization, and supervision of the Nursing Department for the El Camino College and El Camino College Compton Center. Prior to joining the staff at El Camino, Ms. Morris was the Dean of Student Learning in the Health, Natural Sciences, and Human Services Division at El Camino College Compton Center. Her responsibilities include management of the overall quality and integrity of the Child Development Program, the Child Development Center (CDC), the Nursing Department, and other assigned areas. Ms. Morris is responsible for assuring that the programs maintain compliance with regulatory boards to which the Child Development, Child Development Center, and Nursing programs report.

Prior to being hired as a Dean of Student Learning (Health, Natural Sciences, and Human Services Division), Ms. Morris served the CCCD as Director of Nursing. In this capacity, Ms. Morris provided the administration, management, organization, and supervision of the Nursing Department. She also provided academic and administrative leadership of the Associate Degree Nursing, Vocational Nursing, and the Nursing Assistant/Home Health Aide educational programs. Additionally, Ms. Morris serves as Assistant Director of Nursing for El Camino College and Compton Center.

Ms. Morris is an active member of the Council of Black Nurses, the National Council of Black Nurses, American Nurses Association, the Association of California Community College Administrators, and the Historically Black Colleges and University National Strategy Advisory Board.

EL CENTRO COLLEGE

Pattricia Burnett

Most Helpful to Students, Business

ELIZABETHTOWN COLLEGE

Paul Gottfried

Best Researcher/Scholar, Political Science

"Dr. Gottfried is the most widely published scholar in the department and the college. He has written hundreds of articles and about 10 books. His work is reviewed, cited and commented on internationally. He is, by far, the most prominent scholar in the history of the college."

EUREKA COLLEGE

Marygrace Yale Kaiser

Most Helpful to Students, Psychology

After 10 years on the faculty at the University of Miami in Coral Gables, Florida, my family and I moved back to Central Illinois in 2011. I was very excited to find a position teaching at Eureka College! I currently live in Tremont with my husband and two boys.

FAIRFIELD UNIVERSITY

Anna-Maria Aksan

Most Helpful to Students, Economics

I am a development economist studying demography and health with a focus on sub-Saharan Africa. Current work focuses on distinguishing between the roles of disease-caused morbidity and mortality on population growth, human capital accumulation and lifelong health, and the consequences these have for economic development. Some recent research investigates the link between population health and economic development locally, specifically the issue of food deserts in Bridgeport, CT. I am an assistant professor at Fairfield University in Fairfield, Connecticut teaching undergraduate economics.

Edward Deak

Best Overall Faculty Member, Economics
Best Teacher, Economics
Best Overall Faculty Member, Business

William Mazariegos

Best Researcher/Scholar, Economics

Originally from Guatemala, William F. Vásquez Mazariegos is an associate professor at Fairfield University specializing in sustainable development of Latin America. His academic credentials include four master degrees, and a Ph.D. in Economics from the University of New Mexico. He has worked as a consultant for the International Food and Policy Research Institute (IFPRI), the United Nations Development Program (UNDP), the United Nations Economic Commission for Latin America and the Caribbean (ECLAC), and the Central American Institute of Fiscal Studies (ICEFI). Dr. Vásquez has implemented household and community surveys in Brazil, Guatemala, Mexico, and Nicaragua. Dr. Vásquez enjoys working with students in field research projects in Latin American countries.

Carl Scheraga

Best Researcher/Scholar, Business

Carl A. Scheraga is Chair of the Department of Management in the Dolan School of Business and Professor of Business Strategy and Technology Management. His fields of research and teaching include transportation and international logistics, global strategic management, cross-cultural management, and the management of technology and innovation. Scheraga has published numerous articles in Transportation Research Series A, Transportation Research Series E, Journal of Transportation Management, Transportation Journal, Journal of the Transportation Research Forum, Journal of Public Policy and Marketing, Technology in Society: An International Journal, Journal of Banking and Finance, Global Business and Finance Review, Journal of Investing, Management International Review, International Journal of Advertising, and International Review of Economics and Finance. He also has published chapters in such volumes as Japanese Direct

Investment in the United States: Trends, Developments and Issues, International Financial Market Integration, Neuroeconomics and the Firm, and On-Line Readings in Psychology and Culture.

Joan Van Hise

Best Teacher, Business
Most Helpful to Students, Business

Dr. Joan L. Lee (formerly Van Hise) is Professor of Accounting at the Charles F. Dolan School of Business at Fairfield University in Fairfield, Connecticut where she teaches financial and managerial accounting, business ethics and accounting ethics. She received a PhD in Accounting from New York University and is licensed as a CPA in New York. She also earned an M.B.A. in Finance as well as a B.S. in Accounting, both from Fordham University. Dr. Lee has published numerous articles in the areas of accounting education and accounting ethics. Dr. Lee was named Alpha Sigma Nu Teacher of the Year 2008 at Fairfield University and was the recipient (with Dr. Dawn Massey) of the 2001 American Accounting Association Innovation in Accounting Education Award. In 2014, she received the Robert J. Spitzer, S.J. Award from the Colleagues in Jesuit Business Education for her service as Board President, Board member and Editor of Journal of Jesuit Business Education. Dr. Lee is Past President of Colleagues in Jesuit Business Education and a member of the American Accounting Association and the American Institute of Certified Public Accountants.

FAYETTEVILLE STATE UNIVERSITY

Angela Taylor

Best Researcher/Scholar, Criminal Justice

Dr. Taylor received her M.A. and Ph.D. degrees in criminal justice from Rutgers, The State University of New Jersey. Her research areas are violence, drugs-crime linkages, drug use measurement, and health disparities and crime. Prior to coming to FSU, Dr. Taylor was a project director at the National Development Research Institutes, Inc., where she participated in research on quality-of-life policing, the accuracy of self-reported substance use, and sweat testing of drugs, among other topics.

FERRIS STATE UNIVERSITY

Paula Hagstrom

Best Teacher, Health Sciences and Nursing

"She is amazing with her students!"

Greg Vanderkooi

Best Overall Faculty Member, Education

Dr. Gregory P. Vander Kooi is a Professor at Ferris State University, College of Education and Human Services, Department of Criminal Justice in Big Rapids MI. His current responsibilities include faculty member, Deputy Director of the Ferris State University's Law Enforcement academy and the Graduate Coordinator of Criminal Justice Masters program. As a retired Michigan State Police Post Commander, he is acutely aware of the training needs for both recruit schools and in-service training. Dr. Vander Kooi earned his Ph.D. from Western Michigan University in Educational Leadership with a focus on adult learning strategies.

FITCHBURG STATE UNIVERSITY

Christine Devine

Best Overall Faculty Member, Nursing

Teresa Finn

Most Helpful to Students, Nursing

"She has changed the sophomore level for the better of all students and faculty!"

Joseph McAloon

Best Overall Faculty Member, Business Administration

"best to relate to students"

Barbara Powers

Best Teacher, Health Sciences and Nursing

Deborah Stone

Best Researcher/Scholar, Nursing

Deborah earned her Bachelor of Science in nursing degree from Texas Woman's University in 1985. In 1995, Deborah obtained Sexual Assault Nurse Examiner (SANE) training from the Attorney General's Office in Texas. She was one of the first three nurses in Texas certified to perform sexual assault nurse examiner assessments on both adult and pediatric patients. In 2006, she was inducted into the Nursing Honor Society, Sigma Theta Tau International, Eplison Beta Chapter. She received her Master of Science in Forensic Nursing degree from Fitchburg State College in 2006. In 2007 she returned to Fitchburg State College as a new instructor in the undergraduate-nursing program. After her first year of working as an educator, she knew this was her passion, therefore in 2009 she joined the doctoral program in Nursing Health Promotion at the University of Massachusetts in Lowell. Her area of research is dating violence among adolescents. Ms. Stone's primary areas of practice are nursing education and emergency nursing. I aspire to share my passion for nursing not only with my patients, but also with my students; for I believe that educating others, bearing witness, and learning from those that demonstrate the best and worst of humankind is truly one of life's privileges.

FLORIDA AGRICULTURAL AND MECHANICAL UNIVERSITY

Reeder

Best Teacher, Business

Saundra Drumming

Best Overall Faculty Member, Business

Eisenhower Etienne

Best Researcher/Scholar, Business

Doctor of Philosophy (Business Administration), 1982, University of Western Ontario Master of Business and Administration (Business Administration), 1975, University of Western Ontario Bachelor of Science (Management), 1974, University of the West Indies. Associate Professor, Supply Chain/Total Quality Management

Mazhar Islam

Best Overall Faculty Member, Finance
Best Researcher/Scholar, Finance

Dr. Mazhar M. Islam is Professor of Finance in the School of Business and Industry at Florida A&M University. Professor Islam has over thirty years of academic experience in a variety of multicultural settings in higher education. Professor Islam received his Master's and Ph.D. from Vanderbilt University, U.S.A. He is the Editor-in-chief of three international refereed journals: American Journal of Finance and Accounting; Global Review of Business and Economic Research; and International Journal of Business Forecasting & Marketing Intelligence. Professor Islam is the founding Chair and CEO of the Global Academy of Business and Economic Research and has organized numerous international conferences. He has published over seventy scientific papers in international refereed journals. Professor Islam has presented over hundred research papers at national and international conferences. Among numerous awards for his teaching and research, professor Islam is the recipient of U.S. Rotary International Teaching Award, Irvin Distinguished Paper Award, and the most prestigious Fulbright Senior Research Award.

William Bill Ravenell

Most Helpful to Students, Business

FLORIDA ATLANTIC UNIVERSITY

Sharon Darling

Most Helpful to Students, Education

Charles Dukes

Best Researcher/Scholar, Education
Best Teacher, Education

Dr. Dukes is an Associate Professor in the Department of Exceptional Student Education. He received his doctorate from Florida International University in Miami, Florida. Dr. Dukes teaches classes in Applied Behavior Analysis, Classroom Management, and Inclusive Education for General Educators. Dr. Dukes has been a member of TASH (formerly know as The Association for Persons with Severe Handicaps) and the Council for Exceptional Children (CEC) since 1999. Charles is active in TASH, serving on the review board for the research journal, Research and Practice for Persons with Severe Disabilities (RPSD). In addition to work on the national level, Charles also co-facilitates a support group for adults with Asperger's Syndrome for both students attending FAU as well as adults from the local community. His research interests include: positive behavioral interventions (including culturally responsive interventions), autism, inclusive education, social capital of individuals with severe disabilities, sexuality issues for individuals with severe disabilities, and transition to adulthood for individuals with developmental disabilities.

Peggy Goldstein

Best Overall Faculty Member, Education

Dr. Goldstein Specializes in early childhood special education, language disabilities and teacher education. She collaborates extensively with both the local school district and community organizations providing teacher training, program evaluation, and research. Current research interests include defining and evaluating quality in early childhood programs, strategies to teach vocabulary and concept development, and effective practices for inclusion.

Cynthia Wilson

Best Researcher/Scholar, Education

FLORIDA ATLANTIC UNIVERSITY - TREASURE COAST

Mara Schiff

Best Researcher/Scholar, Criminal Justice

FLORIDA COASTAL SCHOOL OF LAW

Christopher Roederer

Best Researcher/Scholar, Law

"Excellent scholar and colleague"

FLORIDA GULF COAST UNIVERSITY

Christine Andrews

Best Researcher/Scholar, Accounting
Best Teacher, Accounting
Most Helpful to Students, Accounting

CHRISTINE P. ANDREWS, Associate Professor of Accounting. D.B.A. in Accounting, Cleveland State University (1998). CPA. She is a member of the American Accounting Association. Research interest in: environmental accountability and accounting information systems.

Charles Daramola

Best Teacher, Occupational Therapy & Community Health

Sarah Fabrizi

Best Overall Faculty Member, Occupational Therapy & Community Health

Sarah Fabrizi received her Bachelor of Science (BS) and Masters of Health Science (MHS) from the University of Florida. She has more than 10 years of experience working in adult rehabilitation and pediatric NICU, Early Intervention, PPEC, outpatient, and private practice as an Occupational Therapist. Dr. Fabrizi's research areas of interest include: early intervention in pediatrics, play and playfulness, and community playgroups for caregivers and their children.

Susan Gregitis

Most Helpful to Students, Health Science

Doug Morris

Best Overall Faculty Member, Health Science

"Doug is knowledgeable and dedicated to students. He is an extraordinary teacher and active in faculty affairs for the advancement of the College."

Doug joined Florida Gulf Coast University in May of 1999. He was born and raised in India and lived in Asia until he graduated from high school. Doug spent four years in the U.S. Army following high school. After his departure from the Army, he attended college at Duquesne University in Pittsburgh, Pennsylvania, where he completed a bachelor's degree in Health Science and a master's degree in Occupational Therapy. Doug was employed as an occupational therapist at Allegheny General Hospital, a large urban medical center, where he had experience in acute neurological and orthopedic rehabilitation. He served as the occupational therapist on a multi-disciplinary treatment team at one of the nation's largest outpatient spina bifida clinics. Doug has provided occupational therapy services to homecare clients in Pennsylvania and geriatric residents at skilled nursing facilities in Pennsylvania and Florida. Prior to moving from Pittsburgh, he served as an adjunct faculty member to the occupational therapy program at his alma mater.

Since earning his PhD in Health Science in 2007, Doug has continued to research the role of spirituality in occupational therapy practice, the topic of his published dissertation. He has also studied the construct of "quality of life" in the geriatric population, and is currently researching the impact of Alzheimer's caregiving on the community-dwelling caregiver.

Susan Okon

Best Overall Faculty Member, Medicine

Sue blew into Florida during Hurricane Charley in 2004. She earned her undergraduate degree in Occupational Therapy from Boston University and her master's degree in School Psychology and doctoral degree in Educational Psychology from American International College (AIC) in Springfield, MA. Sue served on the occupational therapy faculty at AIC from 1996–2004. Sue also completed a Certificate of Advanced Graduate Study (C.A.G.S.) in Educational Psychology at AIC. She is licensed as an occupational therapist and certified as a school psychologist in Florida. Her professional areas of expertise include psychosocial and pediatric occupational therapy and service learning in community settings.

FLORIDA INTERNATIONAL UNIVERSITY

Tim Birrittella

Best Overall Faculty Member, Marketing
Best Teacher, Marketing

Prior to joining Florida International University, Mr. Birrittella was on the faculty of the Marketing Department at Berkeley College in New Jersey. He also taught courses at the International Fine Arts College in Miami. In 1995, Mr. Birrittella founded Zacko Sports, Inc., a bicycle helmet manufacturer. After two years, he negotiated the sale and merger of Zacko Sports, Inc. to PTI Sports, Inc., the second largest manufacturer of bicycle accessories in the United States. Mr. Birrittella continued working with PTI Sports, Inc. as a marketing consultant until 2000. He is the main retail Professor for the Retail Certificate Program, as well as the faculty advisor for the FIU student chapter of the American Marketing Association (AMA). At the AMA's 2011 & 2006 International Collegiate Conference in Orlando, the AMA@FIU chapter received the Top Collegiate Chapter of the Year Award, which ranks the chapter first in the United States. Currently, the AMA@FIU chapter ranks #1 (and in the Top 10 (a position held for the past 10 years.) Mr. Birrittella is currently a consultant and former Senior VP, Marketing, of Baby Abuelita Productions, a toy company designed to preserve Hispanic heritage.

John Clark

Best Overall Faculty Member, International Studies
Best Researcher/Scholar, International Studies
Best Teacher, International Studies
Most Helpful to Students, International Studies
Best Researcher/Scholar, Public Affairs, Political & Policy Sciences
Best Overall Faculty Member, Public Affairs, Political & Policy Sciences

John F. Clark is Professor in the Department of Politics and International Relations at Florida International University (Miami). He was Chairperson of the Department of International Relations at FIU from August 2002 to December 2008. He specializes in the state-society relations of African polities and the international relations of sub-Saharan Africa. He is co-editor of Political Reform in Francophone Africa (1997), editor of The African Stakes of the Congo War (2002), and author of The Failure of Democracy in the Republic of Congo (2008). He has also published over forty articles and book chapters, including articles in African Affairs, the Journal of Democracy, the Journal of Modern African Studies, Comparative Studies in Society and History, African Security, and the Africa Spectrum. During the 1999–2000 academic year, he was a Fulbright lecturer and research scholar at Makerere University in Kampala, Uganda, and he has made seven research trips to the Republic of Congo and the Democratic Republic of Congo since 1990. He is currently revising the Historical Dictionary of Congo, which will appear in its fourth edition in 2013. His next research project is on the intersection of religious practice and politics in Central Africa. Please also see http:/news.fiu.edu2011/02/fiyou-john-f-clark/for an FIU profile of Professor Clark.

Erica Gollub

Best Teacher, Epidemiology

Tatiana Kostadinova

Best Researcher/Scholar, Public Affairs, Political & Policy Sciences

Her research and teaching interests include political institutions with a special emphasis on electoral systems and reform, East European democratic transition, and comparative public policy.

Sumit Kundu

Best Researcher/Scholar, Management

"Highly published–others on the list are not researchers or are no longer at FIU"

Dr. Sumit K. Kundu is the James K. Batten Eminent Scholar Chair in International Business in the College of Business Administration at Florida International University.

Dr. Kundu is the Academic Director of the Masters of International Business program [2009–present] and served as the Faculty Director of the Executive MBA program [August 2004–June 2008] and Ph.D Coordinator [August 2004–June 2009].

Dr. Kundu has taught several international business courses at both the graduate and undergraduate levels at Florida International University, Saint Louis University, State University of New York, Northeastern University, and Rutgers University. His extensive international experience includes teaching at Chulalongkorn University (Thailand), City University of Hong Kong (China), Saint Louis University Madrid Campus (Spain), and the Indian Institute of Management. Dr. Kundu has been the recipient of several teaching awards namely, Best Professor in Professional MBA program (2010), Best Professor in International MBA program (2007), Best Professor award in Masters in International Business program (December 2012, May 2012, 2011, 2006), Outstanding Teacher in Executive MBA program (2004), and Best Course award in Masters in International Business program (2012). Dr. Kundu was named Outstanding Graduate Teacher of the Year (2003), Teacher of the Year for Executive Masters in International Business program (2003), and Teacher of the Year for the full-time MBA program (2003) at Saint Louis University.

On the research front, Dr. Kundu sits on the editorial board of several premier journals, Journal of International Business Studies, Management International Review, Global Strategy Journal, Journal of World Business, International Business Review, Journal of International Management, Thunderbird International Business Review, and Journal of Teaching in International Business. Dr. Kundu has published several articles in prestigious journals, namely, Journal of International Business Studies, Journal of Management Studies, Management International Review, Journal of World Business, Journal of International Management, Journal of International Marketing, Journal of Business Research, Journal of Business Ethics, Journal of World Business, Journal of International Management, International Business Review, Leadership Quarterly and Journal of Small Business Economics. He has served as a Chair and member on many dissertation committees. Dr. Kundu has presented numerous papers in the Academy of International Business Academy of Management, and Strategic Management Society conferences. He served as the track chair for its annual conference in Stockholm, Sweden in July 2004; Milan, Italy in June 2008; San Diego, California, USA in June 2009 and Nagoya, Japan in June 2011. Dr. Kundu organized the Junior Faculty Consortium at the annual Academy of International Business 2006 conference in Beijing, China and the Doctoral Consortium at the annual Academy of International Business 2012 conference in Washington DC. He has also served as the president and program chair for the Midwest Academy of International Business Conference in 2003 and 2002.

His corporate experience includes cash flow management at Phillips Petroleum PLC and international and domestic marketing at Fedders Lloyd PLC, and India Foils PLC. Dr. Kundu has been a consultant to several multinational corporations – Novartis, MasterCard International, Ingersoll Rand-Hussmann International, Boeing, and CPI-Sears Portrait Studio.

Paul W Miniard

Best Researcher/Scholar, Marketing

Dating back to 1978, Dr. Miniard has published in the leading journals in his field during each of the past five decades. His research has appeared in journals such as the Journal of Advertising, Journal of Advertising Research, Journal of Consumer Psychology, Journal of Consumer Research, Journal of Marketing, Journal of Marketing Research, and many others. His current research interests include consumer response to retailers' price comparisons, how the impact of marketing materials on Hispanic bilinguals depends on whether these materials use English or Spanish, the influence of consumer avatars in online shopping environments, and how the accessibility of a brand extension's parental heritage shapes consumers' product evaluations. He co-authored six editions of a consumer behavior textbook over a 20 year timespan. He is a member of the American Marketing Association, Association for Consumer Research, and Society for Consumer Psychology.

Nancy Rauseo

Most Helpful to Students, Marketing

John Stack

Most Helpful to Students, Public Affairs, Political & Policy Sciences

A.B., Stonehill College M.A., University of Denver Ph.D., University of Denver J.D., University of Miami

Professor Stack holds a joint appointment as Professor of Political Science and Law. He is Director of the Jack D. Gordon Institute for Public Policy and Citizenship Studies at Florida International University, and Director of the Ethnic Studies Certificate Program. He twice served as chair of the Department of Political Science, and has been editor of the Florida International University Press

FLORIDA STATE COLLEGE AT JACKSONVILLE

Brad Biglow

Best Overall Faculty Member, Medicine
Most Helpful to Students, Medicine

Hey. Thanks for taking a peek at my life! A lot of the time instructors forget to tell the world a little about themselves leading to students remarking about knowing absolutely nothing about what their professor does other than teach courses. Here's a quick rundown on my life and interests (feel free to fast-forward, er...scroll, through the dull parts):

I was born in the Northwoods of Wisconsin with a window view of Lake Superior. For those of you who think it gets cold in North Florida and are lucky to have seen snowflakes perhaps once in your life, I recall many a day shoveling snow into piles taller than I was and thinking it was warm when it would hit zero (since moving to Florida, my cold threshold has lowered to about 50 degrees). Northern Wisconsin was home to a lot of nature, but it also had its problems. Rather than development, it suffered from economic decline and depopulation. It is now one of those places where Wal-Mart has replaced nearly every other business in the area. Despite these growing problems of outsourcing of union labor to foreign countries and growing poverty, I am proud to call such an atmosphere "home." Why? Well, for starters, it is in just that area that made me realize how close people can be to nature, and why anthropology was right for me. Within about 60 miles of my hometown there were several Indian reservations, all bands of the Chippewa (Anishnabe or Ojibwe) Indian people. I went to school with Indians, I played with Indians, and I saw the discomfort they possessed in uncertain economic times when they were combating a system that just didn't understand what it meant to be "indigenous" (native). While my Indian friends dropped out of school, I always wondered "why?" at the time. In hindsight,

I can look back and see public school textbooks that said nothing about the accomplishments or history of the local Indian peoples - nothing about their heritage, religion, values. Throw in lack of economic opportunities or community development, and you have an equation for rampant poverty, alcoholism and domestic violence. So how did people escape this? Well, the youth moved away to opportunities elsewhere: Duluth, Minneapolis, mega-cities to those of us from a town of 8,000. And I did the same.

I went to the University of Wisconsin, Eau Claire where I was first introduced to Anthropology. It was rather a fluke, actually, as I had intended to become a chemist, but the fear of living in isolation in a laboratory for hours on end eventually lost out to more creative endeavors. Soon I was taking one, then two, then three, then . . . oops, every Anthropology course that was offered (archaeology and physical), minoring in that and also in Sociology. Spanish became my main subject as there was no major in Anthropology at UWEC. I was particularly entranced by Latin American civilizations and cultures, though languages in general also interested me. This was, however, to be put on hold for a while.

Carolyn Keister

Most Helpful to Students, Nursing

"Carolyn is always helpful to the students. She is in charge of the Florida Student Nurses Association."

Lara Michel Moses

Best Teacher, Psychology

Wayne Singletary

Best Teacher, Law
Best Overall Faculty Member, Law

FLORIDA STATE UNIVERSITY

Joseph Calhoun

Best Teacher, Economics
Best Teacher, Business

Joseph P. Calhoun is a Lecturer in the Department of Economics and the Assistant Director of the Stavros Center for the Advancement of Free Enterprise and Economic Education at Florida State University. He currently teaches large principles of economics classes with annual enrollment of nearly 3,000 students. To enhance student learning and manage multiple large sections, he is a heavy user of technology both in and outside the classroom. He is also well known around campus for showing video clips to illustrate concepts and inject humorous breaks. In addition to leading workshops for the Stavros Center, he has presented teaching ideas and technology at several national conferences. He is also responsible for training and mentoring new graduate student teaching assistants for their initial teaching responsibilities at FSU.

A graduate of Illinois State University with a BS in finance and economics and of DePaul University with an MBA, he began his career in the health care industry before leaving to become an economist. His Ph.D. is from the University of Georgia.

Dr. Calhoun has received numerous teaching awards including the Outstanding Graduate Student Teaching Award at the University of Georgia, the Undergraduate Teaching Award at FSU, and three times received the Service Excellence Award for Teaching from Phi Eta Sigma at FSU. He also won first place in the Economics Communicators Contest in 2008 which was cosponsored by the Association of Private Enterprise Education and Market Based Management Institute. He currently resides in Tallahassee, FL with his wife and four daughters.

FOOTHILL COLLEGE

Barbara Shewfelt

Most Helpful to Students, Education

Barbara Shewfelt teaches in both the Biology Department and the Dance Department at Foothill College. As an undergraduate, she earned a bachelor's degree from UC Santa Barbara as a double major in both disciplines, and went on to complete a master's degree in dance from the Tisch School of the Arts at New York University and a master's degree in biology from Stanford University with a research specialization in neuroscience.

Barbara also spent five years as a competitive ballroom dancer. She has taught at Foothill for 15 years, offering popular classes in biology, ballet, stretching and social dance.

FRANCISCAN UNIVERSITY OF STEUBENVILLE

Regina Boerio

Best Overall Faculty Member, Psychology
Best Teacher, Psychology
Most Helpful to Students, Psychology

Tiffany Boury

Best Researcher/Scholar, Education

"Dr. Boury has proven to be a sound researcher and excellent teacher."

Kaybeth Calabria

Best Overall Faculty Member, Education
Best Researcher/Scholar, Education
Best Overall Faculty Member, Education

"Dr. Calabria is a sound researcher and excellent teacher."

Education B.A. in Philosophy - Bethany College (Graduated Summa cum Laude) M.A. in Education - Ohio State University Ph.D. in Education - University of Pittsburgh Vocational Certification (Non-degree graduate program) - Kent State University Professional Experience College Professor - West Liberty State College (West Liberty, West Virginia) Adjunct Professorships - University of Pittsburgh, Jefferson County Community College Vocational Special Education Instructor/Job Training Coordinator - Jefferson County Joint Vocational School School Psychologist - Jefferson County School District, PSI Associates, Inc., Toronto

City Schools, Madison County Schools, London, Ohio Additional Experience Includes: Professional Services in the Field of Mental Retardation and Developmental Disabilities, Teaching Experiences with Children, Adult Education Professional Memberships Association for Young Childhood Education International Council for Exceptional Children National Association for the Education of Young Children The Association for the Severely Handicapped Awards and Honors 1997: Certificate of Recognition by the Ohio Association of Supervisors and Work-Study Coordinators 1998: The Eastern Ohio Regional Education Service Center: Excellence in Education Award 1998: The Ohio Department of Education: The Franklin B. Walter Outstanding Educator Award Certifications State of Ohio, Department of Education - School Psychologist, Provisional (lapsed) State of Ohio, Department of Education - Education of the Handicapped (K–12), Specific Learning Disabled (K–12), and Transition to Work

Mary McVey

Most Helpful to Students, Education
Most Helpful to Students, Education

"Dr. McVey goes above and beyond to help her students"

Prior to joining Franciscan University, Dr. McVey served a 9-year tenure in the Noble Local School System in Southeast Ohio. She was recognized as an outstanding classroom teacher being awarded as a Martha Holden Jennings Scholar.

Dr. McVey has a research interest in issues related to educational leadership, particularly teacher education programs, effective teaching, mentoring and teacher induction. She has presented on these issues at the local and state levels.

During Dr. McVey's tenure in education, she has served as a Praxis III evaluator and on various local and state leadership committees.

Susan Poyo

Best Teacher, Education
Best Teacher, Education

"Prof. Poyo is an excellent teacher as demonstrated by her evaluations"

Stephen Sammut

Best Researcher/Scholar, Psychology

Dr. Stephen Sammut received a B.Pharm from Monash University in Victoria, Australia and a Ph.D. in neuroscience from the University of Malta in Malta, Europe. His research interests lie in the utilization of behavioral models and combined experimental techniques to investigate the interaction between the endocrine, immune and nervous systems and their role in CNS development, functioning and psychopathology. His scientific career experience has been broad and has included experience in a number of animal models of psychiatric disorders including depression, schizophrenia, Parkinson's disease and psychostimulant-induced drug sensitization. Moreover, he has utilized various in vivo and in vitro techniques including electrophysiology, electrochemistry, reverse-microdialysis, individually or in combination with each other, in order to investigate questions related to behavior, cellular activity, neurotransmitter release, and how these are altered in psychiatric diseases in brain regions of interest. He has authored and co-authored several papers in leading scientific journals related to the research he has conducted and also presented his work at various conferences and institutions nationally and internationally. His scientific career has also given him the unique experience of having a leading role not only in the daily running of the laboratory in the capacity of lab manager including the overseeing of the budget and laboratory spending but also in being very actively involved and responsible for the original setting up of laboratories and equipment. Dr. Sammut is currently an Assistant Professor of Psychology at Franciscan University of Steubenville, OH, where he teaches and is also involved in efforts to initiate behavioral research geared at increasing the research opportunities for students in behavioral animal research pertaining to post-natal depression and drug abuse.

FRANKLIN UNIVERSITY

Matt Barclay

Best Researcher/Scholar, Business

"Matt Barclay is an excellent business course designer and I great researcher and a gifted scholar."

Dr. Matt Barclay is a member of the Instructional Design Faculty at Franklin. He is particularly interested in instructional design principles for practice, learning and cognition, improving online instruction, adult learning, human performance improvement, and multimedia for instruction. Dr. Barclay holds a Ph.D. in Instructional Technology & Learning Sciences from Utah State University, a master's degree in Instructional Systems Technology from Indiana University, and a bachelor's degree in Family Sciences from Brigham Young University.

Bruce Campbell

Best Overall Faculty Member, Finance

"Outstanding faculty member and department chair individual."

Bruce Campbell currently serves as both the MBA Program Chair and the Financial Management Program Chair at Franklin University. Bruce began his career as an International Banking Officer for Pittsburgh National Bank (now PNC Financial) in Pittsburgh. He later served as a Vice President at PNC's office in Cleveland after working as a manager at the bank's former Hong Kong affiliate, where he was responsible for corporate business development throughout Asia.

Prior to joining the Franklin faculty, Bruce served as Academic Dean at the International Christian University-Kyiv, an affiliate of National Economics University in Kyiv, Ukraine. More recently, he led MBA Programs at both Myers University and the University of Phoenix's campus in Cleveland.

Bruce earned his Ph.D. in Finance, with a minor in International Economics, from Kent State University, his Master of Business Administration in Finance and International Business from the University of Washington, his Master of Arts in Economics from Cleveland State University, his Master of Arts in Slavic Languages and Literatures from Indiana University, and his Bachelor of Arts in Russian from the University of Oregon. He also holds professional certification as a Certified Treasury Professional.

Ernest Massie

Best Overall Faculty Member, Economics
Best Overall Faculty Member, Business

Bruce Ramesy

Best Overall Faculty Member, Marketing

"Bruce is focused and persistent."

Bruce Ramsey serves as the Chair of the Undergraduate Marketing Program of Franklin University in Columbus, Ohio. He holds Masters degrees in Communications and Business Administration from Indiana University and Ohio University respectively. His primary academic interest is in the development and application of innovative marketing and communication tactics. Mr. Ramsey conducts seminars and workshops on such topics as Relationship Marketing, Fun in the Workplace, and Internal Marketing. He has been a Director of Marketing for a hospital, has developed intensive management training seminars, authored web sites, produced over 150 instructional videotapes, and provided business and marketing consulting services.

Gary L. Sigrist

Most Helpful to Students, Business
Best Teacher, Business

"Gary goes out of his way to support his students. He does this by being responsive, providing excellent feedback and providing real world examples in his teaching."

Michael Tanner

Best Teacher, Economics

FROSTBURG STATE UNIVERSITY

Tom Hawk

Most Helpful to Students, Business

"in addition to being an outstanding teacher, he provides extensive and helpful feedback to all students on each assignment and project."

Paul Lyons

Best Overall Faculty Member, Business

"Logical with a great sense of humor. Published, creative, smart and easy to work with."

Jodi Nichols

Best Teacher, Education

Evan Offstein

Best Overall Faculty Member, Business

"All-around top performer, team player, student advocate, intelligent, fierce, passionate and ethical."

Todd Rosa

Most Helpful to Students, Education

Todd Rosa was born and raised in Western Massachusetts. After serving four years in the U.S. Army, he earned a BA in History from the university of Massachusetts at Amherst and a PhD in history from George Washington University in Washington, DC. While finishing his dissertation, Todd entered secondary-level teaching through the Baltimore City Teaching Residency program. He taught United States History and American Government at Dr. Samuel L. Banks High School, a zoned neighborhood school located in one of the tougher neighborhoods in the city. He joined the educational professions faculty at FSU in 2008. His research interests include Cold War United States foreign relations (particularly during the Carter Administration), the history of American education, and the politics of education.

Kim Rotruck

Best Overall Faculty Member, Education

FULLERTON COLLEGE

Jennifer Combs

Best Researcher/Scholar, Education

Jeffrey Hendrix

Best Teacher, Physical Ed

Jeff Hendrix is a master jazz and tap teacher with an M.F.A. in Dance from U.C.Irvine. His classes offer solid technique and cover a broad range of styles. Performance credits include "A Chorus Line", "Tap" (the movie), "Good Morning America" (with Rita Moreno), "Guys and Dolls", "Anything Goes", and "Oklahoma". Choreography and teaching at Luigi's (N.Y.C.), Svetlova Dance Centre, Bermuda Ballet, Ballet Russe de Montreal, Columbo (Zurich), Schutz Ballet (Vienna), Maui Academy of Performing Arts, and three tours of Japan. Musicals choreographed include "Baby", "H.M.S. Pinafore" and "Joseph . . .".

Jeff is on staff at Mt.SAC, Santiago Canyon and Golden West colleges. He also teaches acting, musical theatre, salsa, swing, yoga, videography, graphics, animation and computer dance.

Rolando Sanabria

Best Overall Faculty Member, Counseling

Peter Snyder

Best Overall Faculty Member, Education

GALVESTON COLLEGE

Pat Perry

Best Overall Faculty Member, Nursing
Best Researcher/Scholar, Nursing
Best Teacher, Nursing
Most Helpful to Students, Nursing

Edward Stout

Best Researcher/Scholar, Health Sciences and Nursing
Best Teacher, Health Sciences and Nursing
Most Helpful to Students, Health Sciences and Nursing
Best Overall Faculty Member, Health Sciences and Nursing

GEORGE MASON UNIVERSITY

Linda Chrosniak

Most Helpful to Students, Psychology

"Linda goes above and beyond to ensure the success of her honors students"

Dr. Chrosniak has been the director of the Honors Program in Psychology at GMU since 2003. She received a B.S. degree in psychology from the University of Texas at Dallas and her Ph.D. in experimental psychology from George Washington University. Her areas of research interest include implicit and explicit memory, cognitive aging, stress and health. More recently her interests have included the interaction effects of stress and cognitive processes and their effect on health and health behaviors.

Her published work has appeared in journals such as the Journal of Experimental Psychology: Learning, Memory and Cognition, Psychology and Aging, Journal of Geochemical Exploration, Journal of Personality and Social Psychology and Military Medicine.

She was nominated for a University Teaching Excellence Award in 1992 and 1999 and was a recipient of the University Teaching Excellence Award from George Mason University in 2000. She also received the Teacher of the Year Award from the GMU chapter of Psi Chi, the National Honor Society in Psychology in 2003 and 2005. She was the recipient of the BIS Outstanding Faculty Mentor Award in 2010.

Desmond Dinan

Best Researcher/Scholar, Public Policy

Bio:
Desmond Dinan is Professor of Public Policy; Director, International Commerce and Policy Program and holds the Jean Monnet Chair in European Public Policy at George Mason University. He has been an adviser to the European Commission in Brussels and a Visiting Fellow at the Netherlands Institute for International Relations, The Hague. He was a Visiting Professor at the College

of Europe, Bruges campus (1998–2000) and the College of Europe, Natolin campus (2005–2008). His research interests include the historiography of European integration; the history of the European Union (EU); institutions and governance of the EU; enlargement of the EU; regional integration in the context of globalization; and prospects for global governance. He has written several books on the EU and, since 1999, has written the article on institutions and governance for the Journal of Common Market Studies' highly regarded annual review of developments in the EU. He has received the George Mason University Student Government "Professor of the Year" Award for Excellence in Instruction and the School of Policy, Government, and International Affairs' annual teaching award. Professor Dinan is a native of Ireland and a permanent resident of the United States.

Walter E. Williams

Best Teacher, Economics
Best Teacher, Business

GEORGE WASHINGTON UNIVERSITY

Lisa Delpy Neirotti

Best Teacher, Business
Most Helpful to Students, Business
Best Overall Faculty Member, Business

Lisa Delpy Neirotti, Ph.D. Associate Professor, School of Business, The George Washington University.

Lisa Delpy Neirotti has been a professor of sport, event, and tourism management at the George Washington University for twenty-four years. In this time, Dr. Delpy Neirotti has established a strong academic program at both the undergraduate and graduate levels and serves as the Director of the Masters of Tourism Administration program and is the lead Sport Management faculty for MBA, MTA, and BBA students. Dr. Neirotti also serves on the faculty of the International Olympic Committee's Executive Masters In Management of Sports

Organizations (MEMOS) which takes her around the world. Furthermore, Dr. Neirotti helped to develop the Certified Sports Administrator educational program for the National Council of Youth Sports, the Sport Philanthropy Certificate., and the GW Green Sports Scorecard that helps to educate, motivate, and evaluate the sustainability of sport facilities, organizations, and events.

In addition to her responsibilities at the university, Dr. Delpy Neirotti works with a number of sport organizations and professional teams to conduct economic, spectator, and market research studies including Olympic Games, World Cup, BNP Paribas Open, Citi Tennis Open, Marine Corp Marathon, Rock and Roll Marathon, Cherry Blossom Ten Mile Race, Army Ten Miler, Military Bowl, Washington Wizards, Washington Capitals, Washington Nationals, and Washington Redskins. She also consults in the area of Sport for Development for Non-Governmental Organizations (NGOs).

As a pioneer in the field of sports tourism, Dr. Delpy Neirotti founded the annual TEAMS: Travel, Events, and Management in Sports conference. Since 1997, TEAMS serves to define, develop and expand the fast growing field of Sports Tourism. Numerous organizations such as Marriott International, State Tourism offices of Maryland, Idaho, Missouri, Ohio, and Ministries of Tourism in St. Martin and Belize have commissioned her to conduct tourism assessments and help create strategic plans and marketing campaigns related to sports and event tourism. She also worked on a USAID economic development project for Montenegro focused on sport tourism.

Dr. Delpy Neirotti co-authored The Ultimate Guide to Sport Event Management and Marketing and serves on the editorial board of SportsTravel magazine. She also is a member of the Women's Sport Foundation (WSF), Up2Us, and Council for Responsible Sports advisory boards as well as Vice-President of the DC Chapter of Women In Sports and Events (WISE).

Prior to arriving at The George Washington University, Dr. Delpy Neirotti traveled to 56 countries around the world studying the development and organization of the Olympic Movement. Since 1984, she has attended 17 consecutive Olympic Games, 4 World Cups, and hundreds of other major sport events as a consultant, volunteer or researcher. In 1994, she served on the World Cup host committee in Washington, DC.

Born and raised in California, Dr. Delpy Neirotti received her undergraduate degree from California Polytechnic State University, San Luis Obispo; a MS in Sport Management from George Mason University, Fairfax, VA; and a Ph.D. in Sport Administration from the University of New Mexico in Albuquerque. Her doctoral dissertation was on the organizational structure and effectiveness of the U.S. national sport governing bodies. She resides in Silver Spring with her husband and two children.

Tonya Dodge

Best Overall Faculty Member, Psychology
Best Teacher, Psychology
Most Helpful to Students, Psychology

Dr. Dodge's work began by investigating factors that affect decisions to use performance enhancing substances (e.g., Anabolic Steroids, creatine) that are used in the physical activity or sports domain. This interest has expanded into the academic domain as well (e.g., prescription stimulant misuse). Exploring decisions to use performance enhancing substances led her to question how engaging in physical activity may make one more susceptible to misuse some substances (e.g., Anabolic Steroids, alcohol use), but this is not necessarily the case for other drugs (e.g., marijuana and tobacco). Some of her current work focuses on identifying psychophysiological processes that can explain these differences.

Jodi Ganiban

Best Researcher/Scholar, Psychology

William Handorf

Most Helpful to Students, Business
Best Teacher, Business

Derrick Heggans

Best Overall Faculty Member, Sports Management
Best Researcher/Scholar, Sports Management
Best Teacher, Sports Management
Most Helpful to Students, Sports Management

Derrick Heggans • Received Bachelor of Arts from Duke University in 1992.
 • September 1992–April 1994 worked at global advertising agency, D'Arcy, Masius, Benton, and Bowles, Inc. ("DMB&B"), helping to manage Cadillac Motor Car Division's sponsorship of the Sr. PGA Tour and various other golf activities.
 • After leaving DMB&B, served as a Client Manager/NBPA Player Agent in the basketball division of Advantage International, now "Octagon" for 3 ½ years.
 • Graduate of George Washington University's School of Law (Evening Division).
 • May 1998 began working at NFL Properties, the marketing and licensing division of the National Football League in the Legal & Business Affairs Department in Staff Attorney role primarily serving as business affairs counsel for Corporate Sponsorships, Marketing and Special Events.
 • May 2001 moved to Office of the Commissioner of the National Football League as Asst.Counsel for Broadcast Operations and Policy. Chief responsibilities include assisting Broadcasting Department, NFL Films, and NFL Enterprises in negotiating, structuring and administering various League media contracts; advising clubs on broadcast issues; managing League participation in copyright tribunals in the United States, Canada and elsewhere; serving

as chief business affairs counsel for implementation of Super Bowl, Pro Bowl and other NFL run special events. Served on Super Bowl Policy Committee.
- Served as Chief Operating Officer for Collegiate Images, licensing agency for archived collegiate sports media images from March 2004–March 2005.
- July 2005, joined the Legal Department of AOL, handling a variety of transactional matters, including sports content deals.
- December 2007, became General Manager of AOL's Sports Channel, overseeing all business and programming operations. In 2008, was named by Sports Business Journal as one of the "20 Most Influential in Sports Digital Media". Left AOL in March of 2009.
- In March 2009, named to Sports Business Journal 2009 "Forty Under 40" list of most dynamic young executives in business of Sports under the age of 40.
- In March 2009, formed Heggans and Company Enterprises (H.A.C.E.) as consultancy on sports and media related matters. Some current clients include "T-Time Productions"-Serves as a consultant and Executive Producer on "Third and Long" documentary about the history of African American Professional Football Players, former WTA tennis player, Chanda Rubin, and XCO SportsLink, a sports focused smart phone mobile solutions developer.
- Served as Adjunct Lecturer at George Washington University from 2006–2010.
- In April 2010, joined The Wharton School at the University of Pennsylvania as Managing Director of the Wharton Sports Business Initiative.
- Resides in Northwest, Washington, D.C. with wife, Tanya and son, Timothy.

George Jabbour

Best Teacher, Finance

Dr. George Jabbour - MSF Program Director and Professor of Finance.

Dr. Jabbour is the Associate Dean for Executive Education, Professor of Finance, and Director of the Masters in Finance Program at The George Washington University School of Business for both DC and China cohorts. He taught risk management, financial engineering, investments, corporate finance, financial markets, options, and advanced financial management. Dr. Jabbour has been a consultant conducting professional training for the World Bank Group in Washington DC (USA), Vienna (Austria) and Islamabad (Pakistan), for The International Training Banking Center in Budapest (Hungary), the Center of Excellence in Management in Kuwait, and for Kipco Asset Management Group (KAMCO) in Kuwait, and Renmin University (China). He was a Visiting Professor at Franklin College in Lugano (Switzerland), Dongseo University (South Korea), and Sorbonne University in Paris (France). Dr. Jabbour has several publications in professional refereed journals. He co-authored "The Option Trader Handbook" for trade adjustments. His research papers were presented in Orlando (Florida), New Orleans (Louisiana), Palm Beach (Florida), San Francisco (California), Cincinnati (Ohio), New York (New York), Washington DC, Baltimore (Maryland), Las Vegas (Nevada), Paphos (Cyprus), Acapulco (Mexico), Istanbul (Turkey), Rhodes (Greece), Salt Lake City (Utah), Cancun (Mexico), Chicago (Illinois), Sydney (Australia), Dallas (Texas), Paris (France), and Bangkok (Thailand). He also participated in seminars in Korea, Romania and China. He is affiliated with the Financial Management Association, American Finance Association, Eastern Finance Association, Southern Finance

Association, Middle East Economic Association, and International Association of Financial Engineers. In addition, he is a member of the editorial advisory board of the Journal of Business and Economics Research in the USA, and GITAM Review of International Business in India. He is member of the Educational Committee of International Association of Financial Engineers. He also served as member of the Advisory Board of the Professional Risk Managers International Association. One of his co-authored papers was listed on the SSRN's Top Ten download list for "Risk Management Recent Hits". Four of his co-authored papers won the best paper awards. He has also received several Teaching Excellence Awards. He participated in the Forum for Democracy and Development in Doha (Qatar), in the Education without Borders and the Festival of Thinkers in Abu Dhabi and Dubai (United Arab Emirates.) In addition to his academic career, Dr. Jabbour is very active in equity and derivatives trading. He was a Senior Financial Analyst at Federal Home Loan Mortgage Corporation and a consultant to several American corporations and international organizations. He is Managing Director of Global Asset Management Inc., President of Global Finance Associates, Inc; Advisor to the CEO of NGP V Real Estate Fund, and a Board member of Energy Research Solutions. He has a Ph.D. in Finance and Investments, an MBA in Finance, a BBA in Marketing, and BS in Mathematics.

Scheherezade Rehman

Best Researcher/Scholar, Business
Best Teacher, Business

Robert Stoker

Best Teacher, Political Science

"This professor has an uncanny capacity to connect with students so they want to learn. He does not just convey information but gets his students to think about political science and the issues before it. He has a remarkable talent for the presentation of balanced material. Bob is a real credit to GWU."

Professor Stoker's research and teaching interests are in the field of public policy. His first book, Reluctant Partners, analyzed the difficulties of policy implementation in a context of diffuse authority. His second book, When Work is not Enough, focused on policies and programs to assist low-income working families. His current research focuses on urban revitalization. He is presently completing a book on the Empowerment Zone program and participating in a cross-national study of neighborhood revitalization initiatives. Stoker regularly teaches courses relating to policy analysis methods and the politics of the policy making process. He occasionally teaches undergraduate courses relating to social policy problems.

Susan Wiley

Most Helpful to Students, Public Affairs, Political & Policy Sciences

Professor Wiley's primary research interests are American political behavior and domestic public policy. Along with her teaching duties in Political Science, she serves as department undergraduate coordinator, directs the undergraduate internship program, and teaches quantitative methods for political managers in the Graduate School of Political Management.

GEORGE WASHINGTON UNIVERSITY LAW CENTER

Michael Abramowicz

Best Researcher/Scholar, Law

Michael B. Abramowicz specializes in law and economics, spanning areas including intellectual property, civil procedure, corporate law, administrative law, and insurance law. His research has been published in the California Law Review, Columbia Law Review, Cornell Law Review, Harvard Law Review, Michigan Law Review, New York University Law Review, Stanford Law Review, University of Chicago Law Review, Virginia Law Review, Yale Law Journal, and many others. He has also published a book, Predictocracy: Market Mechanisms for Public and Private Decision Making, with the Yale University Press.

Before coming to GW, Professor Abramowicz served as an Assistant and then Associate Professor at George Mason University School of Law. Professor Abramowicz has also served as a Visiting Assistant Professor at Northwestern University School of Law and as a Visiting Associate Professor at the University of Chicago Law School.

Professor Abramowicz graduated summa cum laude from Amherst College, where he majored in economics and served as Editor-in-Chief of the campus newspaper. After spending a year as a research assistant at the Federal Reserve Board, he attended Yale Law School, where he served as Executive Editor of the Yale Law Journal and as a Co-Director of the landlord tenant clinic. After law school, he clerked for the Honorable Patrick E. Higginbotham of the U.S. Court of Appeals for the Fifth Circuit.

William Kovacic

Best Teacher, Law

Before joining the Law School in 1999, Professor Kovacic was the George Mason University Foundation Professor at the George Mason University School of Law. From January 2006 to October 2011, he was a member of the Federal Trade Commission and chaired the agency from March 2008 to March 2009. He was the FTC's General Counsel from June 2001 to December 2004. In 2011 he received the FTC's Miles W. Kirkpatrick Award for Lifetime Achievement.

Since August 2013, Professor Kovacic has served as a Non-Executive Director with the United Kingdom's Competition and Markets Authority. From January 2009 to September 2011, he was Vice-Chair for Outreach for the International Competition Network. He has advised many countries and international organizations on antitrust, consumer protection, government contracts, and the design of regulatory institutions.

At GW, Professor Kovacic has taught antitrust, contracts, and government contracts. He is co-editor (with Ariel Ezrachi) of the Journal of Antitrust Enforcement. His publications since returning to GW in 2011 include "Good Agency Practice and the Implementation of Competition Law" in European Yearbook of International Economic Law (Christoph Hermann ed. 2013); "Antitrust in High-Tech Industries: Improving the Federal Antitrust Joint Venture" in George Mason Law Review (2012); "Behavioral Economics: Implications for Regulatory Agency Behavior" in Journal of Regulatory Economics (2012) (with James Cooper); "Competition Agency Design: What's on the Menu?" in European Competition Journal (2012) (with David Hyman); "Plus Factors and Agreement in Antitrust Law" in Michigan Law Review (2011) (with Robert Marshall, Leslie Marx & Halbert White); "Ensuring Integrity and Competition and Public Procurement Markets: A Dual Challenge for Good Governance" in The WTO Regime on Government Procurement: Challenge and Reform (Sue Arrowsmith & Robert Anderson, eds. 2011) (with Robert Anderson & Anna Caroline Mueller); "The International Competition Network: Its Past, Current, and Future Role" in Minnesota Journal of International Law (2011) (with Hugh Hollman); "The William Humphrey and Abram Myers Years: The FTC from 1925 to 1929" in Antitrust Law Journal (2011) (with Marc Winerman); Professor Kovacic also is co-author (with Andrew Gavil & Jonathan Baker) of Antitrust Law in Perspective: Cases, Concepts and Problems in Competition Policy (2d ed. 2008) and Antitrust Law & Economics in a Nutshell (5th ed. 2004) (with Ernest Gellhorn & Stephen Calkins).

Gregory Maggs

Best Teacher, Law
Most Helpful to Students, Law
Best Overall Faculty Member, Law

Gregory E. Maggs joined the George Washington University Law School faculty in 1993. He is a a Co-director of the law school's National Security and U.S. Foreign Relations LLM program. He was the Interim Dean of the law school from 2010–2011 and from 2013–2014 and the Senior Associate Dean for Academic Affairs from 2008–2010. He teaches mainly in the areas of commercial law, constitutional law, contracts, and counter-terrorism law, and he has written extensively on these subjects. By vote of the graduating class, he received the law school's Distinguished Faculty Service Award in 1997, 1998, 2004, 2005, 2011, 2012, 2013, and 2014. In 2012, the university gave him the George Washington Award for outstanding service.

Professor Maggs is a graduate of Harvard College and Harvard Law School. He was a law clerk for Justices Clarence Thomas (1991–1992) and Anthony M. Kennedy (1989–1990) of the U.S. Supreme Court and for the late Judge Joseph T. Sneed of the U.S. Court of Appeals for the Ninth Circuit (1988–1989). He also taught for two years as an assistant professor at the University of Texas School of Law. His other past experience includes service as a special master for the U.S. Supreme Court, a consultant to Independent Counsel Kenneth Starr in the Whitewater Investigation, and an assistant to Robert H. Bork in private practice and research. He is a member of the Advisory Board for the Heritage Foundation's Center for Legal and Judicial Studies and a member of the American Law Institute.

Professor Maggs is a colonel in the U.S. Army Reserve, Judge Advocate General's Corps. He was commissioned in 1990, and has been assigned as a trial or appellate military judge since 2007. He is a graduate of U.S. Army War College, the Military Judge Course, the Command and General Staff Officer Course, the Judge Advocate Officer Advanced and Basic Courses, the Air Assault School, and the Infantry Weapons Specialist Course. He was called to active duty in 2007–2008. In 2002, he received the Judge Advocates Association's Outstanding Career Armed Services Attorney Award.

Todd Peterson

Most Helpful to Students, Law

Professor Peterson joined the Law School faculty in 1987. Earlier, he had been a partner at the Washington, D.C., firm of Ross, Dixon & Masback, where he specialized in commercial litigation. Professor Peterson also served as an attorney adviser in the Department of Justice's Office of Legal Counsel, which is responsible for providing advice to the attorney general and White House on constitutional law issues. He began practice in the fields of administrative and commercial litigation at the D.C. firm of Crowell & Moring.

Professor Peterson has served as a consultant to the National Commission on Judicial Discipline and Removal and as co-chair of the D.C. Circuit Special Committee on Race and Ethnic Bias. From 1997 to 1999, Professor Peterson returned to the Office of Legal Counsel to serve as deputy assistant attorney general. Professor Peterson teaches civil procedure, federal courts, and separation of powers. He writes principally in the areas of separation of powers and the federal judicial system.

Edward T. Swaine

Best Overall Faculty Member, Law
Best Researcher/Scholar, Law

Professor Swaine teaches and writes in the areas of public international law, foreign relations law, international antitrust, and contracts. He is the co-author of Foreign Relations and National Security Law: Cases, Materials and Simulations (4th ed. 2011) (with Franck, Glennon, and Murphy) and has published work in the American Journal of International Law, Columbia Law Review, Duke Law Journal, Harvard International Law Journal, Stanford Law Review, University of Pennsylvania Law Review, Virginia Journal of International Law, William and Mary Law Review, and Yale Journal of International Law, among others. He has consulted on matters involving treaty law, antitrust, intellectual property, and international litigation and arbitration.

Professor Swaine joined the GW faculty in 2006, after serving for one year as the Counselor on International Law at the U.S. Department of State. His previous academic appointment was as a tenured professor at the Wharton School, with a secondary appointment at the University of Pennsylvania Law School. Before entering academia, Professor Swaine practiced law at the Brussels office of Cleary, Gottlieb, Steen and Hamilton, where his work focused on European Community law and antitrust, and served as a member of the civil appellate staff at the U.S. Department of Justice. He is a graduate of Yale Law School, where he was the editor-in-chief of the Yale Law Journal, and Harvard College.

At GW, Professor Swaine has served as the Senior Associate Dean for Academic Affairs and as Director of the Competition Law Center. He is a member of the Executive Council of the American Society of International Law and co-chair of the International Law in Domestic Groups interest group, and member of the Advisory Committee on Public International Law for the U.S. State Department.

GEORGETOWN UNIVERSITY LAW CENTER

Peter Edelman

Most Helpful to Students, Law

Peter Edelman is the Carmack Waterhouse Professor of Law and Public Policy at Georgetown University Law Center, where he teaches constitutional law and poverty law and is faculty director of the Georgetown Center on Poverty and Inequality. On the faculty since 1982, he has also served in all three branches of government. During President Clinton's first term, he was Counselor to HHS Secretary Donna Shalala and then Assistant Secretary for Planning and Evaluation.

Professor Edelman has been Associate Dean of the Law Center, Director of the New York State Division for Youth, and Vice President of the University of Massachusetts. He was a Legislative Assistant to Senator Robert F. Kennedy and was Issues Director for Senator Edward Kennedy's Presidential campaign in 1980. Earlier, he was a Law Clerk to Supreme Court Justice Arthur J. Goldberg and before that to Judge Henry J. Friendly on the U.S. Court of Appeals for the Second Circuit. He also worked in the U.S. Department of Justice as Special Assistant to Assistant Attorney General John Douglas.

Professor Edelman's most recent book, So Rich So Poor: Why It's So Hard to End Poverty in America, was published by The New Press in May 2012. He previously wrote Searching for America's Heart: RFK and the Renewal of Hope, which was published by Houghton-Mifflin in January 2001. He also co-authored Reconnecting Disadvantaged Young Men, which was published by the Urban Institute in 2006, and is the author of many articles on poverty, constitutional law, and issues about children and youth. His article in the Atlantic Monthly, entitled "The Worst Thing Bill Clinton Has Done," received the Harry Chapin Media Award.

Peter Edelman has chaired and been a board member of many organizations and foundations. He is currently chair of the District of Columbia Access to Justice Commission and the National Center for Youth Law, and formerly board chair of the American Constitution Society for Law and Policy and the Public Welfare Foundation, board president emeritus of the New Israel Fund, and a board member of the Center for Law and Social Policy, the Center for American Progress Action Fund, and a half dozen other nonprofit organizations.

He has been a United States-Japan Leadership Program Fellow, was the J. Skelly Wright Memorial Fellow at Yale Law School, and has received numerous honors and awards for his work. He grew up in Minneapolis, Minnesota.

Michael Gottesman

Best Overall Faculty Member, Law

Professor Gottesman served as an adjunct professor at the Law Center from 1978–88, and joined the faculty as a full-time professor in 1989. Specializing in the fields of labor law, constitutional law, and civil rights, Professor Gottesman practiced with the Washington, D.C., firm Bredhoff and Kaiser from 1961–88, and has argued numerous cases in the U.S. Supreme Court. From 1977–81, he served, by appointment of President Carter, on the Judicial Nominating Commission for the District of Columbia, reviewing hundreds of candidates for vacancies on the U.S. Court of Appeals and the U.S. District Court for the District of Columbia. Professor Gottesman is a member of the Board of Trustees of the Lawyers' Committee for Civil Rights Under Law, a member of the American Academy of Appellate Lawyers, and a member of the Law Committee of the American Association of University Professors.

Martin Lederman

Best Teacher, Law

Robin West

Best Researcher/Scholar, Law

Professor West came to the Law Center from the University of Maryland Law School, where she taught from 1986–1991. She has been a visiting professor at the University of Chicago and Stanford Law Schools. She also taught at Cleveland-Marshall College of Law at Cleveland State University, from 1982–1985. Professor West has written extensively on gender issues and feminist legal theory, constitutional law and theory, jurisprudence, legal philosophy, and law and literature.

GEORGIA COLLEGE

Nicole Declouette

Best Teacher, Education

GEORGIA GWINNETT COLLEGE

Dovile Budryte

Best Researcher/Scholar, Political Science
Best Teacher, Political Science

Dovile Budryte, Ph.D, is a professor of political science at Georgia Gwinnett College. Her areas of interest include democratization, gender studies and nationalism. She was a 2004 National Endowment for the Humanities grant recipient, 2000–01 Carnegie Council on Ethics and International Affairs (New York) Fellow, and a 1998–99 Fellow at the College for Advanced Central European Studies at Europa University Viadrina, Frankfurt (Oder), Germany. Her publications include articles on minority rights and democratization in the Baltic states and three books, Taming Nationalism? Political Community Building in the Post-Soviet Baltic States (2005), Feminist Conversations: Women, Trauma and Empowerment in Post-Authoritarian Societies (co-edited with Lisa M. Vaughn and Natalya T. Riegg, 2009) and Memory and Trauma in International Relations: Theories, Cases and Debates (co-edited with Erica Resende, 2013).

Education
Doctorate – International Studies – Old Dominion University Master of Arts – International Studies – Old Dominion University Bachelor of Arts – Communications Studies in Journalism – Vilnius University, Lithuania Academic Interests.

Democratization Minority rights/gender studies Distinctions.

University System of Georgia Regents' Teaching Excellence Award, FY 2015 Outstanding Faculty Teaching Award, Georgia Gwinnett College, 2014 Nominee, the Association for the Advancement of Baltic Studies 2006 book prize Ann Austin Johnston Outstanding Faculty Member Award (the highest award for teaching at Brenau University), 2004 Greek Council Faculty Member of the Year (the highest award for campus service and teaching at Brenau University; selected by students), 2004 Professional Affiliations.

International Studies Association Association for the Advancement of Baltic Studies.

Arlenda Peterson-Murphy

Best Overall Faculty Member, Physical Education

Mary Saunders

Best Overall Faculty Member, Management

Dr. Saunders has held teaching positions at Mercer University, the University of Georgia-Gwinnett, Georgia State University, the Medical University of South Carolina and Miami Dade Community College. She has taught both graduate and undergraduate courses in business administration, health care policy and administration, and nursing. Her professional nursing experiences have included positions as a pediatric staff nurse, a public health staff nurse, a public health nursing supervisor and as a district director of the Visiting Nurse Association of Atlanta.

GEORGIA HIGHLANDS COLLEGE

David Mathis

Best Overall Faculty Member, Physical Ed

"David works above and beyond any job title!!! I worked with him for years and he puts a smile on your face. He is the most willing to help, non complaining, giving, caring individual I've ever known!"

GEORGIA PERIMETER COLLEGE - DECATUR

Mary Mattson

Best Overall Faculty Member, Education

Professor Teacher Education Ph.D. English Education Georgia State University

Janet Orr

Most Helpful to Students, Economics

"Dedicated to her students. Readily available Helpful to colleagues when needed Volunteers her service to the department"

GEORGIA STATE UNIVERSITY

Paul Alberto

Best Researcher/Scholar, Education

Dr. Paul Alberto, interim dean of Georgia State University's College of Education since 2012, has been appointed dean of the college following a national search.

Alberto, a Regents' Professor in Intellectual Disabilities, has led the college's adoption of a seven-year strategic plan that sets a strong course for the future in research, student recruitment and learning, and collaboration.

MEDIA CONTACT Angela Turk Director of Communications College of Education 404-413-8114 aturk@gsu.edu "Paul Alberto is a highly respected researcher with a proven record of excellent leadership," said Provost Risa Palm. "I am confident Dr. Alberto will take the College of Education to the next level in its reputation and prominence." Alberto received his bachelor's degree in sociology from Hunter College, a master's degree in mental retardation from Fordham University and a doctor's degree in severe disabilities from Georgia State.

After graduating from Georgia State in 1976, he joined the College of Education faculty as an assistant professor of special education. He was later promoted to associate professor and then full professor. In addition to teaching special education courses, he was director of the university's Bureau for Students with Multiple and Severe Disabilities program, chaired the university's Senate executive committee for eight years and co-chaired the university's strategic planning committee.

Toby Bolsen

Best Teacher, Political Science

Professor Bolsen's research focuses on the study of political behavior, preference formation, political communication, experimental methods, and U.S. energy policy and climate change. He has published in outlets such as the Annals of the American Academy of Political and Social Science, American Journal of Political Science, American Politics Review, Political Behavior, Political Communication, Public Opinion Quarterly, International Journal of Press/Politics, Journal of Experimental Political Science, and the Journal of Communication.

The National Science Foundation has supported Professor Bolsen's work, and, in 2011, he was the recipient of two Best Dissertation awards from organized sections of the American Political Science Association. He also received an award for Best Paper presented on a Political Psychology panel at the annual meeting of the American Political Science Association in 2011. In 2013, Professor Bolsen received the Franklin L. Burdette/Pi Sigma Alpha Best Paper Award for the best paper presented at the annual meeting of the American Political Science Association. He received his Ph.D. in Political Science from Northwestern University in 2010.

Henry Carey

Best Researcher/Scholar, Political Science

Henry F. (Chip) Carey is Associate Professor of Political Science at Georgia State University, where he has been based since 1998. He has published many books and articles on international law, human rights and comparative democratization. He received his Ph.D. from Columbia University in 1997. His most recent books are: Understanding International Law Through Moot Courts: Genocide, Torture, Habeas Corpus, Chemical Weapons and the Responsibility to Protect (2014); European Institutions, Democratization, and Human Rights Protection in the European Periphery (2014); Trials and Tribulations of International Prosecution (2013); Privatizing the Democratic Peace: Policy Dilemmas of NGO Peacebuilding (2012); and Reaping what you Sow: A Comparative Examination of Torture Reform in the United States, France, Argentina, and Israel (2012).

Ryan Carlin

Best Researcher/Scholar, Political Science
Best Researcher/Scholar, Public Affairs, Political & Policy Sciences

Michael Evans

Most Helpful to Students, Political Science

Michael Evans is an Assistant Professor of Political Science at Georgia State University. He received a BA from Western Washington University (WWU) in 1999, majoring in Politics, Philosophy and Economics, and an MA from WWU in Political Science in 2001. In 2009, he received a PhD in Political Science from the University of Maryland. His exam fields at Maryland were American Government and Politics, Philosophy, and Public Policy.

Sarah Gershon

Best Teacher, Political Science

Dr. Sarah Allen Gershon is an Associate Professor of Political Science at Georgia State University. She received her Ph.D. from Arizona State University in 2008. Her research focuses primarily on the incorporation of traditionally underrepresented groups (including women, and racial and ethnic minorities) into the American political system. In seeking to explain the challenges faced by these groups, Dr. Gershon's work emphasizes the role of communication, campaigns and political attitudes. Her work combines research from multiple fields and relies on a diverse set of methods, including content analysis, experimentation, surveys and in-depth interviews. Dr. Gershon's research has been published in several academic journals, including Political Communication; Political Research Quarterly; the Journal of Women, Politics and Policy; Social Science Quarterly and the Journal of Politics.

Charles Hankla

Best Teacher, Public Affairs, Political & Policy Sciences

Charles R. Hankla is associate professor of political science at Georgia State University in Atlanta. He received his PhD in 2005 from Emory University, and he also holds degrees from Georgetown University and the London School of Economics. In the spring and summer of 2013, he was a visiting scholar at Sciences Po Lille in France.

Beverly Langford

Best Teacher, Marketing
Most Helpful to Students, Marketing

Langford, who serves as Program Coordinator for the Business Communication Programs in the Department of Marketing, has been involved in the Business Communication Programs of the Robinson College of Business since its early development. She participated in developing strategies and courses for this initiative and designed both undergraduate and graduate courses that she also teaches.

Langford is author of The Etiquette Edge: The Unspoken Rules for Business Success, published by AMACOM Press. She is also co-author of Communication Skills and Strategies: Guidelines for Managers at Work. She is President of an Atlanta-based corporate communications consulting firm whose clients include Southern Company, St. Joseph's Hospital, ista, Jamestown, YKK North America, Bennett Thrasher, Leadership North Fulton, and TenCate Protective Fabrics.

Langford is a member of the Association of Business Communicators. She is also a member of the Association of Professional Communication Consultants and has completed her certification as a Certified Course Designer/Developer.

Carrie Manning

Best Overall Faculty Member, Political Science
Best Overall Faculty Member, Public Affairs, Political & Policy Sciences

Dr. Carrie Manning is Professor and Chair of the Department of Political Science at Georgia State University in Atlanta, Georgia. She is the author of three books and more than two dozen journal articles and contributions to edited volumes. Her work on comparative democratization and post-conflict politics has been published in such journals as Comparative Politics, Journal of Democracy, Party Politics, Studies in Comparative

International Development, and Democratization, among others. Her most recent book, Costly Democracy: Building Democracy after Civil War (Stanford University Press, 2012), coauthored with Christoph Zurcher and others, seeks to explain success and failure in post-conflict democratization using a cross-regional comparison of nine countries. Her second book, The Making of Democrats: Elections and Party Development in Post-War Bosnia, El Salvador and Mozambique (Palgrave MacMillan, 2008) examines the impact of participation in electoral politics over a ten-year period on former armed opposition groups turned parties in each of these cases. Her first book, The Politics of Peace in Mozambique (Praeger 2002), traces the dual process of war termination and democratization in that country from 1992–2000. She lived in southern Africa from 1994–1998 and served as Country Director for the National Democratic Institute in Angola in 1997–98. She has conducted seminars in civil-military relations in more than a dozen African countries. Manning holds a PhD in Political Science from the University of California, Berkeley (1997), an MPA from the Woodrow Wilson School of Public and International Affairs at Princeton University (1991), and a BA from Wesleyan Univeristy (1986).

Sean Richey

Best Overall Faculty Member, Political Science
Best Researcher/Scholar, Political Science

I am an Associate Professor in the Department of Political Science at Georgia State University. I teach and research political communication and political behavior, with a specialization in voting behavior, political discussion networks, social capital, and political participation.

In 2013 to 2014, I was on a Fulbright Scholar Fellowship in Japan at the University of Tokyo and Japan Women's University.

Denish Shah

Best Overall Faculty Member, Marketing
Best Researcher/Scholar, Marketing

"Great researcher. Great teacher."

Shah's research mainly focuses on enhancing the role and effectiveness of marketing in firms. His research has earned best paper awards from the Journal of Marketing, MSI, and Journal of Service Research. His research for Prudential was adjudged as a finalist for the INFORM's Practice Prize award. Shah's research has been published in top academic and practitioner journals such as the Marketing Science, Journal of Marketing Research, Journal of Marketing, Harvard Business Review, and Sloan Management Review. His doctoral dissertation earned three awards.

Amy Steigerwalt

Most Helpful to Students, Political Science
Most Helpful to Students, Public Affairs, Political & Policy Sciences

Amy Steigerwalt is an Associate Professor of Political Science at Georgia State University. Her research focuses on the federal judicial selection process, as well as the role of courts as institutions and the differing influences on court operations and decision-making. She has published three books to date, including Battle Over the Bench: Senators, Interest Groups and the Politics of Courts of Appeals Confirmations (UVA Press, 2010), as well as articles in journals such as Journal of Politics, Political Research Quarterly, and Justice System Journal. Her most recent book is The Puzzle of Unanimity: Consensus on the U.S. Supreme Court, coauthored with Pamela Corley and Artemus Ward (Stanford Press, 2013). Dr. Steigerwalt, along with Drs. Corley and Ward, was recently awarded the 2013 Hughes-Gossett Award for the best article published that year in the Journal of Supreme Court History.

Jonathan Wiley

Best Researcher/Scholar, Real Estate

Professor Wiley teaches undergraduate and graduate courses on Real Estate Development and Real Estate Finance and Mortgage Banking at Georgia State University. Professor Wiley has published a number of articles on commercial real estate, investment, valuation, development, brokerage, and sustainable real estate. Many of these publications appear in leading academic real estate journals including Real Estate Economics, Journal of Real Estate Finance and Economics and Journal of Real Estate Research. Professor Wiley has presented his research at the annual meetings of premier industry organizations including the American Real Estate and Urban Economics Association, the American Real Estate Society, and the Financial Management Association.

Prior to joining the faculty at Georgia State University, Professor Wiley held faculty positions at Clemson University and the College of Charleston. In Charleston, he also served as director of the Carter Real Estate Center. In addition, Professor Wiley has served as a consultant to real estate developers and investors on numerous projects throughout the region.

GEORGIAN COURT UNIVERSITY

Ashley Elmore

Best Teacher, Business
Best Overall Faculty Member, Business

Professor Elmore is currently in her 2nd year as a marketing and business professor at Georgian Court University. Prior to relocating to New Jersey, Ashley Elmore taught marketing education in Fairfax County Public Schools for seven years. Elmore's students competed at international marketing competitions where they earned many accolades. Professor Elmore was recognized by WJLA Channel 7 News as DC, Maryland and Virginia's Working Woman of the Year for her unique and high energy approach to teaching.

Cathleen McQuillen

Best Overall Faculty Member, Business
Best Researcher/Scholar, Business
Most Helpful to Students, Business

GETTYSBURG COLLEGE

Kevin Wilson

Best Overall Faculty Member, Psychology

GONZAGA UNIVERSITY

Don Hackney

Best Researcher/Scholar, Business

"Deep interest in his subject – looks beyond the obvious to find those obscure but extremely interesting facts that make the subject more interesting and connect the present with the past to complete our understanding of how we became who and what we are."

Hackney, D. D. (2012). Local Labor Markets, Employment Distributions and Consumer Bankruptcy Filings: Evidence from Eastern Washington. Journal of Accounting and Finance, 12(5): 133–139.

Personal Bankruptcy and Tax Debt: An Examination of the Usefulness of Chapter 13 in Managing IRS Claims, (Donald D. Hackney) Journal of Legal Technology Risk Management, Fall 2010, Issue 5.1.

Medical Expense and Bankruptcy, (Donald D. Hackney) Atlantic Economics Journal, 2009, 38:121–122.

Do Health Care Debts Affect Bankruptcy Chapter Filing Choices, (Donald D. Hackney) Atlantic Economic Journal-Anthology, 2009.

Kent Hickman

Best Overall Faculty Member, Business

"Extremely intelligent, greatly creative. Friendly, personable, a great sense of humor. Helpful to all who ask for help. A degree of common sense that surpasses the common."

Dr. Hickman teaches Principles of Finance as well as courses in Valuation, Mergers and Acquisitions, Sustainable Business, and Corporate Governance. He has held appointments as a Teaching and Research Fellow at Aston University in Birmingham, United Kingdom, the Audi Guest Professor at the Catholic University of Eichstaett-Ingolstadt, Germany, and Visiting Professor at the Graduate School of Business in Rouen, France. He is the co-author of two textbooks, Principles of Finance and Managerial Finance, his refereed publications appear in Journal of Financial Economics, Financial Management, The Sustainability Review, Journal of Portfolio Management, Economics Letters, Journal of Economic Behavior and Organization, Journal of Applied Corporate Finance and Business Horizons, among others. He enjoys biking and fishing and watching the Zags.

GOODWIN COLLEGE

Lois Daniels

Most Helpful to Students, Health Sciences and Nursing

Lois graduated from SUNY at Buffalo (BSN) and Yale University (MSN, CNM) schools of nursing. She has functioned in a variety of roles including teaching in LPN, ADN, RN-BSN, MSN and nurse-midwifery programs. She has been a Nurse Consultant for the State of Connecticut Department of Public Health, where she managed adolescent pregnancy/parenting, infant mortality and maternal mortality programs as well as writing federal grant applications in these areas. She developed an evaluation and approval program for the Centering Healthcare Institute that has invented and pioneered a group model of prenatal care, nationally and internationally. She also assumed a leadership role in the development and implementation of the Associate Degree in Nursing program at Goodwin College. Her hobbies include sewing handcrafted items, creating jewelry and writing poetry.

GOVERNORS STATE UNIVERSITY

Jennifer Groebner

Best Researcher/Scholar, Health Administration

Currently serve as the Undergraduate Program Director and an Instructor in the Health Administration Department and she is the Faculty Advisor for the Health Administration Alumni Association, and Chair of the Bachelor in Health Administration Advisory Board.

(*ref:http://www.govst.edu/uploadedFiles/Academics/Colleges_and_Programs/CHHS/Departments/Health_Administration/Jennifer%20Beth%20Groebner%20docx%20resume%20docx%20GSU.pdf)

Colleen Sexton

Best Overall Faculty Member, Education

"Colleen is very dedicated to maintaining the integrity of the education program at GSU. She is extremely conscientious, caring, and always researching innovative, relevant, and useful technology tools for the classroom. She demonstrates the highest concern for making sure that the future teachers have the best possible preparation for being a classroom teacher."

GRADUATE CENTER - CUNY

David Jaeger

Best Overall Faculty Member, Economics

"Terrific teacher, researcher, and dissertation supervisor."

GRAND VALLEY STATE UNIVERSITY

Mary Bower-Russa

Most Helpful to Students, Psychology

Luke Galen

Best Researcher/Scholar, Psychology

I am currently specializing in the field of the psychology of religion and irreligion. My recent work has examined religious fundamentalism, secularity, prosocial behavior, and social cognition. I have conducted projects focusing on the interpretation of religious versus secular prosociality, the influence of fundamentalism on attitudes toward animal rights and evolution, the formation of interpersonal impressions based on others' religiosity, and the Belief in a Just World. I also study non-religious individuals and secular groups as well as the process of apostasy. I teach courses on counseling and psychotherapy, controversial issues in psychology and human sexuality and have an interest in the intersection of religion and sexuality.

Robert Hendersen

Best Teacher, Psychology

Brian Lakey

Best Overall Faculty Member, Psychology

GRINNELL COLLEGE

Brad Graham

Most Helpful to Students, Economics

Bradley Graham, Assistant Professor, has B.S.E. and J.D. degrees from the University of Iowa, and M.A. and Ph.D. degrees from the University of Colorado at Boulder. His course offerings include empirical methods in economics, law and economics, and macroeconomic analysis. His research interests are law and economics and the economics of innovation.

Mark Montgomery

Best Researcher/Scholar, Economics
Best Teacher, Economics

Jack John Mutti

Best Overall Faculty Member, Economics
Best Researcher/Scholar, Business

HAMPTON UNIVERSITY

Ralph Charlton

Most Helpful to Students, Education

Expertise: Sport Marketing, Sport Management, Higher Education Administration, Social Marketing

Ten years of instruction and research in the areas of sport marketing and management. Emphasis on new/social media and grassroots marketing in the sport industry. PhD in Higher Education Administration supports teaching content in leadership and organizational behavior.

HARPER COLLEGE

John Clarke

Best Overall Faculty Member, Psychology

"Professionalism, great communication skills, easy to access"

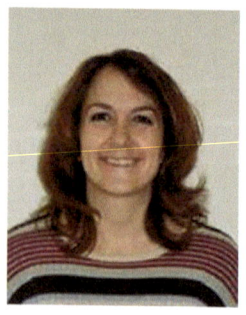

Dawn McKinley

Best Overall Faculty Member, Accounting

Dawn McKinley is a Instructor and Coordinator of the Accounting Department at Harper College. She is a Certified Public Accountant (CPA) and Certified Management Accountant (CMA). She has completed a Masters of Accountancy and Bachelors in Business Administration, Accounting from University of Iowa and also she has completed a Associates in Arts from North Iowa Area Community College.

Julie Nemmer

Best Teacher, Accounting

Beth Nudelman

Most Helpful to Students, Health Sciences and Nursing

Mary Schaefer

Most Helpful to Students, Health Science

Michael Thiry

Most Helpful to Students, Accounting

HARRISBURG AREA COMMUNITY COLLEGE

Carol Niblette

Best Teacher, Education

"Carol has been with HACC for many years and has seen many changes. She has embraced these changes and has always gone the extra mile to serve her students and college. Carol continually updates and evaluates her classes based on changes in education or her own professional development. She holds high expectations but is willing to help any student that is willing to try. Carol is also an excellent advisor, many students have commented on how knowledgeable and helpful she has been. Any students I have in my classes have very positive comments about her as far as interesting classes and projects applicable to real life. Carol also does an excellent job with her practicum students. Carol also helps out often with college events or attends college functions despite the fact she is not a full time employee. I have worked with Carol for the last four years and find her planning and working with her students exceptional. I would like to see Carol receive the recognition due to her for her dedication and creativity so that is why I nominate her."

HARTNELL COLLEGE

Barbara Durham

Best Researcher/Scholar, Nursing

Janeen Whitmore

Most Helpful to Students, Nursing

HARVARD UNIVERSITY

Harvey Mansfield

Best Researcher/Scholar, Political Science

Harvey C. Mansfield, William R. Kenan, Jr., Professor of Government, studies and teaches political philosophy. He has written on Edmund Burke and the nature of political parties, on Machiavelli and the invention of indirect government, in defense of a defensible liberalism and in favor of a Constitutional American political science. He has also written on the discovery and development of the theory of executive power, and has translated three books of Machiavelli's and (with the aid of his wife) Tocqueville's Democracy in America. His book on manliness has just been published. He was Chairman of the Government Department from 1973–1977, has held Guggenheim and NEH Fellowships, and has been a Fellow at the National Humanities Center. He won the Joseph R. Levenson award for his teaching at Harvard, received the Sidney Hook Memorial award from the National Association of Scholars, and in 2004 accepted a National Humanities Medal from the President. He has hardly left Harvard since his first arrival in 1949, and has been on the faculty since 1962.

Robert Paarlberg

Most Helpful to Students, Political Science
Most Helpful to Students, Public Affairs, Political & Policy Sciences

I do most of my research and consulting in the area of international food and agricultural policy, especially in Africa and the developing world. This topic connects me both to my own family history (my father grew up on a farm in Indiana) and to an important current issues in international development: How to help farmers in Africa – most of whom are women – increase their productivity to better feed their families and escape poverty. In the past decade, I have worked in more than a dozen countries in Africa, supported by the Bill and Melinda Gates Foundation, the International Food Policy Research Institute, and the United States Agency for International Development. My 2008 book from Harvard University Press (Starved for Science: How Biotechnology is Being Kept Out of Africa) has a foreword by two Nobel Peace Prize winners, Jimmy Carter and Norman Borlaug. My current research examines the impact of international trade on agricultural land use. In recent years, my students at Wellesley have taken an increased interest in issues of food and farming around the world. They want to know what kinds of food and farm systems can provide not just increased production, but social justice, improved nutrition, and environmental sustainability as well. I address these questions in a senior seminar I teach every year, and in 2010 I will publish a new book from Oxford University Press (Food Politics: What Everybody Needs to Know) based on the materials developed in this seminar. I also teach two large international relations courses every year, one on international economic policy and the other on "theories" of United States foreign policy. In addition, I teach the introductory course in our department, which showcases eight important books written by political scientists, from Machiavelli to the present. My work on international agriculture engages me with a wide variety of audiences beyond the academic world. In the past year I have given talks to the executive leadership of the Mars Candy Company (on cocoa, in Africa), to the Pontifical Academy of Sciences in Rome (on agricultural technology), and to a conference on African farming in Uganda. In addition, I gave testimony in 2009 to the Senate Foreign Relations Committee, on U.S. agricultural development assistance policy. Most interesting to me, however, are the visits I make to farms and farmers in the developing world, where a combination of bad history and bad current policy have held too many people in poverty for too long. In the summer months, between work and travel, I enjoy retreating with my wife Marianne to our place on the coast of Maine, where there is plenty of hiking, biking, swimming, fishing, boating, golfing, and picture taking to do, and where the lobsters always taste good.

Robert Putnam

Best Overall Faculty Member, Political Science
Best Teacher, Political Science
Best Researcher/Scholar, Public Affairs, Political & Policy Sciences
Best Overall Faculty Member, Public Affairs, Political & Policy Sciences

Robert D. Putnam is the Peter and Isabel Malkin Professor of Public Policy at Harvard, where he teaches both undergraduate and graduate courses, and is the 2013–14 Distinguished Visiting Professor at Aarhus University (Denmark). Professor Putnam is a member of the National Academy of Sciences, a Fellow of the British Academy, and past president of the American Political Science Association. In 2006, Putnam received the Skytte Prize, the world's highest accolade for a political scientist, and in 2012, he received the National Humanities Medal, the nation's highest honor for contributions to the humanities. Raised in a small town in the Midwest and educated at Swarthmore, Oxford, and Yale, he has served as Dean of the Kennedy School of Government. The London Sunday Times has called him "the most influential academic in the world today."

He has written fourteen books, translated into twenty languages, including the best-selling Bowling Alone: The Collapse and Revival of American Community, and more recently Better Together: Restoring the American Community, a study of promising new forms of social connectedness. His previous book, Making Democracy Work, was praised by the Economist as "a great work of social science, worthy to rank alongside de Tocqueville, Pareto and Weber." Both Making Democracy Work and Bowling Alone are among the most cited publications in the social sciences worldwide in the last half century.

He consults widely with national leaders, including the last three American presidents, the last three British prime ministers, and the last French president. He co-founded the Saguaro Seminar, bringing together leading thinkers and practitioners to develop actionable ideas for civic renewal. His earlier work included research on comparative political elites, Italian politics, and globalization. Before coming to Harvard in 1979, he taught at the University of Michigan and served on the staff of the National Security Council.

Putnam's most recent book, American Grace, co-authored with David Campbell of Notre Dame, focuses on the role of religion in American public life. Based on data from two of the most comprehensive national surveys on religion and civic engagement ever conducted, American Grace is the winner of the American Political Science Association's 2011 Woodrow Wilson Foundation Award for the best book on government, politics, or international affairs.

He is currently working on one major empirical project: Inequality and opportunity: the growing class gap among American young people and the implications for social mobility.
Courses Fall
DPI-360 Social Capital and Public Affairs: Research Seminar
Research

For a complete list of faculty citations from 2001–present, please visit the HKS Faculty Research Connection. Selected Publication Citations:

Academic Journal/Scholarly Articles Frederick, Carl B., Kaisa Snellman, and Robert D. Putnam. "Reply to Gao et al: Racial Composition Does Not Explain Increasing Class Gaps in Obesity." Proceedings of the National Academy of Sciences of the United States of America 111.22 (June 2014): 2238–2238. Commentary Putnam, Robert D. "Crumbling American Dreams." New York Times, August 3, 2013.

Michael Sandel

Best Teacher, Public Affairs, Political & Policy Sciences

Michael J. Sandel is the Anne T. and Robert M. Bass Professor of Government at Harvard University, where he has taught political philosophy since 1980. His recent book, What Money Can't Buy: The Moral Limits of Markets, takes on one of the biggest ethical questions of our time: What should be the role of money and markets in our society? Sandel's work has been translated into 27 languages. His books include Liberalism and the Limits of Justice (Cambridge University Press, 1982, 2nd edition, 1998), Democracy's Discontent (Harvard University Press, 1996), Public Philosophy: Essays on Morality in Politics (Harvard University Press, 2005), and The Case against Perfection: Ethics in the Age of Genetic Engineering (Harvard University Press, 2007), and Justice: What's the Right Thing to Do? (Farrar, Straus and Giroux, 2009). At Harvard, Sandel's courses include "Ethics, Biotechnology, and the Future of Human Nature," "Ethics, Economics, and Law," and "Globalization and Its Critics." His undergraduate course "Justice" has enrolled over 15,000 students, and was the first Harvard course to be made freely available online (www.JusticeHarvard.org) and on television. A recipient of the Harvard-Radcliffe Phi Beta Kappa Teaching Prize, Sandel was recognized by the American Political Science Association in 2008 for a career of excellence in teaching. He has been a visiting professor at the Sorbonne (Paris), delivered the Tanner Lectures on Human Values at Oxford University, and in 2009 delivered the BBC Reith Lectures. In 2010, China Newsweek named him the "most influential foreign figure of the year" in China. From 2002 to 2005, Sandel served on the President's Council on Bioethics. He is a member of the American Academy of Arts and Sciences, the Board of Trustees of Brandeis University, the Board of Directors of the Institute for Human Sciences (Vienna), and the Council on Foreign Relations. A graduate of Brandeis University (1975), Sandel received his doctorate from Oxford University (D.Phil.,1981), where he was a Rhodes Scholar.

HENRY FORD COMMUNITY COLLEGE

Janice Bartos

Most Helpful to Students, Nursing

"Jan Bartos is very kind and humble to her students. Jan goes above and beyond for the students. She communicates well with the staff and her students. She has a very caring heart and loves her job. She is very creative with her work in her new job role. I highly recommend Janice to receive this recognition."

Eric Gackenbach

Most Helpful to Students, Education

Gwendolyn Pringle

Most Helpful to Students, Counseling
Best Researcher/Scholar, Education
Best Teacher, Education
Best Overall Faculty Member, Education

HOFSTRA UNIVERSITY SCHOOL OF LAW

James Sample

Best Overall Faculty Member, Law

Before joining the Hofstra Law faculty in 2009, Professor Sample served as an attorney in the Democracy Program at the Brennan Center for Justice at New York University School of Law since 2005. Before joining the Brennan Center, he worked on Brian Schweitzer's gubernatorial campaign in Montana and clerked for Judge Sidney R. Thomas of the U.S. Court of Appeals for the Ninth Circuit.

Professor Sample received his J.D. from Columbia Law School in 2003. While at Columbia, Professor Sample was a Harlan Fiske Stone and James Kent Scholar and served as a notes editor on the Columbia Law Review. He has written several publications and legal briefs on the topic of judicial elections, campaign financing and recusal.

Professor Sample regularly comments on voting rights and constitutional issues in leading media outlets, including The Wall Street Journal, The New York Law Journal, Slate.com and The Huffington Post, as well as at national conferences.

Citations to Professor Sample's work on issues related to democracy include in opinions of the U.S. Supreme Court, articles in the Harvard Law Review, Columbia Law Review and other leading journals, as well as in major media, including The New York Times, The Washington Post, The Economist, U.S. News & World Report, the Los Angeles Times, National Public Radio, USA Today and The National Law Journal, as well as in leading blogs and regional outlets throughout the country.

Prior to attending law school, Professor Sample was a three-time Emmy Award winner for his work as a producer with NBC Sports.

HORRY-GEORGETOWN TECHNICAL COLLEGE

Ann Daniels

Best Overall Faculty Member, Health Sciences and Nursing

"As the Department Chair, she is focused on ensuring students are treated fairly and consistently. She has made some faculty moves to ensure quality in the nursing program."

Sherry James

Best Overall Faculty Member, Nursing

HOSTOS COMMUNITY COLLEGE

Nieves Aguilera

Best Researcher/Scholar, Nursing

HOUSTON COMMUNITY COLLEGE (ALL CAMPUSES)

Katherine Abba

Best Overall Faculty Member, Education
Best Researcher/Scholar, Education
Best Teacher, Education
Most Helpful to Students, Education
Best Researcher/Scholar, Education

Best Teacher, Education
Most Helpful to Students, Education
Best Overall Faculty Member, Education

Ibrahim Firat

Best Overall Faculty Member, Business Administration

A native of Istanbul, Turkey, Firat earned his bachelor's degree in History and Mathematics from the University of St. Thomas (Houston, TX). In 2007, Firat co-founded UZO Umbrellas in Houston, TX, and started serving as the International Business Director overseeing all global sourcing, focusing on Europe. His outstanding negotiation and communication skills helped acquire 23 international patents for the company, and in 2009, Firat expanded UZO into the European markets, locating buying partners from Germany, Switzerland, and England. Firat obtained his Master's in Business Administration (MBA) with a concentration in Marketing and Entrepreneurship in 2008 with Summa Cum Laude honors, and became an honorable member of the International Business Administration Society, Delta Mu Delta. In March 2008, as an opportunistic entrepreneur, Firat decided to capitalize on his unique skills, experience, passion for learning, and teaching through his own highly exclusive, individually-catered educational consulting company, Firat Educational Solutions (FES). In 2010, Firat opened his international educational consulting company, Firat Academy, in Turkey. Firat published various papers in nationwide educational journals, spoke at national conferences, and appeared on local and national TV news as an education and entrepreneurship expert. He is an active member of Higher Education Consultants Association (HECA), Independent Educational Consultants Association (IECA), Houston Livestock Show and Rodeo, and is a professor of entrepreneurship and business management at the Houston Community College (HCC). Firat has continuous relationships and activities with the Small Business Administration (SBA) Houston, SCORE Houston, Houston Minority Supplier Development Council (HMSDC), Houston East End Chamber, and Houston Intercontinental Chamber of Commerce. In 2011, he was appointed as the Lead Faculty at HCC for the Goldman Sachs 10,000 Small Businesses initiative. Firat also serves as a member for 10,000 Small Businesses National Curriculum Review Board, and as a Lead Faculty for the National Cohort of the Goldman Sachs 10,000 Small Businesses at Babson College, Boston, MA.

Linda Koffel

Best Teacher, Business Administration

Shafiqullah Rahman

Best Teacher, Business

Tyrone Sharp

Best Overall Faculty Member, Health Sciences and Nursing

Mia Taylor

Most Helpful to Students, Business

Alden Tiggs

Best Teacher, Business Technology
Most Helpful to Students, Business Technology

HOWARD UNIVERSITY

Ben Fred-Mensah

Most Helpful to Students, Political Science

Lorenzo Morris

Best Overall Faculty Member, Public Affairs, Political & Policy Sciences

HUDSON COUNTY COMMUNITY COLLEGE

Siroun Meguerditchian

Best Overall Faculty Member, Culinary Arts

HUDSON VALLEY COMMUNITY COLLEGE

Eileen Mahoney

Best Overall Faculty Member, Education

Lori Purcell

Best Overall Faculty Member, Mortuary Science

ILLINOIS CENTRAL COLLEGE

Kelly Crawford-Jones

Best Overall Faculty Member, Health Science
Best Researcher/Scholar, Health Science
Best Teacher, Health Science
Most Helpful to Students, Health Science
Best Researcher/Scholar, Health Sciences and Nursing
Best Teacher, Health Sciences and Nursing
Most Helpful to Students, Health Sciences and Nursing
Best Overall Faculty Member, Health Sciences and Nursing

Duane Hill

Best Overall Faculty Member, Accounting

Paul Swanson

Best Overall Faculty Member, Business

ILLINOIS VALLEY COMMUNITY COLLEGE

Jill Urban-Bollis

Best Teacher, Psychology

INDIANA STATE UNIVERSITY

Deb Barnhart

Most Helpful to Students, Nursing

"Deb Barnhart has devoted her entire life to the success of our students prior to entering and during the nursing program. She has sacrificed a lot and given her all. Very deserving of this award."

John Conant

Best Researcher/Scholar, Business
Best Overall Faculty Member, Business

Ryan Donlan

Best Researcher/Scholar, educational leadership

"Ryan is engaged in research and presenting research at the local, national, and international levels. He could not do more."

Dr. Ryan A. Donlan, as Assistant Professor in Indiana State University's Department of Educational Leadership, served for twenty years in K–12 education, most of that time as a Superintendent, as well as a school building leader for at-risk students. He is a public speaker and visionary for the future of public education. Dr. Donlan, once a frequent skydiver, today enjoys more conservative, contemporary pursuits, such as cooking, reading, and camping. He has written/co-written Gamesmanship for Teachers: Uncommon Sense is Half the Work and The Secret Solution: How One Principal Discovered the Path to Success, and can be followed on Twitter at www.twitter/ryandonlan or on the Indiana State University's on-line "ISU Ed. Leadershop," where he offers thoughtful commentary each week to school leaders at www.k12edleadershipatISU.blogspot.com.

Susan Eley

Most Helpful to Students, Health Sciences and Nursing

Roseanne Fairchild

Best Researcher/Scholar, Nursing

Shiaw-Fen Ferng-Kuo

Best Researcher/Scholar, Health Sciences and Nursing

Robert Guell

Best Overall Faculty Member, Economics
Most Helpful to Students, Business

Micheal Harmon

Best Teacher, Business

Jeff Harper

Most Helpful to Students, Business

Terry McDaniel

Most Helpful to Students, educational leadership

"Terry does so much for students I would like to give him release time just for advising. It is his gift to the university."

Robert McMahan

Best Teacher, Business

Bobbie Jo Monahan

Best Teacher, Education

"She is the best recruiter and advisor we have, which manifests in the classroom as she brings a passion that transcends the students in the room, and trickles out to potential students, as our students talk about the great experiences they are having with her. She gets them excited about the program and keeps them here with her teaching."

Jin Park

Best Researcher/Scholar, Business

David Robinson

Best Overall Faculty Member, Business

Debra Vincent

Best Teacher, Health Sciences and Nursing

INDIANA UNIVERSITY BLOOMINGTON

Cheryl Hughes

Best Teacher, Business
Most Helpful to Students, Business
Best Overall Faculty Member, Business

Orville Powell

Best Overall Faculty Member, Management
Best Researcher/Scholar, Management
Best Teacher, Management
Most Helpful to Students, Management

Craig Ross

Best Researcher/Scholar, Education

INDIANA UNIVERSITY OF PENNSYLVANIA

James Jozefowicz

Best Teacher, Economics

"Jim is the most helpful to students in and out the classroom."

Dr. James J. Jozefowicz is a professor in the Department of Economics. He has been a member of the Department of Economics since 1999. Dr. Jozefowicz received his Ph.D. degree from the State University of New York at Albany. He holds a B.S. degree in Biology and a B.A. degree in Economics from Marist College. He received the IUP University Senate Distinguished Faculty Award for Teaching in 2005. In 2012, Dr. Jozefowicz was one of 14 economics professors selected for inclusion in the Princeton Review's Best 300 Professors, a guide recognizing top professors in the nation. Dr. Jozefowicz is the faculty co-advisor for the IUP Economics Club, chair of the Department of Economics Student Affairs Committee, chair of the Department of Economics Evaluation Committee, a member of the editorial board of the Pennsylvania Economic Review, and a past president of the Pennsylvania Economic Association. His areas of expertise are econometrics, economic education, industrial organization, managerial economics, and monetary economics.

Dr. Jozefowicz teaches Introduction to Econometrics and Advanced Econometrics as part of the Department of Economics Honors Track. His econometrics students have given presentations of their research papers at the IUP Undergraduate Scholars Forum, the Europe: East and West Undergraduate Research Symposium at the University of Pittsburgh, and conferences held by the Eastern Economic Association, Midwest Economics Association, Southern Economic Association, and Pennsylvania Economic Association. Several of these students have won awards for their papers and presentations, including the Best Undergraduate Student Paper at the 2011, 2012, and 2013 Pennsylvania Economic Association Conferences.

Dr. Jozefowicz has published more than a dozen articles in refereed journals such as Applied Economics, Atlantic Economic Journal, International Advances in Economic Research, International Journal of Applied Economics, Journal of Economic Education, Journal for the Scientific Study of Religion, National Social Science Journal, New York Economic Review, Pennsylvania Economic Review, and Perspectives on Economic Education Research. Many of these journal articles have been co-authored with undergraduate students.

Raymond Pavloski

Best Researcher/Scholar, Psychology

INDIANA UNIVERSITY SOUTH BEND

Steven Gerencser

Best Overall Faculty Member, Political Science
Most Helpful to Students, Political Science
Most Helpful to Students, Public Affairs, Political & Policy Sciences
Best Overall Faculty Member, Public Affairs, Political & Policy Sciences

Neovi Karakatsanis

Best Teacher, Political Science
Best Teacher, Public Affairs, Political & Policy Sciences

Jamie Smith

Best Researcher/Scholar, Political Science
Best Researcher/Scholar, Public Affairs, Political & Policy Sciences

INDIANA UNIVERSITY-PURDUE UNIVERSITY FORT WAYNE

David Dilts

Best Researcher/Scholar, Business

Dr. Dilts is the executive editor of the Journal of Collective Negotiations in the Public Sector. He was appointed the journal's executive editor in 2003 and is only the third person to have ever served as executive editor of the Journal.

Dr. Dilts joined the Indiana University-Purdue University Fort Wayne faculty in 1987 after serving two years at Ball State University and seven years at Kansas State University.

Dr. Dilts has been recognized with an award for teaching by the Graduate Business Council, EMBAT 2004, for research by Indiana-Purdue University - Fort Wayne with the 2004 Outstanding Research Award, and the Irwin Distinguished Paper Award in 1992 by the Midwest Society for Human Resources and Industrial Relations. Dr. Dilts has also been included in several editions of Who's Who Among America's Teachers after having been nominated by students at IPFW.

Dr. Dilts is a member of Omicron Delta Epsilon, Delta Sigma Pi, and Phi Kappa Phi.

John Kessler

Best Teacher, Business

Michael Slaubaugh

Most Helpful to Students, Business

Dr. Michael Slaubaugh received a Ph.D. in accounting from Indiana University in 1992. He joined the faculty of the Richard T. Doermer School of Business faculty in 1995. Professor Slaubaugh previously taught at Ball State University, Indiana University, University of Tennessee, and Ashland University.

Nichaya Suntornpithug

Best Overall Faculty Member, Business

James Toole

Most Helpful to Students, Political Science
Most Helpful to Students, Public Affairs, Political & Policy Sciences

Georgia Ulmschneider

Best Teacher, Political Science
Best Teacher, Public Affairs, Political & Policy Sciences

Michael Wolf

Best Overall Faculty Member, Political Science
Best Researcher/Scholar, Political Science
Best Researcher/Scholar, Public Affairs, Political & Policy Sciences
Best Overall Faculty Member, Public Affairs, Political & Policy Sciences

IRVINE VALLEY COLLEGE

Robert Melendez

Best Overall Faculty Member, Counseling

"Robert goes above and beyond his job description every day. He works hard to advocate for students and counselors alike. He is wonderful with students, is so helpful and informative, is a primary source of assistance for our Honors students, and has improved morale in our department since becoming our chair. Students and colleagues alike are grateful to have Robert to work with!"

Kathryn Milostan Egus

Best Teacher, Health Sciences PE & Athletics

"Knowledgeable, dedicated, passionate about her subject matter, kind to her students and colleagues"

Kathryn Milostan-Egus (Department Chair) is the founder of the Dance Program at Irvine Valley College. She formerly was a member of the companies of Annabelle Gamson

(Dance Solos, Inc.), Senta Driver (Harry), and Beverly Blossom (Blossom and Co.), where she performed in New York City, toured the U.S., and France. She has also performed in the works of other well-known modern dance choreographers such as Lucas Hoving, Douglas Nielsen, Stephen Koester and Mary Anthony. Additionally, she also performed with jazz dance companies, in films, commercials and stage productions; and in concerts for professional theatre, musical theatre and vocal groups. Her professional experience also includes serving as the Artistic Director of SAGA, a Hip Hop dance company in Los Angeles; and manager of the Victoria Marks Performance Company in New York City, where she attained state, local, and federal grants for the company, organized tours, and visits to local schools for an Arts in Education program. Kathryn has been on the faculties of numerous community colleges and universities in Southern California, Illinois and New York; taught at various festivals, including American College Dance Festival; offered workshops for athletes and dance instructors; and been an ACE-certified private trainer in New York City and California. She is certified with Distinction in Pilates and Core Intelligence and is also a GYROKINESIS® Pre-Trainer, for those interested in becoming a certified instructor. Kathryn has most recently been working on statewide career pathways in dance science, dance history, dance and technology and dance performance. In addition, she is the founder and president of InsightHELPS Foundation, Inc. (Healing, Empowering, Learning, Preventions and Science), for which she created the "There is Hope" presentation, designed for various Parkinson Associations, as well as specialized workshops and classes for private clients and their care-givers, making significant progress in their sense of smell and movement abilities.

IVY TECH COMMUNITY COLLEGE: INDIANAPOLIS

Dennis McCrory

Best Teacher, Accounting

J. SARGEANT REYNOLDS COMMUNITY COLLEGE

Kathy Larue

Best Overall Faculty Member, Education

"Kathy is extremely effective as a educator and colleague at Reynolds. She in very helpful to students and other faculty."

Sonya Shaw

Best Teacher, Early Childhood Education

JACKSONVILLE STATE UNIVERSITY

Joseph Akpan

Best Researcher/Scholar, Education

"He published 6 articles in the 2013–2014 academic year."

Janet Bavonese

Best Overall Faculty Member, Education

Misty Cothran

Best Teacher, Psychology

Kyoko Johns

Most Helpful to Students, Education

Kyoko Johns, PhD, an assistant professor of mathematics education, teaches elementary mathematics methods at the credential level and supervises practicum teachers at Jacksonville State University in Jacksonville, AL. Her areas of interest include examining students' thinking processes when solving mathematical problems, investigating social and cultural effects on mathematics education, and exploring ways to incorporate technology in the classroom.

Nina King

Best Teacher, Education

Patricia Lowry

Most Helpful to Students, Education

Cynthia McCarty

Most Helpful to Students, Business

Lynetta Owens

Best Overall Faculty Member, Education

David Palmer

Best Overall Faculty Member, Business

Gina Riley

Best Teacher, Education

Melinda Staubs

Best Researcher/Scholar, Education

Dr. Melinda Staubs holds a B.S. in Elementary Education from the University of Tulsa, and an M.S., Ed.S., and Ed.D. in Elementary Education from The University of Alabama. Prior to joining the faculty at Jacksonville State University as an Assistant Professor of Elementary Education, Dr. Staubs taught elementary and middle school in Oklahoma, Virginia, and Alabama.

Dr. Staubs's educational interests include teacher education, social studies, and cultural diversity. Presently, she is the Elementary Graduate Faculty Chair, Comprehensive Exam Coordinator for the Elementary Master's (MSEd) program, and Program Advisor for Elementary Educational Specialist (EdS) candidates.

James Thomas

Best Overall Faculty Member, Marketing
Best Researcher/Scholar, Marketing
Best Researcher/Scholar, Business

Mike Zenanko

Most Helpful to Students, Education

Mike is the Director of the Instructional Services Unit, Acting Coordinator of the Teaching/Learning Center, and Supervisor of the Multimedia Instructional Laboratories. He assists faculty, staff, and students in the use of the hardware and software. Mike received his M. Ed. from Vanderbilt University in Nashville, Tennessee, and his B.A. from Hendrix College, Conway, Arkansas. Research areas include tutoring and instructional technology. Mike also publishes on the areas of tutoring.

JACKSONVILLE UNIVERSITY

Jim Mirabella

Best Researcher/Scholar, Business
Best Teacher, Business
Most Helpful to Students, Business
Best Overall Faculty Member, Business

Dr. Jim Mirabella has been with Jacksonville University since 1998 when he joined the inaugural ADP faculty as an adjunct and has since become full-time in 2008. He enjoys spending time with his wife, Karen, and his 9-year old son, Sean (whom they homeschool). Jim and his son are members of the International Brotherhood of Magicians and the Fellowship of Christian Magicians, and they both enjoy performing Gospel magic, and Jim likes to use magic when teaching. He has written and directed many plays and enjoys leading the drama ministry at his church, and loves to travel (mostly cruising) as well as playing & watching sports.

Vince Narkiewicz

Best Overall Faculty Member, Marketing
Best Teacher, Marketing
Most Helpful to Students, Marketing

Vince Narkiewicz has over thirty-five years of undergraduate and graduate level teaching experience, and has been a member of the Davis College of Business faculty since the founding of the business college. His major areas of expertise include: market and opportunity analysis, product and new venture development, advertising management, marketing strategy and sales & customer relationship management. Since 1975, in addition to teaching, he has provided consulting services to trade and professional associations, consumer package goods companies in the food and cosmetic industries, as well as to firms in the building products, manufactured housing, and construction industries.

A few of the firms that he has worked with over the years include: The American Association of Advertising Agencies, Campbell Soup Company, Hershey Foods, Redken, The National Frozen Food Association, Armstrong World, Coastal Design & Construction, Deck House, Moffitt Corp., The National Association of Home Builders, Web4Minds, Inc., and at the community level: The Salvation Army of N.E. Florida and as a member of the Board of Directors at Gateway Community Services and First Coast Management Services. Prior to teaching he held marketing management positions in the frozen food industry. He also has an established track record as an investor and managing partner in the start-up and development of new businesses.

Dennis Ratliff

Best Teacher, Accounting

JAMES MADISON UNIVERSITY

Margaret Bagnardi

Best Researcher/Scholar, Nursing
Best Researcher/Scholar, Health Sciences and Nursing

Little more than a year after coming to JMU, nursing professor Margaret Bagnardi is changing patient care in Harrisonburg. In collaboration with Rockingham Memorial Hospital, Margaret empowers nurses to conduct research, implement evidence-based care and evaluate the results — "linking academia and the bedside," she says. Through professional seminars, she is building a dynamic network of information that flows from academia to health care professionals, effectively disseminating and exchanging state-of-the-art information. JMU's nursing students are involved too; they provide promising health care information from professional literature to RMH nurses, who in turn evaluate and use it to improve patient care. "It's creating an excitement. It's not just coming to work and doing a job . . . patients will get the most current research-based care they can get." Margaret who holds degrees in nursing and education — and who works as a nurse at RMH herself — knows this meshing of research and patient care is critical for patients. When asked who inspired her to pursue nursing, Margaret's answer is instantaneous: "Mom. She was a nurse. She taught us to see people as they are, and see the good in everyone.

Andrea Knopp

Best Overall Faculty Member, Nursing
Best Teacher, Nursing
Most Helpful to Students, Nursing
Best Teacher, Health Sciences and Nursing
Most Helpful to Students, Health Sciences and Nursing
Best Overall Faculty Member, Health Sciences and Nursing

JEFFERSON COLLEGE

Amy Kausler

Best Overall Faculty Member, Psychology

Amy Kausler earned an Associate of Arts Degree from Jefferson College, a Bachelor of Arts Degree in Psychology from Saint Louis University, a Master of Arts Degree in Psychology from University of Missouri - St. Louis, and a Ph.D. from the University of Missouri - St. Louis. A graduate of Crystal City High School, Dr. Kausler is a lifelong resident of Jefferson County and enjoys working with students from the community in which she resides. Her interests include reading, traveling, and spending time with her family.

JOHN CABOT UNIVERSITY

Pamela Harris

Most Helpful to Students, Law

"She is an excellent teacher, very devoted."

Pamela Harris studied at Berkeley and Harvard Law School, and clerked at the Italian Constitutional Court. She is the Associate Dean of Academic Affairs and Chair of the Department of Political and Social Sciences at John Cabot University and teaches courses in international law, constitutional law, American government, and political theory. She has written articles on law and literature as well as on religious liberty in the United States.

JOHN JAY COLLEGE OF CRIMINAL JUSTICE

William Gottdiener

Best Overall Faculty Member, Psychology

"The faculty member is an award winning teacher, mentor, researcher and successful leader."

Peter Mameli

Best Researcher/Scholar, Public Administration
Best Teacher, Public Administration

Professor Mameli's current research interests include studying the impacts of globalization processes on public administration, transnational crime and the oversight of public sector organizations. Recent projects have focused on such topics as government surveillance operations, national data protection authorities, international crime statistics, the impact of the Arab Spring on Middle East and North African public sectors, and transnational human trafficking. Some of his articles have appeared in Law & Policy, Crime, Law and Social Change, Critical Issues in Justice and Politics, International Public Management Journal, The Public Manager, and, Human Rights Review. In 2011, Dr. Mameli and his co-authors published, Security and Privacy: Global Standards for Ethical Identity Management in Contemporary Liberal Democratic States.

Professor Mameli has served as the Director of the Protection Management program, Coordinator of the undergraduate Public Administration program, and Director of the Saturday Masters of Public Administration program at different times while at John Jay College. Prior to working in the Department of Public Management, he held positions with the New York City Council, the New York City Board of Education and the New York City Office of Management and Budget.

Judy-Lynne Peters

Best Teacher, Public Administration
Most Helpful to Students, Public Administration

"Enthusiasm, mastery of discipline and willingness to work with students."

JOHN MARSHALL LAW SCHOOL

Susan Brody

Best Teacher, Law

After law school, Susan Brody served as law clerk for the Honorable Lloyd A. Van Deusen of the Illinois Appellate Court. Then she practiced with Beermann, Swerdlove, Woloshin, Barezky, and Berkson, concentrating in family law and appellate practice. Professor Brody joined the John Marshall faculty in 1982, and from 1985–1995, she served as director of the Lawyering Skills Program. She was associate dean for academic affairs from 1995–1998 and associate dean for institutional affairs during 1998–1999. Professor Brody teaches Civil Procedure and Feminist Legal Theory.

Professor Brody is active with the American Bar Association Section on Legal Education and Admission to the Bar, serving as chair or member of numerous site inspection teams as part of the ABA accreditation process. In September 2014, she will serve as chair for the site inspection team to visit Indiana Tech School of Law. She has recently also served on teams that visited Golden Gate University School of Law (2013), Lincoln Memorial University School of Law (2011), Liberty University School of Law (2009), and Western States University School of Law (2009). She has helped prepare site team chairs, law school representatives, and new site evaluators for future site inspections by presenting at ABA-sponsored workshops, most recently in 2010, 2011, and 2012.

Professor Brody also participates in conferences concerning feminist legal theory and storytelling, especially their intersection with teaching law, becoming a skilled lawyer, and understanding issues of diversity. Most recently she presented on these topics at an in-house seminar for the Illinois Attorney Registration and Disciplinary Commission (November 2012) and at Mercer University School of Law, where she served as a panelist on Feminist and Critical Theory (April 2011) and also presented a paper, "Opening the Lens of Otherness Through the Legacy of Virginia Woolf: Stories and Lessons in Feminist Legal Theory," based in part on her article, Law Literature, and the Legacy of Virginia Woolf: Stories and Lessons in Feminist Legal Theory, 21 Tex J Women & Law 1 (2011). Her article, "Twilight: The Unveiling of Victims, Stalking, and Domestic Violence," will be published this fall, 21 Cardozo J. L. & Gender – (forthcoming fall 2014).

Professor Brody serves on the Women's Board of the Museum of Contemporary Art in Chicago and the Board of Youth Services of Glenview/Northbrook.

Kim Chanbonpin

Best Teacher, Law
Best Overall Faculty Member, Law

Kim D. Chanbonpin joined the John Marshall faculty in 2008. Professor Chanbonpin received her bachelor's degree in English literature from the University of California at Berkeley. She earned her JD from the University of Hawai'i at Manoa, William S. Richardson School of Law, graduating cum laude with a certificate in Asian-Pacific Legal Studies. After law school, she was a law clerk to the late Judge John S.W. Lim, Intermediate Court of Appeals in Honolulu. Professor Chanbonpin also earned an LLM, with distinction, and a Certificate in National Security Law at the Georgetown University Law Center. While in Washington, D.C., she was a Short-Term Consultant at the World Bank.

Professor Chanbonpin is a member of the State Bar of California, and has been involved in several pro bono public cases litigating a variety of legal issues, including post-conviction relief, Violence Against Women Act (VAWA) self-petitions, and police brutality claims. In September 2012, she was appointed to a two-year fellowship under the Illinois State Bar Association's (ISBA) Diversity Leadership Council. She sits on the ISBA's Criminal Justice Section Council. Professor Chanbonpin is also currently serving as a Board Member of the Legal Writing Institute and on the Board of Governors for the Society of American Law Teachers (SALT).

Prior to coming to John Marshall, Professor Chanbonpin was a Westerfield Fellow at Loyola University New Orleans College of Law. During her fellowship, she taught National Security Law & Civil Liberties, Legal Research & Writing, and Moot Court (Appellate Advocacy).

Professor Chanbonpin teaches Lawyering Skills, Criminal Law, Torts, Gender Race and Class, and National Security Law. She also taught Introduction to the U.S. Legal System to LLM students in China's State Intellectual Property Office. Her scholarly writing considers redress and reparations law, policy, and social movements in the United States. In a 2011 article, she proposed the Inclusive Model for Social Healing, a new paradigm for understanding reparations projects. This model draws on anti-subordination and narrative principles rooted in LatCrit and Critical Race Theory scholarship, and is a part of the School of the Art Institute Sullivan Gallery's 2012 exhibition, "Opening the Black Box: The Charge is Torture." Her work on the law's power to exclude and to include continues in her 2013 article in the U.C. Irvine Law Review.

She is a contributor to the SALT Law blog, and her scholarly work has appeared in the U.C. Irvine Law Review, the Northwestern Journal of Law & Social Policy and the Mercer Law Review.

Michael Heyman

Best Researcher/Scholar, Law

Ann Lousin

Best Researcher/Scholar, Law

Between college and law school, Ann Lousin studied political science at the University of Heidelberg in Germany. After graduating from law school in 1968, she was a research assistant at the Sixth Illinois Constitutional Convention, where she worked on the drafting of the 1970 Illinois constitution. From 1971 to 1975, she was on the staff of the Speaker of the Illinois House of Representatives, including two years as Parliamentarian of the House.

She has served on several not-for-profit boards and governmental commissions, including a term as Chairman of the Illinois State Civil Service Commission. She is active in the commercial law committees of the American and Chicago Bar Associations, and has been the chair of the CBA Constitutional Law Committee. She has been a leader in other legal organizations, including service as Chair of the Board of Governors of the Armenian Bar Association from 1995 to 1998. She lectures and consults on the Illinois Constitution, general public law issues, and commercial law in the US and abroad. In 2009, she was elected a member of the American Law Institute.

Professor Lousin joined the faculty in 1975. She primarily teaches Sales Transactions.

Mary Nagel

Most Helpful to Students, Law

Following law school, Mary Nagel was a judicial law clerk for Hon. Thomas E. Hoffman on the Circuit Court of Cook County and for Hon. Fred A. Geiger, an appellate justice on the Second District Appellate Court for the State of Illinois. She was an associate at the Chicago offices of Querrey & Harrow and Bollinger, Ruberry & Garvey. From 1999 to 2001, Professor Nagel was an Illinois assistant attorney general, and from 2001 to 2003, she was chief legal counsel for the Illinois Department of Labor.

Most recently, she had been a judicial law clerk for Hon. Bill Taylor and Hon. Barbara J. Disko in the Circuit Court of Cook County, drafting court rulings for motions, evidentiary issues, and trials. She had been an adjunct professor at The John Marshall Law School from 1999 until 2007.

Ralph Ruebner

Best Overall Faculty Member, Law

Ralph Ruebner came to John Marshall with 12 years of experience as an appellate litigator, heading the Elgin and Chicago offices of the State Appellate Defender and representing indigent criminal defendants at all levels of appellate review. He has represented victims of government abuse in a number of countries, including the former Soviet Union and Peru.

Dean Ruebner has published articles and presented papers at various international conferences on human rights topics, and has testified before Congress on human rights conditions in Peru. From 2001–2003, he served as legal director of the Law Consortium for Palestinian Legal Education, a Rule of Law Project of U.S. A.I.D. In that capacity, he also visited Palestinian law schools in the West Bank.

He has drafted legislation in Illinois, making it the first state to criminalize international terrorism. He is the author of Illinois Criminal Trial Evidence, editor/author of Illinois Criminal Procedure, and co-author of Illinois Decisions on Search and Seizure; Illinois Evidence: Illinois Rules of Evidence, Statutes, and Constitution, A Compendium for Criminal Litigation; and faculty co-author of Illinois Judicial Benchbook on Criminal Law and Procedure. Dean Ruebner has served as a faculty member of the National Academy for Judicial Education and as a reporter of judicial conferences in Illinois. He serves as the Reporter of the Supreme Court Committee on Illinois Evidence.

Dean Ruebner joined the faculty in 1981. He founded John Marshall's Criminal Justice Clinic and served as its director for four years, served as Moot Court director for 16 years, and was the law school's Centennial Planning Committee chairperson. He was named associate dean for academic affairs in 2007. He teaches Criminal Procedure, Evidence, and International Human Rights.

Julie Spanbauer

Most Helpful to Students, Law

After graduating from law school, Julie Spanbauer served as a law clerk to Hon. Andrew P. Rodovich, U.S. Magistrate Judge, and Hon. James T. Moody, U.S. District Court Judge. Since joining the John Marshall faculty in 1990, she has published numerous articles in the areas of employment discrimination, constitutional law, and women's issues.

Professor Spanbauer was recently named a Fulbright Senior Specialist Grant Recipient and she taught at the Institute of Technology Law, National Chiao Tung University in Taipei, Taiwan. Professor Spanbauer has organized and served as a panel moderator at diverse conferences involving discrimination issues. She has presented lectures at Trinity College in Ireland and in China. Professor Spanbauer has served as the program director for two programs sponsored by the International Law Institute in Washington, DC.

These programs prepare international LLM students to enter law schools throughout the United States. She has also served as a member of the board of directors of the Friends of Battered Women and their Children, a not-for-profit organization providing counseling, advocacy, and education services for abused women and their children.

From 2004–2008, Professor Spanbauer served as director of the special admissions program at John Marshall, the Summer College for Assessing Legal Education Skills (SCALES). She teaches Employment Discrimination, Contracts, and Lawyering Skills.

JOHNS HOPKINS UNIVERSITY

Louise Schiavone

Most Helpful to Students, Business

"Louise is a consistent advocate for the students and works tirelessly to help bridge any communication gaps non-American students have."

Louise L. Schiavone (Journalism, Columbia University School of Journalism, New York City) joined the Johns Hopkins Carey Business School in 2011. She is a Senior Lecturer with a specialty in communications.

KANSAS STATE UNIVERSITY

Bronwyn Fees

Most Helpful to Students, Education

Craig Harms

Best Overall Faculty Member, Kinesiology
Best Researcher/Scholar, Kinesiology
Best Teacher, Kinesiology
Best Researcher/Scholar, Education
Best Teacher, Education
Best Overall Faculty Member, Education

Ann Knackendoffel

Best Teacher, Education

Sherri Martinie

Most Helpful to Students, Education

"Dr. Martinie is one of the best mathematics educators I've ever seen. She is a mentor to other faculty members as well as early career teachers, plus spends a great deal of time supporting her pre-service teachers. She is also very involved at the state level in addition to teaching at Kansas State University."

B.S. - Education, University of Kansas M.S. - Education, University of Kansas Licensure - Math, 7–12, Kansas State University Ph.D. - Curriculum and Instruction, Kansas State University

Robert Pettay

Most Helpful to Students, Kinesiology

Chwen Sheu

Best Researcher/Scholar, Business

Joe Ugrin

Best Teacher, Business

KENNESAW STATE UNIVERSITY

Lucy Ackert

Best Researcher/Scholar, Business

Hope Baker

Best Teacher, Business

Doug Bell

Best Researcher/Scholar, Early Childhood Education

"Doug has been very successful with multiple research and or journal submissions. His work should be recognized."

Stacy Campbell

Most Helpful to Students, Management

Cristen Dutcher

Best Teacher, Business

Thomas Kolenko

Best Teacher, Management
Most Helpful to Students, Management

Dr. Kolenko has taught on the graduate faculties of the University of Wisconsin–Madison and Wake Forest University before joining Kennesaw State University's Department of Management and Entrepreneurship as an Associate Professor of Management. He currently teaches undergraduate and MBA courses in labor relations, compensation systems, human resource management, and organizational behavior.

Dr. Kolenko has made significant contributions to improving organizational effectiveness as a human resource and organizational change consultant to many local and Fortune 500 firms including Georgia–Pacific, RJR–Nabisco, ITT, General Motors, BellSouth, and Bausch & Lomb. These consulting engagements have focused on human resource systems planning, organizational transformation and change, management development systems, organizational assessment surveys, and leadership development programs.

Dr. Thomas A. Kolenko earned his undergraduate degree at General Motors Institute in Organizational Development, followed by an MBA from Michigan State University in Human Resource Management in 1975. After holding several managerial posts in General Motors Corporation, he earned his Ph.D. from the University of Wisconsin–Madison.

He is an active member of the Greater Atlanta Human Resource Planning Group, Human Resource Planning Society, and Information Management Forum. Professor Kolenko has been active within the Academy of Management since 1978 and the Southern Management Association since 1984. He has presented over 25 scholarly papers in these professional associations in the areas of executive self–management, person–job matching strategies, strategic human resource planning, and college recruiting. His pedagogical interests and presentations have focused on executive development programs, designing effective video learning systems, and executive self–management tools.

He has been active on the boards of the Management Education & Development (MED), Careers, and Managerial Consultation (MC) Divisions of the Academy of Management. He was elected to the office of Division Chair of the 1000+ member MED Division in 1997.

Paul Lapides

Best Overall Faculty Member, Business

Paul D. Lapides is Co-founder and Director of the Corporate Governance Center in the Coles College of Business at Kennesaw State University, where he is a professor of management and entrepreneurship. His research and teaching interests include corporate governance, entrepreneurial finance, management, real estate, and venture creation.

A Research Fellow at the University of Tennessee's Corporate Governance Center, Mr. Lapides has taught at Directors' College at the University of Georgia, and was named an "ATDC Champion" for his work with entrepreneurs at the Advanced Technology Development Center (ATDC), a nationally recognized science and technology incubator, at the Georgia Institute of Technology.

Mr. Lapides received the 2000 Kennesaw State University Distinguished Service Award; has been recognized as one of the leading academic authorities on corporate governance by Corporate Board Member; and is the author or coauthor of more than 100 articles and twelve books, including several best sellers. He is a winner of the International Facility Management Association's Distinguished Author Award for Facility Management.

A frequent speaker at business, professional and academic organizations, his opinions have appeared in literally thousands of publications and on national and local television and radio, including, The Wall Street Journal, Business Week, Forbes, USA Today, CNBC's Morning Call, Bloomberg television and radio, NPR, London's Financial Times, Canada's National Post, Associated Press, Dow Jones, Bloomberg, Reuters, CBSMarketWatch, New York Times, Washington Post, Los Angeles Times, Chicago Tribune, Boston Globe, Atlanta Journal-Constitution, The Deal, Director's Monthly, Corporate Board Member, Trustee, and Directors & Boards.

Mr. Lapides is co-author of "21st Century Governance and Financial Reporting Principles," issued in March 2002 by the Corporate Governance Center at Kennesaw State University. The Governance Principles were endorsed by the Institute of Internal Auditors, presented to the New York Stock Exchange, and sent to members of Congress and the White House.

He was a member of the National Association of Corporate Director's Blue Ribbon Commission on Audit Committees. The Commission developed recommendations for improving audit committee performance, and published, The Report of the Blue Ribbon Commission on Audit Committees: A Practical Guide (1999, 2004).

Mr. Lapides is a member of the board of directors of Sun Communities, Inc. (NYSE: SUI), a real estate investment trust; EasyLink Services International Corporation (NASDAQ: ESIC), a provider of business-to-business e-commerce solutions; and the Board of Directors Network, Inc. (BDN), whose mission is to increase the number of women on corporate boards. He also serves on the advisory boards of the National Association of Corporate Directors (NACD); the Newman Real Estate Institute at Baruch College; Grubhub.com LLC, a lead generation company; and W. Ray Wallace & Associates, Inc., an Inc. 500 company providing training cost recovery services.

His business and consulting experience includes advising hundreds of start-up, growth and mid-market companies, as well as many of America's Fortune 500 companies, managing a diversified $3 billion real estate portfolio, raising more than $1 billion in equity capital for start–up and growth companies, and providing expert witness and litigation services.

A CPA, Mr. Lapides earned a BS with honors in economics from The Wharton School at the University of Pennsylvania and an MBA from New York University. Prior to joining the faculty at Kennesaw State University, he held faculty positions at New York University and Columbia University.

Stuart Napshin

Best Overall Faculty Member, Management
Best Researcher/Scholar, Management
Best Teacher, Management

Nancy Prochaska

Most Helpful to Students, Business

Department of Management & Entrepreneurship Burruss Building 311 470.578.3513 nprochas@kennesaw.edu Linked-In

Research & Teaching Areas - International Experiences - Experiential Education - Co-ops and Internships - Human Resource Management

Recent Consulting - Sany America, Inc. - Habif, Arogeti & Wynne, LLP - Career Services, KSU - Help the Poor - World Foundation

Academic Positions - Corpus Christi State University (Texas A&M University - Corpus Christi) - Texas A&I University (Texas A&M University - Kingsville) - Kennesaw State University

Professional Experience - St. Catherine of Siena Church and School - St. Joseph Catholic School - St. Michael the Archangel LifeTeen Program - Arden Trace Homeowners' Association - Farming Operations

Representative Publications

"Characteristics of Labor Market and Human Resources Management in the Republic of Kazakhstan," with Dr. Bolat Tatibekov and Dr. Janet Adams. Journal of Global Competitiveness. Named Best Conceptual Paper at the 2002 Conference of the American Society for Competitiveness.

"Assessing Interdisciplinary Learning in Theme-Based, One Semester Communities," "with Hoerrner, K., Goldfine, R., Buddie, A., Holler, E., Wooten, B., Collins, C. (2008). Journal of Learning Communities Research(Dec. 2008 Special Issue).

Howard Weinstein

Best Overall Faculty Member, Management

Howard is currently an Entrepreneur in Residence and a Board Member at Coles College, Kennesaw State University, where he teaches Strategic Management, Operations Management and Organizational Behavior. He is the former CEO and owner of Atlanta-based Monarch Wine Company, which he purchased as a leveraged-buyout in 1985 and recently sold to Todhunter International.

Howard earned his bachelors degree from Rensselaer Polytechnic Institute and an MBA from Harvard Business School. He has spent 30 years managing businesses in the food industry. Career highlights include:

President of Mayflower Restaurants- 90 full service units in 20 states. Group Vice President-Kane Miller Corp-responsible for Margarine, Peanut Butter (Superman Brand), vegetable farming, vegetable canning/processing and wine subsidiaries. In addition to being Entrepreneur in Residence, Howard also works as a business consultant in New York and Atlanta.

KENT STATE UNIVERSITY

Andrew Barnes

Best Overall Faculty Member, Political Science

Dr. Barnes's research and teaching interests are in post-communist political economies, the politics of international finance and oil, and the links between markets and democracy. His first book was called Owning Russia: The Struggle over Factories, Farms, and Power (Cornell University Press, 2006), and his articles have appeared in Review of International Political Economy, Problems of Post-Communism, Post-Soviet Affairs, and Comparative Politics, among others.

Daniel Hawes

Best Researcher/Scholar, Political Science

Daniel Hawes' research interests deal with questions related to public policy and public administration, broadly, and substantively focus on education and immigration policy. His research incorporates aspects of public administration, public management, and state and local politics in examining questions of public policy and policy performance. A central theme in his research is a focus on the determinants of public policy outcomes, particularly for disadvantaged groups. A fundamental question that his work has sought to address is: How can government – via policy, structure, bureaucracy or management – better address the inequities we observe in policy outcomes for disadvantaged groups? In doing so, his work has explicitly examined the role of public management, organizational structure, political representation, organizational and external environments on shaping policy outcomes. He has approached this broad question through different theoretical lens – e.g., social capital, representative bureaucracy, rational choice, public management – and in different substantive contexts – e.g. K–12 education, higher education, immigration.

Dr. Hawes' publications have appeared in a number of political science and public policy/administration journals including, Public Administration Review, J-PART, Political Research Quarterly, State Politics and Policy Quarterly, Social Science Quarterly, State and Local Government Review and the American Review of Public Administration.

Tom Hensley

Best Teacher, Political Science

Thomas R. Hensley received his B.A. in 1965 from Simpson College, his M.A. from the University of Iowa in 1967, and his Ph.D. from the University of Iowa in 1970. His primary teaching and research interests are in the field of law and the courts, with specific research interests in civil rights and liberties, Supreme Court decision making, and the implementation and impact of judicial decisions. He is the co-author of The Changing Supreme Court: Constitutional Rights and Liberties with Chris Smith and Joyce Baugh (West/Wadsworth: 1997) and is under contract to write a new book entitled, The Rehnquist Court: Justices, Rulings, Legacies. He has won numerous teaching awards, including being named Ohio Professor of the Year by the Council for the Advancement and Support of Higher Education. In addition to continuing his research on the United States Supreme Court, he is also developing a research agenda on the role of state supreme courts in regard to school funding, with a special focus on the Ohio Supreme Court.

Ratchneewan Ross

Best Researcher/Scholar, Nursing

"Dr. Ross exemplifies all the qualities of a wonderful professor and researcher."

Joshua Stacher

Most Helpful to Students, Political Science

Joshua Stacher is an Associate Professor in the Department of Political Science. Stacher's scholarship focuses on authoritarianism and social movements in the Middle East and North Africa. He is currently working on a book project on Egypt's political transition after Mubarak. In 2012–13, he was a Fellow at the Woodrow Wilson International Center for Scholars.

Stacher is the author of Adaptable Autocrats: Regime Power in Egypt & Syria (Stanford UP, 2012) as well as other peer-reviewed journal articles. He is a regular contributor to and on the editorial board of MERIP's Middle East Report. Stacher has made media appearances and written commentary for NPR, CNN, BBC, Al-Jazeera, Foreign Affairs, Jadaliyya, and The New York Times, among others. He is also a founding member of the Northeast Ohio Consortium on Middle East Studies.

KUTZTOWN UNIVERSITY OF PENNSYLVANIA

J. Frederick Garman

Best Overall Faculty Member, Education
Best Researcher/Scholar, Education

Dr. Garman's extensive academic and professional career enables him to incorporate experiences from diverse environments into Kutztown University's Bachelor of Science in Leisure and Sport Studies degree program. With over 29 years in the college classroom, he demonstrates an ongoing ability to effectively promote student learning within the disciplines of Health, Physical Education and Sport Management. Concurrent with his successful teaching, he pursues a scholarly agenda that resulted in numerous publications. These include authoring the textbooks, Sport and Recreation Operations: Experiential Learning in Sport Management and Exploring Health: A Problem-Based Inquiry, writing multiple publications in refereed, professional journals including Education, The International Electronic Journal of Health Education, The Journal of American College Health, and The Research Quarterly for Exercise and Sport, and presenting at numerous national professional conferences. Additionally, Dr. Garman's professional experiences, outside of the academic environment, enable him to bring a wealth of practical knowledge to the Sport Management classroom. He has successfully functioned as a senior administrator in community recreation, corporate wellness, and cardiac rehabilitation programs with a broad array of responsibilities that included, program development and implementation, budget development and oversight, fundraising, human resources management, facility operations and management, construction oversight, marketing and advertising, and media relations. These collective activities have resulted in consulting opportunities in program evaluation and curriculum design.

LA SALLE UNIVERSITY

Joe Burke

Best Overall Faculty Member, Psychology
Best Teacher, Psychology

Brother Joseph Burke received his B.A. degree in English from La Salle University in 1969, a M.Ed. in Educational Administration from the University of Miami in 1971, and a Ph.D. in Human Behavior from United States International University in 1973. Brother Burke joined the Psychology Department in 1973, and served as Department Chair from 1978–1986. In 1996 he was awarded a yearlong Fellowship in higher education administration from the American Council on Education. He spent that year at the University of Hartford, and then remained there as a dean for the next three years. In 1990 he become Provost at La Salle, and from 1992–1998 served as the University's President. Brother Burke became chair of the Psychology Department again in 2002, an office he still holds. He is the author of a graduate level textbook on psychotherapy, as well as numerous articles and book chapters. He is a licensed psychologist in Pennsylvania, and his current research interests include the history of psychology, clinical hypnosis, and psychological services to adjudicated minors. He serves on the board of directors of several non-profits, including a psychiatric hospital.

Leeann Cardaciotto

Best Researcher/Scholar, Psychology

Dr. Cardaciotto earned her Bachelor's degree in Psychology from Franklin and Marshall in 2000 and her doctorate in Clinical Psychology from Drexel University in 2005. She received a broad-based clinical training, gaining experience with clients across the entire developmental spectrum in many levels of care (e.g., inpatient, partial hospital, outpatient, student counseling center). Currently, her main clinical interests are anxiety disorders and the use of mindfulness- and acceptance-based interventions.

Dr. Cardaciotto's research interests primarily focus on constructs related to psychological well-being. Specifically, she is interested in the study of mindfulness, acceptance, defusion/decentering, and compassion (for self and others). She welcomes both undergraduate and graduate students to share in her research endeavors - there are many ways to become involved (e.g., participant recruitment, data collection, project development, presentation of results in posters/publications). Dr. Cardaciotto and her research team have several ongoing projects related to examining the measurement of mindfulness and related constructs with self-report and non-self-report measures; the differential roles of the two key components of mindfulness, awareness and acceptance, in a variety of outcomes and contexts; and examining mindfulness and self-compassion in the areas of social anxiety and disordered eating. Dr. Cardaciotto also is interested in effective pedagogical practices in the field of counseling education.

Outside of work, Dr. Cardaciotto enjoys being with her family and friends, watching football, reading, spending time in nature, and getting up at 6:00AM for "Boot Camp" classes.

David Falcone

Most Helpful to Students, Psychology

EDUCATION
Ph.D. December, 1981 University of Kentucky
Major: Cognition/Developmental Minor: Perception/Learning M.S. June, 1974 Western Illinois University
Major: General-Experimental Minor: Cognition B.S. June, 1972 University of Dayton Major: Psychology
PROFESSIONAL EXPERIENCE
1987–1999 Chairperson, Department of Psychology, La Salle University, Philadelphia, Pa. 1989–1995 Statistical Consultant. Clients have included: Department of Human Development. Bryn Mawr College Maternity Care Coalition, Philadelphia, Pa. Court Mental Health Clinic, City of Philadelphia Office of Probation and Parole, Philadelphia, Pa. Philadelphia Court Mental Health Clinic, Jules deCruz & Albert Levitt 1986 Promoted to Associate Professor and tenured 1981–1985 Assistant Professor; La Salle University 1980–1981 Instructor; La Salle College 1978–1980 Research Appointment; GROW Inc., Lexington, KY 1978–1979 Instructor; University of Kentucky, Lexington, Kentucky. 1976–1977 Research Associate; David Kay (NIDA Addiction Research Center, Lexington, Kentucky) and Ekkhardt Othmer (VA Hospital, Lexington, Kentucky) 1974–1976 Instructor; Academy of Health Sciences, Baylor University, San Antonio, Texas 1972–1973 Research Appointment: Dayton Board of Education, Dayton, Ohio (Project EMERGE: Title VIII, Sec 807, ESPA).

Preston Feden

Most Helpful to Students, Education

I am a former special education teacher at both the elementary and secondary levels. I have been a member of the La Salle University faculty since 1973. Recently, I was appointed Assistant Director of the University Honors Program. I received the Lindback Award for Distinguished Teaching in 1984 and the Provost's Distinguished Faculty Award in 2001. In addition, I am the Founding Director of the La Salle University Teaching and Learning Center for Faculty Development, and La Salle University's Philadelphia Center. My work has included consulting widely with school districts on applying cognitive science to instructional practice. I have made numerous presentations on student learning and faculty development in both basic education and higher education, all focused on the application of research from cognitive science in promoting deep learning and effective instruction. I co-authored a textbook titled Methods of Teaching: Applying Cognitive Science to Promote Student Learning (2004, McGraw-Hill). My latest article, Teaching without Telling: Contemporary Pedagogical Theory Put Into Practice, was recently accepted for publication in the Journal on Excellence in College Teaching (JECT). In addition to being a member of the education department faculty and Assistant Director of the Honors program, I am also an affiliated faculty member in the American Studies program at La Salle. In that program, I teach a course related to the theme of medicine in America

Kim Lewinsky

Best Teacher, Education

Deborah Yost

Best Researcher/Scholar, Education

An expert in Special Education, Dr. Deborah S. Yost has served as a professor in the Department of Education at La Salle University since 1996. In addition to teaching a variety of graduate and undergraduate level courses, Dr. Yost has continuously served as a chair or member in one or more of the many department and university committees. Her individual and collaborative work has been widely published in academic journals and further includes chapter contributions, a book review, and a collaborative book project currently in press on empowering young writers. Through her regional, national, and international conference presentations, education professionals and those in related fields have gained access to and benefited from her vast wealth of knowledge and expertise. She has also authored and co-authored a host of curriculum writing projects for the Education Department.

LAKE FOREST COLLEGE

Siobhan Moroney

Best Teacher, Public Affairs, Political & Policy Sciences

Paul Otogun

Best Teacher, Political Science

LAKE SUPERIOR STATE UNIVERSITY

Joe Susi

Best Teacher, Medicine

LAMAR STATE COLLEGE AT ORANGE

Sherri Foreman

Most Helpful to Students, Health Science

LAMAR UNIVERSITY

Mary Goodwin

Best Teacher, Nursing

LANCASTER BIBLE COLLEGE

Julia Hershey

Best Overall Faculty Member, Education

LANE COMMUNITY COLLEGE

Marianne Farrington

Most Helpful to Students, Physical Education

"Marianne is there to help everyone . . . student . . . faculty at every turn. She is sweet and friendly, smart and extremely approachable and helpful. I vote for Marianne."

Del Nero

Best Overall Faculty Member, Criminal Justice
Best Researcher/Scholar, Criminal Justice
Best Teacher, Criminal Justice
Most Helpful to Students, Criminal Justice

LANSING COMMUNITY COLLEGE

David Brown

Best Researcher/Scholar, Criminal Justice

LEHMAN COLLEGE

Amod Choudhary

Best Overall Faculty Member, Business

"Professor Choudhary is conscientious about his work and about his interactions with other faculty members. He's generous with his time and always willing to help. He works diligently and always see the project to the end."

Professor Choudhary has taught Introduction to Management, Strategic Management, Seminar in Strategic Management, Introduction to Microeconomics, and Principles of Finance courses at Lehman College. Prior to being an educator, he was an engineer and lawyer. Professor Choudhary's research interests are strategic management, leadership, innovation and finance. His guiding philosophy in terms of teaching is to ensure that students are engaged and master the concepts taught in his courses. He engages students by discussing current events as they relate to the course topics and brings out students' viewpoints during class lectures. He also advises student clubs and participates in various Lehman College committees.

John Cirace

Best Teacher, Business

I am currently teaching Business Law I, Business Law II, and Commercial Transactions. I have also taught Introduction to Macroeconomics, Introduction to Microeconomics, Economic Statistics, and Industrial Organization. I have published 18 articles, the most recent of which is, When are Law and Economics Isomoporhic, 39 Golden Gate University Law Review 183–220 (2009). Outside of work, I like to swim, work out with weights, walk, read nonfiction, and go to movies and museums.

Juan Delacruz

Most Helpful to Students, Business

"Professor Delacruz, always has a scrum of students in his office seeking his advice. He is extremely generous with his time and makes the effort to understand and empathize with students."

Juan J. DelaCruz earned a PhD in Economics (2008) from the New School for Social Research. He is an Associate Professor of Economics and Business. His work is concentrated in the area of health economics with an emphasis on HIV/AIDS and its effects on human capital.

LEWIS AND CLARK COLLEGE

Cliff Bekar

Best Overall Faculty Member, Economics

LIBERTY UNIVERSITY

Stephen Parke

Most Helpful to Students, Criminal Justice

Dr. Parke is a Christian, conservative and attorney. He retired from the United States Army after having served for 21 years, primarily as a supervisory Staff Judge Advocate. His last duty station was as the Staff Judge Advocate for Joint Task Force Guantanamo, Guantanamo Bay, Cuba. Each day he thanks God for the privilege of being able to teach young champions for Christ.

LOCK HAVEN UNIVERSITY

Dan Gales

Best Overall Faculty Member, Health Science
Most Helpful to Students, Health Science

Yvette Ingram

Best Teacher, Health Science

Beth McMahon

Best Researcher/Scholar, Health Sciences and Nursing

The Pennsylvania State University, University Park, PA
   Doctorate in Health Education (Ph.D.) - Magna Cum Laude (3.89), 1993
   East Stroudsburg University, East Stroudsburg, PA
   Masters of Science: Exercise Physiology with concentration in Cardiac Rehabilitation and Primary Prevention - Magna Cum Laude (3.8), 1984
   Bachelor of Science: Health and Physical Education - Cum Laude (3.65), 1980
   Courses Most Commonly Taught: Undergraduate: Professional Field Experience in Health Science, Epidemiology, Foundations of School & Community Health, Community Health Organization Management, Current Health Issues, Women's Health Issues
   Gaduate: Epidemiology in Community and Health Practice, Advanced Field Experience in Health Science, Masters Capstone Project, Contemporary Issues in Health and Healthcare
   Courses Developed:
   Undergraduate: Community Health Organization Management, Professional Field Experience in Health Science, First Year Student Seminar, Women's Health Issues, Current Health Issues, Introduction to Epidemiology, Community Health
   Graduate Courses: Master of Health Science Courses: Grant Development, Epidemiology in Community and Health Practice, Contemporary Issues in Health and Healthcare, Advanced Professional Field Experience, Master Capstone Project
   Programs Developed: Master of Health Science/Health Promotion/Education and Healthcare Management, Program Coordinator
   Professional Presentations:
   McMahon, B.F., Eisner, H., Collins, N.E., Flores, F.L. (2013) "Community Collaborative Effort to Serve Youth"; 2013 Pennsylvania Public Health Association/Pennsylvania Office of Rural Health Annual Conference, Harrisburg, PA, October 16.
   McMahon, B.F., Tennant, A., Bellinger, R.G. (2013) "An Overview of Health Improvement Partnerships in Pennsylvania"; 2013 Pennsylvania Public Health Association/Pennsylvania Office of Rural Health Annual Conference, Harrisburg, PA, October 16.

McMahon, B.F. (2013) "Health Status in Lycoming County, Health and Human Service Annual meeting;" Training session for Leadership Lycoming Class of 2013, Susquehanna Health System, Williamsport; PA, February 14.

McMahon, B.F. (2012) "Women: Our Education is Our Strength" Keynote address, Annual Federal Women's Program/National Women's History Month, United States Penitentiary (USP), Lewisburg, PA., March 21.

McMahon, B.F. (2012) "Health Status in Lycoming County, Health and Human Service Annual meeting;" Training session for Leadership Lycoming Class of 2012, Susquehanna Health System, Williamsport; PA, February 9.

LOMA LINDA UNIVERSITY

Keri Medina

Most Helpful to Students, Nursing

"Keri makes time for her students and helps them review their pathophysiology."

LONE STAR COLLEGE (ALL)

Carol Girocco

Best Overall Faculty Member, Nursing
Best Researcher/Scholar, Nursing
Best Teacher, Nursing
Most Helpful to Students, Nursing
Best Teacher, Health Sciences and Nursing
Best Researcher/Scholar, Health Sciences and Nursing
Most Helpful to Students, Health Sciences and Nursing
Best Overall Faculty Member, Health Sciences and Nursing

Michael Green

Best Overall Faculty Member, Psychology

Michael Green

Best Overall Faculty Member, Psychology

Janet Harris

Best Overall Faculty Member, Vocational Nursing

Heidi McDonald

Most Helpful to Students, Education

"Heidi graduated from Harvard University. She is the Education Chair at LSC-Montgomery. As an Educaton faculty member, Heidi makes sure that each faculty member understands the requirements and executes them."

LONG BEACH CITY COLLEGE

Dan Ripley

Best Teacher, Health Science

I had the pleasure of both attending a California Community College, and subsequently teaching 36 years at three of them. I am married and have three children. I enjoy active hobbies such as mountain biking and backpacking.

John G. Smith

Best Teacher, Education

LONG ISLAND UNIVERSITY - C.W. POST

Kathy Isoldi

Best Overall Faculty Member, Health Science
Best Researcher/Scholar, Health Science
Best Teacher, Health Science
Most Helpful to Students, Health Science

"Kathy is an outstanding resource for students. She is knowledgeable, selfless and most professional."

Dr. Isoldi has been a registered dietitian practicing in the field of nutrition for close to three decades. Her interests include body weight regulation, diabetes management, childhood obesity prevention and the prevention of chronic diseases through diet and physical activity. Dr. Isoldi's recent research investigated the foods and beverages offered and consumed during classroom celebrations at a public elementary school. Her future work will be aimed at identifying and communicating preventive measures that schools and families can take to curb the current childhood obesity crisis.

Barbara Shorter

Best Researcher/Scholar, Health Sciences and Nursing
Best Teacher, Health Sciences and Nursing
Most Helpful to Students, Health Sciences and Nursing
Best Overall Faculty Member, Health Sciences and Nursing

"Dr. Shorter recently developed a food sensitivity questionnaire for Interstitial Cystitis patients that is intended for use by urologists internationally. J of Urol 2014 She also presented on prevention of LUTS in woman at the National Institute of Health, NIDDK, Washington DC."

Barbara Shorter has been teaching at LIU Post for over 20 years. She began as an adjunct professor while raising her two children. She then joined the faculty and served for 15 years (until 2011) as Director of the Undergraduate Nutrition Program. During that time, she was tenured and promoted to Associate Professor.

In 2009, Dr. Shorter was invited to join the faculty of the Smith Institute for Urology, North Shore-LIJ Health System where she actively participates in research as well as counseling urology patients. Dr. Shorter is an internationally known expert on Interstitial Cystitis/Bladder Pain Syndrome(IC/PBS). She was lead author of "Effects of Comestibles on Interstitial Cystitis/Painful Bladder Syndrome" published in the Journal of Urology. Other articles Barbara has authored/co-authored have been published in the British Journal of Urology International, Urology, Nutrition Today, AUA News, Topics in Clinical Nutrition, Bladder Pain Syndrome: A Guide for Clinicians, Springer Press, and Collide: Styles, Structures and ideas in Disciplinary Writing, Pearson Publishing. In addition, she was a peer reviewer for the 2011 AUA Guidelines for the Diagnosis and Treatment of IC/BPS. She recently worked with Pro-Change Behavioral Systems, Inc. and a team of IC experts, to develop an on line patient self-help module based on the Transtheoretical Model of Behavioral Change, for IC patients. Barbara is also a Medical Advisory Board Member of the Interstitial Cystitis Association.

Dr. Shorter has been a guest speaker for the Interstitial Cystitis Association, the Interstitial Cystitis Network, the CW Post/LIU Honors Program, the North Shore – LIJ Health System Men's Health Program, the New York State Dietetic Association, the Long Island Dietetic Association, the North Shore-LIJ Monter Cancer Center and the Dr. Melanie Barton World Talk Radio show. She has presented two Podium Sessions for both the American Urological Association and the Society for Inflammation and Infections in Urology.

Dr. Shorter has been interviewed for numerous articles in the IC Update, Newsday and other Women's magazines.

LORAIN COUNTY COMMUNITY COLLEGE

Karen Joris

Most Helpful to Students, Nursing

Assistant Professor, ADN program Program Coordinator, Paramedic to ADN program
Current member of the Service Learning Advisory Committee Began teaching with Service Learning before the official start of the LCCC program in 2006 Service Learning Courses:
NURS 127 Child Health Nursing Why do you teach with Service Learning?

"The Service Learning project allows the student to begin to see the needs of the community beyond the walls of the hospital. As our students provide a service to a community group, they are able to access & share their knowledge of health promotion and more confident in their abilities to increase the overall wellness of our communities. Project Explanation:

Students completing a Service Learning project in Professor Joris' course may be involved in a variety of community based projects with local agencies (eg. Murry Ridge, Catholic Charities, Haven Center, etc.) focused on identifying and addressing community health needs.

LOS ANGELES MISSION COLLEGE

Christopher Williams

Best Teacher, Psychology

LOS ANGELES SOUTHWEST COLLEGE

Todd Roberts

Best Overall Faculty Member, Health Science

Heidie Tatum

Most Helpful to Students, Health Sciences and Nursing

LOS ANGELES VALLEY COLLEGE

Louis Jones

Best Teacher, Health & Physical Education

"He is very helpful."

LOUISIANA TECH UNIVERSITY

Joe Pullis

Best Researcher/Scholar, Business
Best Teacher, Business
Most Helpful to Students, Business
Best Overall Faculty Member, Business

Alison Wall

Best Overall Faculty Member, Business
Best Researcher/Scholar, Business
Best Teacher, Business
Most Helpful to Students, Business

LOYOLA MARYMOUNT UNIVERSITY

Marta Baltodano

Best Researcher/Scholar, Education

Marta Baltodano is a Professor in the Department of Specialized Programs in Urban Education at Loyola Marymount University. Her research focuses on the corporatization of schools of education, teachers' beliefs on social justice, and interracial conflicts in Los Angeles. Her teaching includes issues of critical educational theory, political economy, globalization, social justice, and qualitative research. Baltodano has served in the Council of Anthropology and Education (CAE) of the American Anthropological Association in different capacities, among them chair of the committee on Anthropological Studies of Schools and Culture, Contributing Editor of CAE for Anthropology News, and Co-Chair of CAE's Mission Committee. She is one of the founding members of the California Consortium for Critical Educators (CCCE), and the founder of AERA's SIG Critical Educators for Social Justice (CESJ). She was the 2008–2009 Program Co-Chair of AERA's Division G (Social Context of Education). Baltodano co-edited the book

"The Critical Pedagogy Reader" (Darder, Baltodano & Torres, 2002, 2008, Routledge) and the compendium "Social Justice, Peace and Environmental Education Standards" (Andrzejewski, Baltodano, & Symcox, 2009, Routledge). She is the author of the book "The Appropriation of Social Justice in Education" (in press, Rowman & Littlefield), and she has published several articles in peer-refereed journals and is currently writing a book about the corporatization of education.

Ellen Ensher

Best Overall Faculty Member, Management
Best Researcher/Scholar, Management
Best Teacher, Management
Most Helpful to Students, Management

Ellen A. Ensher is a Professor of Management at Loyola Marymount University in Los Angeles. Ensher is a national expert in mentoring and careers and co-authored the book, Power Mentoring: How Successful Mentors and Protégés Get the Most out of Their Relationships (Jossey Bass). Ensher has been cited in Money Magazine, New York Times, Forbes, and recently delivered a Ted talk on how to get a mentor. For more information visit www.ellenensher.com.

James Konow

Best Overall Faculty Member, Economics
Best Researcher/Scholar, Economics

Jennifer Offenberg

Best Teacher, Economics

Robert Singleton

Most Helpful to Students, Economics

LOYOLA UNIVERSITY

Charles Corprew

Best Teacher, Psychology

Charles S. Corprew, III joined the Loyola University New Orleans psychology faculty in August of 2011. He received his B.A. in History from James Madison University in 1993. After receiving his M.A. in Education from Norfolk State University in 1997, he taught for ten years within the Virginia Beach Public School System, teaching both American History and Advanced Placement Psychology. During his time with the school system, he also served as the Director of the James Madison University Male Academy. The academy is a three week summer program geared toward the academic and social development of adolescent African American males. It would be his experiences teaching as well as with the academy that motivated him to pursue his doctoral degree. Charles completed his Ph.D. in Psychological Science at Tulane University in the spring of 2011, centering his research on issues of academic achievement and masculinity with adolescent and emerging adulthood populations. Dr. Corprew's current research interest focuses on understanding how the intersection of masculinity and culture impact the academic achievement of adolescent African American males. Additionally, his interest will continue to focus on issues of hypermasculinity within adolescent and emerging adulthood populations. In particular, the function the construct serves for males adopting its characteristics.

Chuck Nichols

Best Overall Faculty Member, Psychology

Chuck Nichols is an assistant professor of psychological sciences at Loyola University New Orleans, specializing in social and personality psychology. He studies a set of interrelated topics that can be summarized as the psychology of human valuing, i.e., the psychology of what and whom people value in their lives. First, he is interested in the different outcomes and goals that people value and how they seek to achieve them, an area of study sometimes called personality processes, which includes motivation, values, self-determination, and identity. Second, he is interested in people's value judgments and beliefs, particularly in the political, moral, and interpersonal domains. These include judgments regarding justice and "the good," as well as judgments regarding the value of individual persons, groups of people, and other living beings in the natural environment. Third, he is interested in the psychology of happiness and well-being (sometimes referred to as positive psychology). His work in this area aims to better understand the true causes of individual and collective well-being, with the ultimate aim of improving them. He is also interested in applying psychology to understand and address social and economic issues (e.g., international development).

He enjoys teaching social psychology, introduction to psychology, research methods, and industrial/organizational psychology, as well as mentoring and collaborating with students on research projects. He was educated at the University of Missouri, Columbia, where he studied political science, history, and management, before finishing his PhD in social and personality psychology in 2011.

Timothy Scanlan

Best Overall Faculty Member, Criminal Justice
Best Teacher, Criminal Justice
Most Helpful to Students, Criminal Justice
Best Teacher, Education
Most Helpful to Students, Education
Best Overall Faculty Member, Education

Professor Timothy P. Scanlan has been an adjunct professor with the Department of Criminal Justice since 2004 where he lectures in the field of forensic science. Before beginning his education into the realm of natural science, he first obtained a Bachelor of Criminal Justice degree from Loyola University New Orleans on August 31, 1998, graduating Magna Cum Laude. He then obtained a Master of Science in Forensic Science degree from Florida International University. His graduate research focused on the corrosive effect of blood on projectiles. He presented the results of this extensive study at the American Academy of Forensic Sciences' Fifty-seventh meeting and at the International Forensic Science Symposium in Taipei, Taiwan. Professor Scanlan is currently pursuing his doctorate degree (Ph.D.) in Public Administration with a specialization in Homeland Security Police and Coordination. His current research is divided between the role of forensic science in homeland security and factors effecting expert witness testimony.

As a Captain with the Jefferson Parish Sheriff's Office, he oversees the Crime Laboratory and Crime Scene Divisions. Professor Scanlan is a court qualified expert in firearms and tool mark examination, bloodstained pattern analysis, crime scene reconstruction, and trace evidence analysis. He has testified in numerous criminal trials and in multiple jurisdictions.

In addition to instructing the FRSC-C100: Introduction to Forensic Science and FRSC-C200: Criminalistics I - Crime Scene courses, he is responsible for overseeing all forensic science researched projects (FRSC-C498). These projects allow students to focus on a specific area of interest while conducting research that advances the field of forensic science. Professor Scanlan has presented his own research nationally and internationally, including for the International Association of Identification, the International Forensic Science Symposium, the American Academy of Forensic Sciences, the Southern Institute of Forensic Science, and the Louisiana Association of Forensic Sciences. He also enjoys guest lecturing for many of the local universities and law enforcement agencies.

LOYOLA UNIVERSITY MARYLAND

Gerald Athaide

Best Overall Faculty Member, Business

Gerard A. Athaide is Professor and Chair of the Department of Marketing in the Joseph A. Sellinger, S.J., School of Business and Management at Loyola University Maryland. He received his Ph.D. and MBA from Syracuse University. His teaching, research, and consulting interests focus on innovation management and new product development. Gerard has taught courses on Innovation Management and New Product Development in the undergraduate, graduate, and Executive MBA programs at Loyola University; he has also taught these courses in Chile, India, and New Zealand. In addition, he leads graduate students on an International Marketing study tour of Chile. He has published several articles on innovation with a focus on the commercialization of technology-based innovations and the determinants of effective brand names for new products.

William Blouch

Most Helpful to Students, Accounting

Frank Izzo

Best Teacher, Accounting

Roger Kashlak

Most Helpful to Students, Business

Roger J. Kashlak is Professor of International Business at the Sellinger School of Business at Loyola University Maryland. He has been with Loyola since 1993. During the past ten years, he has also served as Department Chair and Senior Associate Dean of the Business School, chaired the university's NCAA Cycle 3 Re-certification, and was a member of the university's Executive Governance and Budget Committees. He received a B.S. in Economics from the Wharton School of the University of Pennsylvania, an MBA in International Business from Temple University and a Ph.D. in International Business and Strategy from Temple University.

Dr. Kashlak was a member of the International Business Department at the University of Auckland (2000–2002) and subsequently served as Visiting Professor of International

Business there. He has been a Research Fellow at the Voinovich Center for Leadership and Public Affairs at Ohio University and has been an invited lecturer and developed courses and seminars at institutions including: Thunderbird (The American Graduate School of International Management); Temple University-Japan; Manipal University (India); Katholiek University Leuven (Belgium); Institut Technologi Mara (Malaysia); and, Universidad Jesuita Alberto Hurtado (Chile).

Dr. Kashlak's research has focused on topics such as international reciprocity, international alliances, global expansion in telecom, health care and other industries, global control and corporate governance issues, executive education pedagogies, and comparative analyses of leadership and work attitudes. His research has been published in leading journals such as Journal of International Business Studies, Strategic Management Journal, Management International Review, Journal of Business Research, Long Range Planning and Journal of International Management, and presented at over 60 national and international conferences since 1993. He is a co-author of the book, International Management: Managing in a Diverse and Dynamic Global Environment, published in 2005 by McGraw-Hill, with Chinese editions published through 2011. The American 2nd edition was published in 2009, and the international edition in 2010. The text has been used by universities in 22 countries. He has also served on various editorial boards and as a reviewer for international business and strategic management journals.

Dr. Kashlak's teaching focuses on International Management and Global Strategy at the undergraduate, MBA and Executive MBA levels. He has developed and conducted Executive MBA courses throughout the world including, China, Vietnam, Malaysia, Thailand, South Africa, Chile, Argentina, the Netherlands and the Czech Republic. He is the recipient of Loyola's 1997 Distinguished Teacher Award and other teaching honors from Beta Gamma Sigma, The Sellinger School and Alpha Sigma Nu, the Jesuit Honor Society.

Prior to academia, Dr. Kashlak worked for AT&T-Communications International, where he developed the initial international rate negotiation strategy and was responsible for financial negotiations with host governments and telecom entities throughout the world. Subsequent to that job, he established AT&T-Communication's Italian subsidiary. Dr. Kashlak has continued to be involved in the corporate world through executive education with firms such as AEGON, AT&T, Ericsson, Northrop-Grumman, Lucent Technologies, Motorola, and various sub-Saharan African public and private entities.

Dr. Kashlak sees his most important role as focusing on the next generation, including his children and his students, and the most recent edition of his textbook is dedicated to "the next generation of scholars." He is the father of two sons: Adam, who has degrees from McGill University and Johns Hopkins University and who is currently pursuing a Ph.D. in Mathematics at Cambridge University, and Jake, who graduated from Loyola University Maryland in 2012, and just recently completed his Masters of Philosophy In Theater and Analysis at Trinity College Dublin.

Rick Klink

Best Teacher, Business

Richard R. Klink is a Professor of Marketing in the Joseph A. Sellinger, S.J., School of Business and Management. He received his Ph.D. in marketing from the University of Pittsburgh in 1996. His research interests include new product development and management with a focus on branding issues. His current teaching interests are in the areas of marketing strategy and international marketing. He recently took undergraduate students enrolled in his international marketing course to Europe to visit companies such as BMW, Lloyd's of London, and The Economist.

Joseph Langmead

Best Overall Faculty Member, Accounting

Norman Sedgley

Best Researcher/Scholar, Business

Jalal Soroosh

Best Researcher/Scholar, Accounting

Jalal Soroosh is a professor of accounting at Loyola University Maryland. He received his Ph.D. in accounting from The University of Mississippi and holds a Certificate in Management Accounting (CMA). He has published numerous articles in practitioners and academic journals such as the CPA Journal, Management Accounting, Corporate Controller, The Journal of Business and Economic Perspectives and Review of Business. He also has an extensive teaching experience in various MBA/EMBA programs in China. Jalal brings a good mix of teaching and work experience to the accounting program at Loyola University Maryland. In addition to over twenty years of teaching accounting, he has a variety of practical experiences with Arthur Andersen & Co., Coopers & Lybrand (PWC), C.W. Amos & Company (RSM McGladrey), McCormick, Reznick Fedder & Silverman, and NeighborCare.

MACOMB COMMUNITY COLLEGE (ALL CAMPUSES)

Jennifer Gornicki

Best Researcher/Scholar, Business
Best Teacher, Business
Most Helpful to Students, Business
Best Overall Faculty Member, Business

MADONNA UNIVERSITY

Kelly Rhoades

Best Researcher/Scholar, Health Sciences and Nursing
Best Teacher, Health Sciences and Nursing
Most Helpful to Students, Health Sciences and Nursing
Best Overall Faculty Member, Health Sciences and Nursing

Dr. Rhoades completed her Ph.D. in 1994 at Michigan State University in Family Ecology with an emphasis in Family Therapy. She began working specifically with grieving families within hospice bereavement programs in 1987. Dr. Rhoades focused her doctoral course of study on grieving family dynamics and the healing potential of transformative grief. Dr. Rhoades has been involved in direct care as a bereavement counselor and consultant since 1987, initially working in a part-time position as a hospice Bereavement Coordinator. Dr. Rhoades began teaching at Madonna University in the Hospice and Palliative Studies Department in 1995 and became the Chair in 1999. She was instrumental in helping to create the Master's in Hospice and Palliative Studies in 1996, and her vision to offer graduate course work in Bereavement became a reality in 2000 with the creation of the Post-Master's Bereavement Certificate. As an online and on-campus offering, the certificate prepares students to conduct high-risk assessment and intervention with families where complicated grief and trauma may be involved. Dr. Rhoades enjoys teaching graduate courses in bereavement theory and family systems theory, and supervising graduate students in field placements and research activities. She is also pleased to announce the publication (2009) of a story book about children's grief with useful tips for adults that recently (2013) became part of an electronic book offering within a specialty series for grieving children and families. Dr. Rhoades believes that her greatest accomplishment has been that of becoming a mother and sharing the parenting role of their daughter with her husband, Jeff. Her hobbies include freelance writing and photography, sketching, and walking her English Setter.

MARIST COLLEGE

Sherry Dingman

Best Overall Faculty Member, Medicine

Dr. Dingman was born and raised in Montana. She graduated from the University of Montana.

Caroline Rider

Best Teacher, Business

Elizabeth Teed

Best Overall Faculty Member, Psychology

MARQUETTE UNIVERSITY

H. Richard Friman

Best Researcher/Scholar, Public Affairs, Political & Policy Sciences

H. Richard Friman is Eliot Fitch Professor for International Studies, Professor of Political Science, Coordinator of the Interdisciplinary Major in International Affairs, and Director of the Center for Transnational Justice at Marquette University, Milwaukee, Wisconsin. His current research focuses on the intersection of the licit and illicit global economies, politics of (im)migration and crime, smuggling and advanced industrial countries, trafficking in persons, and drug control policy. He has written/edited seven books including Crime and the Global Political Economy, and is the author of numerous book chapters and journal articles.

MARSHALL UNIVERSITY

Jacqueline Agesa

Best Researcher/Scholar, Business

Linda Spatig

Best Researcher/Scholar, Education

Amanda Thompson

Best Teacher, Accounting

MASSACHUSETTS COLLEGE OF PHARMACY & HEALTH SCIENCE

Stacey Taylor

Best Researcher/Scholar, Administration

Caroline Zeind

Most Helpful to Students, Pharmacy

MCNEESE STATE UNIVERSITY

Gerald Calais

Best Teacher, Education

MEDGAR EVERS COLLEGE

Paul Cox

Most Helpful to Students, Business

"Prof Cox conducts research, deals with students, chooses the appropriate level of work and achieves outcomes in accordance with the objectives."

Veronica Udeoganlaya

Best Researcher/Scholar, Economics

"Dr Udeogalanya constantly interacts with students and helps them to achieve better results as respects presentations and papers. While doing all of this she conducts meaningful research relevant to the needs of our School of Business."

MERCY COLLEGE

Rossi Hasaad

Best Researcher/Scholar, Psychology

"Rossi continues to contribute up to date research which he presents around the world. He has also contributed to the new DSM V and he is called upon to educate the public on issues of epidemiology."

MERIDIAN COMMUNITY COLLEGE

Suzie Gibson

Most Helpful to Students, Psychology
Most Helpful to Students, Medicine

Chad Graham

Best Overall Faculty Member, Psychology
Best Researcher/Scholar, Psychology
Best Teacher, Psychology
Best Researcher/Scholar, Medicine
Best Teacher, Medicine
Best Overall Faculty Member, Medicine

"Has a great amount of knowledge and is very helpful to his students!"

MESSIAH COLLEGE

Stephan Gallaher

Most Helpful to Students, Biblical & Religious Studies

Doug Miller

Best Teacher, Physical Ed
Most Helpful to Students, Physical Ed

METROPOLITAN COMMUNITY COLLEGE

Traci Warren

Best Overall Faculty Member, Nursing

"Traci is a devoted teacher to her students. She continuously challenges her students to be the best that they can be and even better. Her devotion to education is profound. She is friendly but still maintains professionalism. She keeps her students accountable but is compassionate and caring. Traci works hard to be the best she can be for her students and her colleagues. She is a team player. If you need her she will be there, no questions."

METROPOLITAN STATE UNIVERSITY

Grover Cleveland

Best Researcher/Scholar, Accounting
Best Teacher, Accounting
Best Researcher/Scholar, Business

Robert Sausen

Most Helpful to Students, Business
Best Overall Faculty Member, Business

Adrianne Slaymaker

Best Overall Faculty Member, Accounting
Most Helpful to Students, Accounting
Best Teacher, Business

Glen Spielmans

Best Researcher/Scholar, Psychology

MIAMI DADE COLLEGE (ALL)

Yamina Alvarez

Best Teacher, Nursing

Franklin Aviles

Best Teacher, Business

Ruth Barrow

Best Researcher/Scholar, Business
Best Teacher, Business

I believe that my role as a faculty member is to help students to develop the skills and knowledge needed for a successful long-term career in business. My goal is to link what students learn in the classroom to the real business world. Through my general business courses, students will develop critical thinking skills that will enable them to become effective lifelong decision-makers and problem solvers.

James Cavalaris

Best Overall Faculty Member, Business

Rene Choy

Best Teacher, Economics

Ana Corrales

Best Overall Faculty Member, Economics
Best Teacher, Economics
Most Helpful to Students, Economics

Miss Corrales brings experience in economic research, project management, strategic/financial planning, and corporate communications to MDC's School of Business. Most recently, she was the director of research and client services at an economic development consulting firm. Miss Corrales served as executive director of a bi-national chamber of commerce and was a business analyst for a Fortune 500 energy company.

Professor Corrales joined the MDC faculty in 2005 and has been teaching in the Honors College and Virtual College since 2006.

In January 2010, Professor Corrales was appointed to the City of Miami's Blue Ribbon Committee on Economic Initiatives by Mayor Tomas Regalado.

Raul De La Cruz

Most Helpful to Students, Psychology

Denise Eccles

Most Helpful to Students, Nursing

Daniel Estape

Best Teacher, Health Sciences and Nursing

Olubisi Faoye

Best Overall Faculty Member, Health Science

Yvonne Garner

Best Overall Faculty Member, Nursing
Best Researcher/Scholar, Nursing
Best Teacher, Health Sciences and Nursing
Best Overall Faculty Member, Health Sciences and Nursing

Mari Kimsey-Ortega

Best Teacher, Health Sciences and Nursing

Morris Knapp

Best Researcher/Scholar, Economics
Best Researcher/Scholar, Business

Maria Mari

Most Helpful to Students, Business

Constance Miller

Best Overall Faculty Member, Nursing
Best Researcher/Scholar, Nursing

Jorge Monserrate

Most Helpful to Students, Health Sciences and Nursing

Marlene Morales

Best Overall Faculty Member, Education

With its proven history of advocacy for quality instruction, Miami Dade College consistently fulfills a leading role as a contributor to the improvement of nationwide teaching standards and practices.

Continuing this strong tradition, Miami Dade College alumna and Associate Professor of Science Education Dr. Marlene Morales was recently invited to participate in the Teacher Preparation Task Force of the American Federation of Teachers (AFT-CIO) in Washington, D.C.

As part of the task force, comprised of higher education and K–12 leaders, Dr. Morales was charged with examining the AFT report "Building a Profession: Strengthening Teacher Preparation and Induction" and providing recommendations to bring teacher education programs into the 21st century. Her contributions will shape future generations of teachers throughout the nation and ensure that every child in America has access to quality teaching.

Shaping the Future "I'm very excited to be a part of this task force," said Morales. "It really has the wonderful potential to influence policy on teacher preparation around the country."

Morales has worked extensively with the state in the area of teacher preparation and certification. At MDC, she has been an integral player in the science education program, developing and revising curriculum while ensuring that it meets and exceeds state standards. Morales credits MDC's high-flying reputation for facilitating this opportunity to influence education policy at the national level.

MDC Makes an Impact "The AFT thinks very highly of MDC, and I believe that is why we were invited to come to the table and present our ideas," Morales said.

Dr. Susan Neimand, director of MDC's School of Education, said she was proud that Morales could represent the College at such a noteworthy event. "Her experience makes her a perfect candidate for this work," she said. "Dr. Morales' input into this initiative will ensure that teacher education programs around the country maintain rigor and high quality.

Edwin Nieves

Most Helpful to Students, Health Sciences and Nursing

Yolanda Nitti

Most Helpful to Students, Nursing

"Always available to help students."

Rebecca Sanchez

Best Overall Faculty Member, Education

Giannina Santos

Best Researcher/Scholar, Health Sciences and Nursing
Best Teacher, Health Sciences and Nursing

MICHIGAN STATE UNIVERSITY

Michael Kasavana

Best Researcher/Scholar, Business

"39 MSU years in various capacities."

Michael L. Kasavana, Ph.D., is the NAMA-endowed professor in hospitality business and holds CHTP (Certified Hospitality Technology Professional) certification. He continues research into the current and near future developments of electronic commerce, information technology, and transaction processing methodology relative to the hospitality industry. Dr. Kasavana has authored or co-authored six books and a host of academic and industry journal articles. In addition, he has also created a series of online instructional materials and software products.

Dr. Kasavana has been actively exploring TCP/IP application (Internet, Intranet and Extranet) opportunities for various aspects of the hospitality industry including virtual cash transactions, application service providers, e-procurement processes, and effective web site design. He recently completed an innovative research project dealing with the impact of technology on industry productivity and competitive advantage and is credited with coining the phrases "V-Commerce", "V-Engineering", and "Menu Engineering".

Dr. Kasavana is the recipient of the MSU Distinguished Faculty Award. He has been inducted into the HFTP International Technology Hall of Fame and was the first recipient of a Distinguished Achievements Award from FS/TEC. He earned BS (Hotel, Restaurant and Travel Administration), MBA (Finance), and Ph.D. (Management Information Systems) degrees from the University of Massachusetts-Amherst.

MIDDLE TENNESSEE STATE UNIVERSITY

Terri Tharp

Best Teacher, Education

"Terri takes the time to learn about each student she teaches and instills in the students the importance of their future career."

Jane Williams

Best Overall Faculty Member, Education

MIDDLESEX COMMUNITY COLLEGE

Stephen O'Leary

Best Teacher, Accounting

MIDLANDS TECHNICAL COLLEGE

Patrica Shahbahrami

Most Helpful to Students, Health Sciences and Nursing

"Mrs. Gates takes a lot of time with students in the Pre Health programs and tries to accomodate special requests when possible. She wants students to become successful in their chosen fields and works hard to help them achieve their goals."

MIDSTATE COLLEGE

Leah Grebner

Best Researcher/Scholar, Health Sciences and Nursing

Leah Grebner, PhD, RHIA, CCS, FAHIMA has 25+ years of experience in the Health Information Management career field. Her diverse background includes acute care inpatient and outpatient, home health, skilled nursing, physician office, and consulting in a variety of settings. She enjoys sharing her passion for this career field, not only through teaching, but also by volunteering with professional organizations at the regional, state, and national levels.

Leah is currently a member of the AHIMA Consumer Engagement Practice Council. Her greatest area of passion in the Health Information Management field is provision of education to the public regarding personal health records and health literacy. This has also inspired her to pursue her dissertation research study related to promoting health literacy. Leah enjoys being actively involved in professional organizations, giving presentations, and writing for both professional and layperson audiences. She has published two textbooks, "Medical Coding: Understanding ICD-10-CM and ICD-10-PCS" in February 2012, and "Ethics Case Studies for Health Information Management" in 2008.

Leah was awarded the Literary Legacy Triumph Award by the American Health Information Management Association in 2012 and she delivered her first international presentation, "Promoting Student Success and Engagement Through Mentoring, Professional Association Involvement, and Networking" at the 17th Congress of the International Federation of Health Information Management Associations in Montreal, Quebec, Canada in May 2013.

Joanna Holly

Best Overall Faculty Member, Health Science
Best Researcher/Scholar, Health Science
Best Teacher, Health Science
Most Helpful to Students, Health Science
Best Teacher, Health Sciences and Nursing
Most Helpful to Students, Health Sciences and Nursing
Best Overall Faculty Member, Health Sciences and Nursing

MILLERSVILLE UNIVERSITY

Justin Garcia

Best Researcher/Scholar, Anthropology

Richard Kerper

Best Overall Faculty Member, Education

Tim Trussell

Best Overall Faculty Member, Anthropology

"Operates a phenomenal archaeology field school that prepares students for real-life job training and experience with archaeological excavation and recovery, processing, and interpretation of data. Is highly valued and appreciated by Anthropology majors and minors."

MINNESOTA STATE UNIVERSITY

Amy Hedman

Best Overall Faculty Member, Health
Best Teacher, Health
Most Helpful to Students, Health

MIRAMAR COLLEGE

Darrel Harrison

Best Overall Faculty Member, Law
Best Overall Faculty Member, Law

Darrel Harrison is a graduate from Western Sierra School of Law. He is the program director and tenured professor for the paralegal program.

Darrel teaches Introduction to Paralegalism (Legl 100a), Legal Procedures (Legl 100b), Legal Research (Legl 105), Legal Communications (Legl 110), Bankruptcy Law (Legl 160), Law Office Management and Technology (Legl 140), and Business Law (Buse 140)

MISSION COLLEGE

Yolanda Coleman

Best Overall Faculty Member, Education

Dennis Haze

Best Researcher/Scholar, Education

Chidhood skateboarding afficionado. Hates ALL cell phones. Favorite restaurant, Harey Carey's in Chicago.

Theresa Lawhead

Most Helpful to Students, Counseling
Most Helpful to Students, Education

"Prof. Lawhead is an outstanding teacher, counselor and invaluable member of the Student Services dept at Mission College. Ms, Lawhead brings her industry experience along with her expertise in STEM majors to engage and encourage students."

Theresa Lawhead is a Counselor and Instructor at Mission College. She joined Mission College in 1993, became a Counselor in 1996 and began teaching "Careers and Life Styles" in 2000. Theresa was also a College Relations Manager at Amdahl Corporation, and an Engineering Career Counselor and Employers Relations Specialist at San Jose State University.

MISSISSIPPI STATE UNIVERSITY

Chris Codling

Best Teacher, Education

Jianxia DU

Best Researcher/Scholar, Education

David Morse

Best Overall Faculty Member, Education

MISSISSIPPI UNIVERSITY FOR WOMEN

Twila Alpe

Most Helpful to Students, Education

Sue Jolly-Smith

Best Researcher/Scholar, Education

"Dr. Jolly-Smith has an extensive history in education and consistently demonstrates knowledge and practices that support staff development, collaboration and collegiality."

MISSOURI SOUTHERN STATE UNIVERSITY

Stu Dunlop

Best Overall Faculty Member, Business

J. Chris Moos

Most Helpful to Students, Business

Dr. Moos has over 25 years of business experience prior to teaching. He has held the positions of CFO, COO, CEO and Consulting for Domestic, International and Multi-National Corporations. Dr. Moos has served in the U. S. Navy (Officer Leadership & Management), and taught English as a Second Language (ESL). He is also a Commander in the Navy (Retired). Dr. J. Chris Moos began his teaching career as an adjunct instructor at MSSU in 2003. He was quickly offered a full time position and has served as the Department Head of Marketing and Management for the Plaster School of Business.

Dr. Moos currently serves on the Budget Advisory Committee, as the Director for the International Trade and Quality Center, and as the adviser for the International Business Club at MSSU. He is one of approximately 1,030 individuals worldwide who are Certified Global Business Professionals. Dr. Moos has also been a member of the Strategic Planning Committee, a Co-Director for the USAID Grant, and a committee member for the Egypt, Brazil, China, and Russia themed semesters. In 2007 he was awarded MSSU's Outstanding International Education Teacher and in 2009 he received the Faculty Research Award.

MISSOURI STATE UNIVERSITY

Melissa S. Burnett

Best Researcher/Scholar, Marketing
Best Teacher, Marketing

Edward Chang

Most Helpful to Students, Finance

Sherry Cook

Best Overall Faculty Member, Marketing

Susan Crain

Best Researcher/Scholar, Finance
Best Overall Faculty Member, Business

John Kent

Best Researcher/Scholar, Marketing

Wayne Mitchell

Best Researcher/Scholar, Psychology

"Dr. Mitchell has a long-term, very productive research program, and is widely known for his contributions to the literature. He has trained many students who have gone on to PhD programs. He has volunteered thousands of hours to colleagues, and to students supervised by other faculty."

Peter Richardson

Best Researcher/Scholar, Business

"Outstanding teacher and researcher."

Christina Simmers

Best Overall Faculty Member, Marketing
Best Teacher, Marketing
Most Helpful to Students, Marketing
Best Teacher, Business
Most Helpful to Students, Business
Best Overall Faculty Member, Business

MISSOURI UNIVERSITY OF SCIENCE AND TECHNOLOGY

Bih-Ru Lea

Best Overall Faculty Member, Business
Best Researcher/Scholar, Business
Best Teacher, Business
Most Helpful to Students, Business

Dr. Bih-Ru Lea earned a bachelor's degree in accounting from Tamkang University in Taiwan, a Master of Professional Accountancy from Clemson University, and a Ph.D. in Industrial Management from Clemson University.

Dr. Bih-Ru Lea devotes her research efforts to interdisciplinary research projects aimed at addressing integration issues and providing insights to improve business performance in the areas of enterprise resource planning (ERP), supply chain management (SCM), accounting information systems, manufacturing information systems, information technology, and business strategy. Dr. Lea has performed funded researches for the National Science Foundation, British Telecom, Monsanto Company, and University of Missouri System Research Board. Dr. Lea collaborated and supervised numerous information technology related student projects with major companies such as UPS, GE, Humana, CNET Networks, Inc., Yum! Brands, and Kindred Healthcare as well as non-profit organizations such as Metro United Way, the Jefferson County Public Schools, and the Louisville/Jefferson County Metro Revenue Commission. Dr. Lea previously served as an Assistant Research Director of UPS Center for Worldwide Supply Chain Management for Logistics and Distribution Institute at the University of Louisville, where she coordinated research projects in the areas of ERP, SCM, and manufacturing performance with her industrial experience and knowledge.

Dr. Lea has served as the SAP Campus Coordinator for the University of Missouri-Rolla since 2004, is currently the director of the ERP Center, and is on the editorial board of Production and Operations Management Journal, Journal of International Technology and Information Management, International Journal of Electronic Finance, and International Journal of Business and Systems Research. Dr. Lea is a regular reviewer for many prestigious academic journals and conferences including International Journal of Production Economics, The Engineering Economist Journal, International Journal of Production Research, Decision Sciences Institute (DSI) conference, Conference of Association of Information Systems (AIS), and Conference of International Association of Information Systems (IACIS).

Prior to coming to Missouri University of Science & Technology (formerly University of Missouri- Rolla), Dr. Lea was an assistant professor at University of Louisville and Anderson College. Dr. Lea has taught classes at both the undergraduate and graduate levels as well as to traditional and non-traditional/adult students. To prompt active learning, through hands-on projects and real-world experience, Dr. Lea coordinated constructions of several state-of-the-art multi-media interactive classrooms and brought in many company projects and software grants for Anderson College, Clemson University, and the University of Louisville. Furthermore, Dr. Lea has extensive experience in curriculum design and planning in the area of computer information systems. In the past few years, Dr. Lea helped to establish a new Computer Information Systems concentration for Anderson College in South Carolina and played a major role in creating a web-based Computer Information Systems curriculum for the University of Louisville in Kentucky. Dr. Lea's other experience include as an research assistant at Clemson Apparel Research Center (CAR) where she participated in various simulation modeling and system analysis projects; a research and teaching assistant at Clemson University, and a research Assistant at Accounting Research & Development Foundation of Taiwan.

Eun Park

Best Researcher/Scholar, Business
Best Teacher, Business
Most Helpful to Students, Business
Best Overall Faculty Member, Business

MOHAWK VALLEY COMMUNITY COLLEGE

Arthur Friedberg

Best Overall Faculty Member, Business
Best Researcher/Scholar, Business
Best Teacher, Business
Most Helpful to Students, Business

Arthur Friedburg

Best Researcher/Scholar, Business
Best Teacher, Business
Most Helpful to Students, Business
Best Overall Faculty Member, Business

MOLLOY COLLEGE

Mary Fassetta

Best Overall Faculty Member, Nursing

Debra R Hanna

Best Researcher/Scholar, Nursing

I am completing a phenomenological study of the lived experience of moral distress for former child protection workers. I had hoped to learn if there are gender-specific differences in the experience of moral distress. I have found differences that are important to understand.

I also have underway a mailed survey of nurses' work to reposition patients to prevent pressure ulcers. The survey is called Learning about Turning. The results will help inform us about the realities about nursing labor involved in repositioning patients in today's high tech work environments.

Rose Schecter

Best Overall Faculty Member, Health Sciences and Nursing

Margaret Whelan

Best Overall Faculty Member, Nursing

MONMOUTH UNIVERSITY

Andrea Hope

Best Teacher, Health

"Very engaging teacher."

Laura Kelly

Best Teacher, Health Sciences and Nursing

"Students love her."

Joe Mosca

Most Helpful to Students, Business

David Paul

Best Overall Faculty Member, Business

For a business academic, Dr. Paul has a somewhat eclectic background. After receiving his B.Sc. in Chemistry and Mathematics from Hampden-Sydney College and his D.D.S. from the Medical College of Virginia, he practiced general dentistry for more than 20 years. During this time, he served 8 years on the Board of Directors of the Virginia Academy of General Dentistry, and 12 years as Assistant Director/Director of a hospital-based Dental General Practice Residency. While practicing dentistry full time, he attended Old Dominion University, where he received his M.B.A. in Marketing. While practicing dentistry part-time, he earned an M.A. in Economics and a Ph.D. in Marketing from Old Dominion, while simultaneously serving as president of his local dental society.

Gene Simko

Best Teacher, Business

Michaeline Skiba

Best Overall Faculty Member, Marketing

Dr. Michaeline Skiba teaches both undergraduate and graduate courses within the department, and graduate courses in the MBA program – Healthcare Management Track. Prior to her academic appointment, Dr. Skiba had a 20+ year business career during which she held senior management positions within three Fortune 500 companies and where she designed, developed, and delivered both line and staff management marketing and management materials. These programs centered on professional development topics that included pharmaceutical marketing strategy (pre-launch), healthcare symposia and colloquia (for CME and CPE credit), journal supplements, market research (focus groups and telephone-based interviews), professional communications, and a variety of management techniques.

MONROE COMMUNITY COLLEGE

Gary Thompson

Best Teacher, Criminal Justice

Professor Thompson has been on the faculty since 1988. He holds a B.A. in Political Science from St. John Fisher College and a Master of Public Administration from the State University of New York. Prior to becoming a full-time faculty member, he served as an adjunct for six years at the State University of New York College at Brockport and as an adjunct at Monroe Community College. He is retired after twenty years of service from the Monroe County's Sheriff's Department where he served as patrol sergeant, planning and research officer, police trainer and SWAT commander. He also is a police instructor at the Criminal Justice and Public Safety Training Center. Prof. Thompson has been awarded the Teaching Excellence Award by the National Institute of Organizational and Staff Development. Prof. Thompson also serves on the College Compact for Service Learning Advisory Committee.

MONTCLAIR STATE UNIVERSITY

William Batkay

Best Researcher/Scholar, Political Science

I am now officially retired from the university, but am happily continuing teaching as a plain ordinary adjunct. In the spring 2015 semester, I am teaching International Organizations–POLS203. My office hours for Spring 2015 are W F 9–9:45 AM. The best way to contact me is by the email address above; by phone it is through the Political Science and Law Office number: 973-655-4238.

I've taught a wide variety of courses since coming to MSU in fall 1971, with an initial focus on American government and politics and comparative politics. Relatively soon, I branched out in to areas studies courses that I introduced into the curriculum: Western European politics, Eastern European politics and the politics in first the Soviet Union and then in post-Soviet states, primarily Russia. Along the way, I was asked to step in and teach the basic course in international relations, which I have continued to do. To that I added international organizations, which I also love.

As part of my past responsibilities, I served as department chair for two terms in the 1980's and more recently have served and continue to serve as deputy chair for political science and as coordinator of departmental academic advisement. I seem to have been particularly good at advisement, in large part because of the web of helpful campus contacts I built up just by being here for so many years.

My attached CV has the details of my research and publications. Here I'll just note that initially I was interested in Hungarian politics in the 1920s, but then became fascinated with Jewish religious politics in Israel. More recently, I have been drawn to issues concerning the nation-state as a type of political arrangement. Along the way, I became experienced at writing book reviews on a variety of academic topics.

Jennifer Bragger

Best Overall Faculty Member, Psychology

"What is wonderful about Prof Jennifer Bragger is that she is balanced in terms of teaching, research, and service. She takes teaching very seriously and her students see and appreciate

this, she conducts top notch research studies and regularly presents and publishes (also with her students, by the way), and she does a tremendous amount of service to the university."

My name is Jennifer DeNicolis Bragger. I graduated from Temple University in 1997. I have a Ph.D. in Industrial Organizational Psychology. Industrial Organizational Psychology is the study of behavior in the work place and deals with issues such as job satisfaction, organizational commitment and how to select individuals for given positions in an organization. I teach graduate and undergraduate courses in Industrial and Organizational Psychology as well as undergraduate courses in Leadership, Personnel Psychology, Organizational Psychology and Teaching of Psychology. I conduct research on stereotype in the job interview, effects of leave taking on performance appraisal ratings, work family conflict, leadership, person-organizational fit and work family conflict. I am married with four young children.

Marie Cascarano

Best Teacher, Health Science

"dedicated, non-judgmental with students."

Julie Farnum

Best Overall Faculty Member, Anthropology
Best Researcher/Scholar, Anthropology
Best Teacher, Anthropology
Most Helpful to Students, Anthropology

Research interests include modeling health processes through the examination of activity patterns, diet, ecology, genetics, occupational stress and pathologies, and social inequalities. Dr. Farnum works with a multi-disciplinary team on the North Coast of Peru and colleagues at the American Museum of Natural History in New York City where she studies human remains from pre-Hispanic cultures that span thousands of years. Some of the results from this project were published in the prestigious journal, Current Anthropology.

Dierdre Glenn Paul

Best Researcher/Scholar, Education

Dr. Dierdre G. Paul started her career as a public school teacher and a Communication Arts Teacher Trainer.
After more than 20 years in academia, she is now full professor at Montclair State University's College of Education and Human Services. She has authored three books, four book chapters and multiple journal articles. Dr. Paul's current research interests include new and digital literacies, the technological pitfalls of PARCC testing, and literacy trends for African American children/youth.

Dierdre also served as department chairperson and a statewide union leader. She first served as Grievance Officer for AFT Local 1904, then Acting President, and President, in which capacity she aggressively negotiated contracts for her members. She became the first African American woman to hold the office of Executive Vice President of the Council of New Jersey State College Locals and was later elected Executive Vice President-Higher Education of American Federation of Teachers New Jersey.

In 2002, Dr. Paul received the distinction of being appointed to the Governor's Early Literacy Task Force.

Katia Goldfarb

Most Helpful to Students, Education
Best Overall Faculty Member, Education
Best Researcher/Scholar, Education
Best Teacher, Education

Scholarly Interests and Specialties
  Latino immigrant families and school relationships Under-representation of ethnic minority families in family journals Reviews.
  NCFR, Conference Proposals, Program Reviews American Educational Research Association (AERA), Conference Proposals Leadership Positions in Professional Organizations.
  National Council on Family Relations 2004–2006 Chair-Elect, Ethnic Minority Section National Council on Family Relations 2006–2008 Chair, Ethnic Minority Section.

Laurence Greenberg

Best Teacher, Early Childhood Education

Nancy Lauter

Most Helpful to Students, Early Childhood Education

John McGinnis

Most Helpful to Students, Business

http://business.montclair.edu/file/478

Farahmand Rezvani

Best Teacher, Economics

Moschos Scoullis

Most Helpful to Students, Economics

Nancy Tumposky

Best Overall Faculty Member, Education

Ana Maria Villegas

Best Researcher/Scholar, Education

Dr. Ana María Villegas is Professor in the Department of Secondary and Special Education and Director of the Doctoral Program in Teacher Education and Teacher Development at Montclair State University. She teaches courses in teacher education policy and research, culturally responsive teaching, and sociocultural perspectives on teaching and learning. She has a Ph.D. in curriculum and teaching from New York University. Prior to joining MSU in September 1996, she was a Senior Research Scientist with the Division of Education Policy Research of Educational Testing Service, a position she held for eight years. She has also taught at the University of Colorado-Denver, Department of Language, Literacy, and Culture. Dr. Villegas began her career as a teacher in the NYC Public Schools.

Vincent Walencik

Best Teacher, Education
Most Helpful to Students, Education

MONTGOMERY COUNTY COMMUNITY COLLEGE (ALL)

Barbara May

Best Researcher/Scholar, Education

MOTT COMMUNITY COLLEGE

Paula Harris

Most Helpful to Students, Health Science

"Paula addresses multiple learning styles in the classroom setting, offers assistance outside of class time, and seeks professional development offerings to improve her own skills/performance. She puts great effort in providing additional resources for students: websites, sample work, narrated PowerPoints, and videos. She is warm and approachable with her students. She is a great asset to Mott Community College."

MOUNT SAINT MARY COLLEGE

Priscilla L Sagar

Best Researcher/Scholar, Nursing

MT. SAN ANTONIO COLLEGE

Michelle Boyer

Best Overall Faculty Member, Nursing
Most Helpful to Students, Nursing
Best Researcher/Scholar, Health Sciences and Nursing

Denise Condra

Most Helpful to Students, Nursing

Melanie Diederichs

Best Overall Faculty Member, Education

Amirk Dua

Best Teacher, Business

"He is very dedicated to student learning."

Lance Heard

Best Overall Faculty Member, Criminal Justice

Professor Lance Heard's Law Enforcement career spanned from 1990 to 2008. As a 10 year sergeant, his assignments included community relations, detective supervisor, and patrol. As an officer, his specialty assignments ranged from D.A.R.E. to beach patrol and investigations, including crimes against persons, financial crimes, and juvenile cases. Prior to transferring to the Hermosa Beach Police Department, he began his career at LAPD and worked in the Harbor, Wilshire, and Pacific area stations.

An army veteran, Lance Heard served six years active duty in New York, Texas, Germany, and Korea. He was commissioned as a 2nd Lieutenant into the Air Defense Artillery in 1982 and left the Army as a Captain in 1988. He served primarily as a surface to air missile officer with the PATRIOT missile system.

Professor Heard has been at Mt SAC since 2008.

Paul Jefferson

Best Overall Faculty Member, Criminal Justice

Rasool Masoomian

Best Overall Faculty Member, Economics
Best Researcher/Scholar, Economics
Best Teacher, Economics
Most Helpful to Students, Economics
Most Helpful to Students, Business
Best Researcher/Scholar, Business
Best Teacher, Business
Best Overall Faculty Member, Business

"He is very serious about student's success."

Don Meredith

Best Teacher, Criminal Justice

While the results of the November elections are in, one race has yet to be settled involving a Mt. SAC adjunct professor. Don Meredith, who has taught criminal justice at Mt. SAC for more than five years, is vying for Los Angeles County Sheriff in a field with three other candidates, including the incumbent, Sheriff Lee Baca. The primary election is set for June 6, 2006. A career law enforcement officer, Meredith says he would "bring a new vision and new direction to the Sheriff's Department—to make the county safer while working within approved budget resources. This vision includes stopping the early release of prisoners—many of whom are currently serving only 10% to 25% of their sentence in county jail. This practice has no deterrent effect to those who commit crimes of violence and theft." Meredith believes he is uniquely qualified for the job, because he says he is decisive, nonpartisan and able to work with the region's diverse communities. Using his expertise in street gang intervention and community policing, Meredith said he would work in partnership with communities, citizens, and agencies to address quality-of-life issues that contribute to escalating gang violence. Meredith is currently a lieutenant and commander of the Community Policing Unit for the Police Department of Glendale, the county's third largest city and one of the nation's safest. As a graduate of the FBI National Academy, the USC School of Public Administration Delinquency Control, and the USC School of Engineering Aviation Safety Program, Meredith treasures his teaching responsibilities here at Mt. SAC. He takes pride in the fact that many of his students have graduated and are enjoying careers in law enforcement, including at least five who are serving at Glendale P.D. During his 33-year career with Glendale PD, Meredith has earned over 60 commendations from the public as well as law enforcement agencies. In addition, he has received departmental awards for bravery, community service and three campaign ribbons for major tactical operations.

Sarah Plesetz

Best Overall Faculty Member, Health Sciences and Nursing

Full-time professor

Curtis Simon

Best Teacher, Political Science
Most Helpful to Students, Political Science

Kirk Smith

Best Teacher, Criminal Justice

Tina Ziolkowski

Most Helpful to Students, Health Science

"Tina is the Program Director for both Paramedic and EMT programs as well as being a primary instructor. She spends countless hours of her own time helping students with both classroom and personal issues. She is incredibly dedicated and deserves this award!"

MUSKEGON COMMUNITY COLLEGE

Chris Hain

Best Teacher, Criminal Justice

NASHVILLE STATE COMMUNITY COLLEGE

Tammy Ruff

Best Overall Faculty Member, Psychology

NATIONAL UNIVERSITY

Thomas Doyle

Best Overall Faculty Member, Business

"Tom has served at all levels of the university. In addition to being a stellar teacher, Tom's service to the university, school, department, and program are exemplary."

Ron Germaine

Best Teacher, Education

"Ron is dedicated to the success of each of his students. As the Program Lead for the MAT program Ron works to ensure program and course integrity."

Dr. Ron Germaine is currently Professor in the School of Education at National University, San Diego, CA and Program Lead for the Master of Arts in Teaching Program. Ron received a BA in geography from the University of British Columbia, an MEd in educational administration from the University of Victoria, and an EdD in educational leadership from the University of San Diego. Ron has published articles in the Journal of Research in Innovative Teaching and the Association of Institutional Research, and serves on the advisory board of the Character Development Center, University of San Diego.

Robyn Hill

Most Helpful to Students, Education

"Robyn's dedication to student success is evidenced by her time and energy. She is tireless in ensuring student success."

Peter Serdyukov

Best Researcher/Scholar, Education

"Peter has served as Founder and Editor of the Journal of Research in Innovative Teaching and has tirelessly worked to increase the quality of research for faculty at NU."

NATIONAL-LOUIS UNIVERSITY

Donna Wakefield

Best Overall Faculty Member, Education

"Dr. Wakefield is my colleague. In the summer of 2013 I taught my first online course at N-LU. Donna went out of her way to

provide me guidance in navigating the tech issues. What stands out above that is—-during the last few weeks of this course I was spending a lot of time with my mom who was very very ill. Donna knew I would be at the hospital in the middle of the night using hospital computers–sometimes successfully and sometimes no access. During this time, Donna helped me grade assignments and even entered my final grades. I have NEVER forgotten how much she reached out to assist — I never asked for help . . . but boy did I need it! My mom did pass during this time–and because of Donna, I was able to be with mom. Donna goes above and beyond not only for her students but for her colleagues"

NEUMANN UNIVERSITY

William Hamilton

Most Helpful to Students, Psychology

Etsuko Hoshino-Browne

Best Overall Faculty Member, Medicine
Best Researcher/Scholar, Medicine

Sheila Keller

Most Helpful to Students, Nursing

Colleen McDonough

Best Researcher/Scholar, Psychology
Best Teacher, Psychology
Best Overall Faculty Member, Medicine

"Is always working on some sort of research, collaboration, etc. Is able to produce several manuscripts in a year's time despite her heavy load. Always thinking of ways to assess things that are important for us to know."

NEW JERSEY CITY UNIVERSITY

William Craven

Most Helpful to Students, Business

Marilyn Ettinger

Best Overall Faculty Member, Business

NEW MEXICO STATE UNIVERSITY

Joan Crowley

Best Researcher/Scholar, Criminal Justice
Best Researcher/Scholar, Education

Dr. Joan E. Crowley received her doctorate in Social Psychology from the University of Michigan in 1979. Prior to joining the faculty of NMSU in the Fall of 1989, Dr. Crowley held a senior research position with the Center for Human Resource Research at the Ohio State University and with the Institute for Social Science Research at the University of Alabama. She taught as

Adjunct Assistant Professor of Psychology at both the Ohio State University and at the University of Alabama. In 1991, she was elected to the three-year presidential cycle of the Southwest Association of Criminal Justice Educators. Her current interests are in the areas of communities and criminal justice, family violence and child abuse, juvenile delinquency, substance abuse, historical criminology, and research methods.

Henry Dimatteo

Most Helpful to Students, Education

Hank DiMatteo received his MPA from the University of Texas at El Paso and his MBA at NMSU. He is completing his Ph.D. in Educational Management and Development at NMSU. For the past 28 years, Hank has been a police officer with the El Paso, Texas Police Department. He is presently a Lieutenant and is responsible for the night operations in Central El Paso. He is also a graduate of the FBI National Academy. His research interests are in police administration, school violence, ethics, and force issues.

R. J. Maratea

Best Researcher/Scholar, Criminal Justice
Best Researcher/Scholar, Education

Dr. R. J. Maratea is an assistant professor of CJ at NMSU. He received his B.A. in political science from Syracuse University, a M.S. in Justice Studies from Arizona State University, and a Ph.D. in sociology at the University of Delaware. His dissertation expands upon social problems theory by examining the ways in which claims-makers use the Internet to distribute problem claims and attract support for their issues. Dr. Maratea's areas of specialization include mass communication, social problems, deviant and criminal subcultures, and criminological theory.

Rory Rank

Best Overall Faculty Member, Criminal Justice
Best Teacher, Criminal Justice
Most Helpful to Students, Criminal Justice
Best Teacher, Education
Best Overall Faculty Member, Education

"He works hard, is passionate about the subject he teaches and he cares about the students."

Mr. Rory L Rank has worked for the New Mexico Public Defender Department since 1992. He is currently the supervisor of the Juvenile Division in the Las Cruces NM office. He has been very proactive in designing and implementing diversion programs for juveniles the 3rd Judicial District. As a practicing attorney for over 30 years, he ensures excellence in juvenile defense and promotes justice for all children.

NEW YORK CITY COLLEGE OF TECHNOLOGY

Nicholas Manos

Best Teacher, Dental

"Prof. N. Manos has been the most helpful and most knowledgable teacher in the Department of Restorative Dentistry for over 20 years. He is a role model to students and faculty alike."

NEW YORK COLLEGE OF PODIATRIC MEDICINE

Steven Goldman

Best Overall Faculty Member, Medicine
Best Researcher/Scholar, Medicine
Best Teacher, Medicine
Most Helpful to Students, Medicine
Best Overall Faculty Member, Medicine

DVA - New York Harbor: - Chief of Podiatry - Site Director of Surgical Service - Director of Podiatric Medical Education
    NYCPM - Associate Clinical Professor of Podiatric Medicine
    American Board of Podiatric Medicine - President Elect - Former Treasurer and VP
    CPME - Site Examiner representing ABPM

NEW YORK LAW SCHOOL

Doni Gewirtzman

Best Teacher, Law

Edward Purcell

Best Researcher/Scholar, Law

Edward A. Purcell Jr. is the Joseph Solomon Distinguished Professor at New York Law School and one of the nation's foremost authorities on the history of the United States Supreme Court and the federal judicial system. In 2013 he received the "Outstanding Scholar Award" from the American Bar Foundation.

His most recent book, Originalism, Federalism, and the American Constitutional Enterprise: A Historical Inquiry (Yale University Press, 2007), examines the original structure and subsequent operations of the federal system and refutes the widely accepted belief that the founding fathers crafted a careful constitutional balance of power between the states and the federal government. The book argues that there was no clear agreement among the founders regarding the "true" nature of American federalism, nor was there any consensus on a "correct" line dividing national authority from state authority. The book maintains that even if there had been some such true "original" understanding, the elastic, dynamic, and underdetermined nature of the constitutional structure would have made it impossible for subsequent generations to maintain any such specific balance.

Professor Purcell's previous book, Brandeis and the Progressive Constitution: Erie, the Judicial Power, and the Politics of the Federal Courts in Twentieth-Century America (Yale University Press, 2000), examines how the Erie case provides a window into the legal, political, and ideological battles over the federal courts in the 20th century and also offers an in-depth study of Supreme Court Justice Louis D. Brandeis's evolving constitutional jurisprudence. The book has been hailed by reviewers as a work destined to occupy an important place in the constitutional-historical canon. It was awarded an American Bar Association Silver Gavel Certificate of Merit, the Supreme Court Historical Society's Triennial Griswold Prize, and the Association of American Law School's Coif Triennial Award.

Professor Purcell first became interested in legal and constitutional issues when he was studying 20th century American intellectual history at the University of Wisconsin, where he began work on legal realism and its relationship to democratic theory and on the rise of totalitarianism during the 1930s. After completing his Ph.D., he taught American history at the University of Missouri and, while there, took a year's leave of absence to serve as Charles Warren Fellow in American Legal History at Harvard Law School. "My research on legal realism had made me realize that I was dealing with issues I didn't fully understand," he recalls, "and I knew I needed some legal training." At Harvard he sat in on first-year classes, which he found "fascinating," and finished his first book, The Crisis of Democratic Theory: Scientific Naturalism & the Problem of Value (University Press of Kentucky, 1973), which was awarded the Frederick Jackson Turner Prize by the Organization of American Historians.

Returning to the University of Missouri, he became a tenured professor in the department of history but found himself increasingly drawn to legal subjects and decided to return to Harvard to complete his legal education. "While I was in law school as a full-time student," he remembers, "I concluded that, if I was really going to understand law and the legal system, I had to go into practice for a time."

After graduation, Professor Purcell joined Paul Weiss Rifkind Wharton & Garrison LLP and remained there from 1980 to 1989, periodically taking leaves of absence to work on what was to become his second book, Litigation & Inequality: Federal Diversity Jurisdiction in Industrial America, 1870–1958 (Oxford University Press, 1992). "By 1988 I realized that I wasn't going to finish the book if I remained a practicing lawyer," Professor Purcell says. "Then, I got a call from New York Law School asking me to teach Federal Courts and Civil Procedure—exactly what I was writing about. Both the timing and the subjects were perfect." The law school environment has provided the "intellectual crackle" he loves. "When a law school class goes really well," he believes, "it's a fantastic experience." In addition to teaching Civil Procedure and Federal Courts, he also teaches courses on Civil Rights Law and Complex Litigation.

Professor Purcell was born in Kansas City, Missouri, where he grew up and attended Rockhurst College. Subsequently, he received an M.A. in American history from the University of Kansas before attending the University of Wisconsin. Prior to law school, Professor Purcell also taught at the University of California, Berkeley and at Wellesley College. He has served on the Board of Directors of the American Society for Legal History and the Community Law Offices of the Legal Aid Society, and he is currently a member of the Advisory Board for the Federal Judicial Center's Narrative History of the Federal Courts and the Board of Academic Advisors of the Institute of Constitutional Studies at George Washington Law School. He has been active in the Association of the Bar of the City of New York on its committees for Housing Court, the Legal Needs of the Poor, and the President's Committee on Implementing the Report of the Pro Bono Housing Court Project. In 2004 he was elected Program Chair of the Federal Courts Section of the Association of American Law Schools, served as the Section's Chair in 2005–06, and continues to serve on the Section's Board of Directors. In addition to his books, he has published widely in law reviews (including the Virginia Law Review, the University of Pennsylvania Law Review, the Harvard Civil Rights/Civil Liberties Law Review, and the UCLA Law Review) and history journals (including the American Historical Review, the Journal of American History, the Law and History Review, and the American Quarterly). He has also contributed chapters to several books, including International Law in the U.S. Supreme Court (Cambridge University Press, 2011), the Cambridge History of Law in America (Cambridge University Press, 2008), Progressive Lawyering, Globalization and Markets: Rethinking Ideology and Strategy (William S. Hein & Co., 2007), and Private Law and Social Inequality in the Industrial Age: Comparing Legal Cultures in Britain, France, Germany, and the United States (Oxford University Press, 2000).

James Simon

Best Researcher/Scholar, Law

NEW YORK UNIVERSITY

Mark Alter

Best Overall Faculty Member, Teaching & Learning
Best Overall Faculty Member, Education

Alter is a Professor of Educational Psychology at New York University and was the founding chair of the Department of Teaching and Learning and served as Chair for 14 years. He has an extensive record of publications, national and international workshops and funded grants in the field of teacher education and special education. In 2005, he was granted A Fulbright Senior Specialist award to Vietnam, in 2007 was awarded The NYU Distinguished Teaching award and in 2011 was named one of the 50 Most Influential Professors in Education. In May 2011 Alter and Pradl published in Education Week: Ending Three-Card Monte in Teacher and argued that teacher education programs must have a direct connection with student learning outcomes. Alter has an extensive background in the classroom, as well as a PhD from Yeshiva University in special education, a focus that gives his teaching a unique context.

Analia Keenan

Best Overall Faculty Member, Dental

"Analia Keenan is a caring individual that will mentor any faculty member to help them advance to higher achievements. She is dedicated to her work and succeeds at best for the role modelling dynamics at NYUCD. I wish her the best of luck for this award, since she deserves it!"

NEW YORK UNIVERSITY SCHOOL OF LAW

J. Steines

Best Overall Faculty Member, Law
Best Researcher/Scholar, Law
Best Teacher, Law
Best Teacher, Law
Best Overall Faculty Member, Law

John P. Steines, Jr. received a degree in engineering from General Motors Institute in 1971 and worked for General Motors

Corporation before receiving his J.D. from Ohio State University College of Law in 1974 (summa cum laude, Order of the Coif). Steines practiced law for three years in Grand Rapids, Michigan and then, as a Gerald L. Wallace Scholar, received an LL.M. in Taxation from New York University in 1978. He became an instructor at NYU and two years later joined the permanent faculty. Steines's teaching includes classes in basic personal and corporate income tax for J.D. students and a wide variety of LL.M. courses in corporate and partnership taxation, consolidated tax returns, tax accounting, international aspects of U.S. taxation, and tax policy. Formerly counsel for many years to Weil, Gotshal & Manges and since 2004 to Cooley LLP (formerly Kronish, Lieb, Weiner & Hellman), Steines's practice focuses on corporate, partnership, and international tax matters. He frequently testifies as an expert in U.S. and foreign tax or tax-related controversies. A former editor-in-chief of the Tax Law Review, his scholarship includes a casebook, International Aspects of U.S. Income Taxation, and several articles on corporate, partnership, and international issues. He was author of the Bittker and Eustice treatise on Federal Income Taxation of Corporations and Shareholders from October 2011 through February 2013.

NORFOLK STATE UNIVERSITY

Benie Marshall

Best Researcher/Scholar, Nursing
Most Helpful to Students, Nursing

Antoinette McCray

Best Overall Faculty Member, Nursing

Ellis Siegel

Best Teacher, Nursing

NORTH CAROLINA STATE UNIVERSITY

Joseph Brazel

Best Researcher/Scholar, Business

I am currently a professor in the accounting department at North Carolina State University where I teach classes in auditing at the undergraduate and graduate levels. I completed my Ph.D. degree at Drexel University, where I defended my dissertation on May 25, 2004. My dissertation investigated auditor decision-making in an enterprise resource planning (ERP) system environment. Prior to starting my doctoral studies, I worked as an audit manager for the firm of Deloitte and Touche in the Tri-State (NY/NJ/CT) and Mid-Atlantic (Philadelphia office) regions and I am a certified public accountant (inactive) in the state of Pennsylvania. My prior educational experiences include a Bachelor of Arts with a double major of accounting and economics from Muhlenberg College in 1994 and an MBA from Drexel University that I completed in 2003. My research interests include judgment and decision-making in auditing, audit review methods, professional skepticism, fraudulent financial reporting, fraud detection, brainstorming, nonprofessional investor/auditor use of fraud red flags, relations between financial and nonfinancial measures, enterprise systems, audit delay, accelerated financial reporting, internal controls over financial reporting, and executive compensation structure.

I grew up in Carlisle, PA, the son of Dr. Joseph and Barbara Brazel. I also have two younger sisters Kate and Elizabeth. While growing up in south central PA, I acquired my hobbies that include fly fishing, playing tennis, jogging, canoeing, and mountain biking. I also enjoy watching my Philadelphia Eagles and going to Wolfpack football games at Carter-Finley Stadium. My wife Kyla (also a former Pennsylvanian), daughters Abigail Grace (born May 18, 2004) and Caroline Olivia (born June 21, 2007), and I reside in Raleigh, NC. Here is a family portrait (from a couple of years ago).

Lee Craig

Best Overall Faculty Member, Business

Shannon Davis

Best Teacher, Management
Best Teacher, Business

Donovan Favre

Most Helpful to Students, Business

NORTH CENTRAL TEXAS COLLEGE

Jeremy Dawson

Most Helpful to Students, Political Science
Most Helpful to Students, Public Affairs, Political & Policy Sciences

Richard Huckaby

Best Teacher, Political Science
Best Teacher, Public Affairs, Political & Policy Sciences
Best Overall Faculty Member, Public Affairs, Political & Policy Sciences

NORTH LAKE COLLEGE (ALL CAMPUSES)

Gabriel Bach

Best Overall Faculty Member, Political Science

Dr. Gabriel Bach earned a law degree from the University of Strasbourg in his native France. He also holds a Master of Business Administration degree from the University of Dallas and a doctorate from Tulane University.

Rebecca Escoto

Most Helpful to Students, Psychology

NORTHAMPTON COMMUNITY COLLEGE

Cori Doughty

Best Teacher, Criminal Justice

"Professor Doughty goes above and beyond in her classes. She has the resources to bring in speakers and working knowledge of her classes."

NORTHEAST IOWA COMMUNITY COLLEGE

Molly Brandel

Best Teacher, Anatomy

"Molly Brandel has taken the credo of intensity and creativity of teaching Anatomy and Physiology to a new level. She interacts with students on so many levels and impacts their lives in so many ways that are impossible to measure. Molly not only understands the material so well that her instruction seems effortless, but she has created innovative and engaging approaches to teaching Anatomy and Physiology that defy traditional paradigms of learning. She understands that her role as an instructor is a catalyst and not a purveyor of information, and she takes very seriously the personal interactions that allow her to make those connections with her students. In a small cadre of instructors here at NICC, in several fields of study, that I consider to be exceptional; she stands alone."

NORTHEASTERN ILLINOIS UNIVERSITY

Eleni Makris

Most Helpful to Students, Education

Education Ph.D. The University of Chicago, Chicago, Illinois Educational Psychology Dissertation: Educational Resilience: Mediating Factors of Adolescents' Adversity Advisor: Mihaly Csikszentmihalyi, Ph.D.
    M.Ed. Loyola University Chicago, Chicago, Illinois Counseling Psychology
    B.A. Loyola University Chicago, Chicago, Illinois

Erica Meiners

Best Researcher/Scholar, Education
Best Researcher/Scholar, Education

Erica R. Meiners teaches, writes and organizes in Chicago. She has written about her ongoing labor and learning in anti-militarization campaigns, educational justice struggles, prison abolition and reform movements, and queer and immigrant rights organizing, in Flaunt It! Queers organizing for public education and justice (2009 (with Therese Quinn), Right to be hostile: schools, prisons and the making of public enemies (2007) and articles in Radical Teacher, Meridians, AREA Chicago and Social Justice. Her work in the areas of prison/school nexus; gender, access and technology; community-based research methodologies; and urban education, has been supported by the US Department of Education, the Illinois Humanities Council and the Princeton Woodrow Wilson Public Scholarship Foundation, among others. Follow her work at http://homepages.neiu.edu/~ermeiner/

NORTHEASTERN STATE UNIVERSITY

Dilene Crockett

Best Researcher/Scholar, Business

NORTHERN ARIZONA UNIVERSITY

Laura Bounds

Most Helpful to Students, Health Science

Laura Bounds has been a member of the Northern Arizona University (NAU) Health Sciences faculty since 2002. Prior to NAU, she taught at Texas A&M University (TAMU) in College Station, Texas. Laura earned her Doctorate from NAU in Curriculum and Instruction with an emphasis in Higher Education. She earned both her Master of Science in Health Education and her Bachelor of Business Administration in Accounting from TAMU. Laura is the lead author of the textbook: Health and Fitness: A Guide to a Healthy Lifestyle, which is now in its fifth edition and has been adopted for use at multiple universities. Laura currently serves as the lead faculty and program coordinator for the Health Sciences Department's online (Extended Campuses) programs. Her past roles at NAU and TAMU included coordinating large credit-based physical activity programs that serve the entire student body.

Ellen Larson

Best Teacher, Health Science

Ellen Larson has been a member of the Northern Arizona University (NAU) Health Sciences faculty since 2000. Prior to NAU she taught physical education, health education, and adaptive physical education in a southeastern Alaskan community. She also served as a health education consultant for the Alaska Department of Education. Ellen has established strong working relationships with various schools and health organizations in Flagstaff, which provide opportunities for her students to gain experience in the community setting.

Debby McCormick

Best Overall Faculty Member, Health Science

Debby McCormick has been a member of the Northern Arizona University Health Sciences faculty since 1999. Prior to coming to NAU, she taught at the University of Texas at San Antonio. She has also taught at Lamar University and Texas State University. Debby earned her Ph.D. from Texas A&M University in Health Education in 1995. Her personal and academic interests are in social, emotional, and spiritual aspects of health and well-being. Debby has additional degrees in Health and Kinesiology (M.S. from Baylor University, Waco, Texas) and Physical Education (B.S. from University of Mary-Hardin Baylor, Belton, Texas). She has co-authored two health textbooks.

Priscilla Sanderson

Best Researcher/Scholar, Health Science

Priscilla R. Sanderson, Ph.D., CRC grew up on the Navajo Reservation in Shiprock, New Mexico. She is an Associate Professor with the Health Sciences Department, College of Health and Human Services, Northern Arizona University, Flagstaff, Arizona. She received her Ph.D. in Special Education and Rehabilitation Counseling from the University of Arizona, her M.S. in Psychology with an emphasis in Rehabilitation Counseling, and BA in Psychology from Southwestern College, Winfield, Kansas. In 2008, she completed her postdoctoral fellowship with the Arizona Cancer Center, College of Medicine, Tucson, Arizona. Her postdoctoral research was colo-rectal cancer screening, knowledge, attitudes, and beliefs on the Navajo Reservation. Her mentors included Dr. Elena Martinez, Dr. Nicolette Teufel-Shone, and Dr. Neil Weinstein. Currently, she teaches in public health area and Co-PI for the Center for American Indian Resilience (CAIR), an NIH/NIMHD P20 center and the first P20 on Northern Arizona University campus. Her mentor, Dr. Teufel-Shone is a CAIR Co-PI with the University of Arizona's Mel and Enid Zuckerman College of Public Health. Diné College is a collaborator with the CAIR NAU and UA team. She is also past chair for Native Research Network, Inc. and serves on the Board of Directors. She is also a mentor with CAIR undergraduate students and Langston University Rehabilitation Research and Training Center on Research and Capacity Building tribal college junior faculty. Her research interests include cancer prevention, resilience, public health, vocational rehabilitation, and disability rehabilitation among American Indians.

NORTHERN ILLINOIS UNIVERSITY

Giovanni Bennardo

Best Overall Faculty Member, Anthropology
Best Researcher/Scholar, Anthropology

Professor Bennardo's area of specialization is linguistic and cognitive anthropology. His primary geographic focus is Oceania, in particular, Western Polynesia, the Kingdom of Tonga, where he conducted extensive fieldwork. Research and interests are interdisciplinary; he brings together linguistic, psychological, and anthropological perspectives to cognitive science. His doctoral and current research focuses on the linguistic, cognitive, and cultural representations of spatial relationships. His specific interest is the investigation of intra-modular and inter-modular conceptual structures or cultural models. He's taught at UCLA, U. of Missouri, Columbia and College of Charleston, South Carolina. He currently teaches courses in linguistic anthropology, cognitive anthropology, and cognitive science.

Jenny Parker

Best Overall Faculty Member, Physical Education

"Dr. Jenny Parker is a dedicated faculty member, who demonstrated excellent teaching, organizational skills, and application of the new knowledge into the field. Excellent communication skills and leadership among faculty and students of the Kinesiology and Physical Education Department, College of Education and Northern Illinois University."

NORTHERN VIRGINIA COMMUNITY COLLEGE

Dahlia Henry-Tett

Best Overall Faculty Member, Physical Education

Nicole Mancini

Most Helpful to Students, Education

"For over five years, I have the privilege of working alongside Ms. Mancini in the Health and Physical Education department at Northern Virginia Community College- Manassas Campus, where she serves as a full-time faculty member. Throughout her tenure at the college- adjunct to now fulltime-, Nicole has shown herself to be a dedicated professional. She constantly exhibits the high level of planning, coordination and communication skills necessary to accomplish the tasks associated with this position. Her endeavors, not only include the duties assigned, but are also replete with instances in which she provides pivotal support to other departments; assists with special projects; and facilitated significant learning experiences for students. With her unique combination of talents, dynamic personality, educational background and passion for her students she has been a valuable addition to this ever growing department. At all times I have found her to be creative, self-motivated, goal oriented, dependable, organized and efficient. Not only is Nicole's work ethic noteworthy, but she impresses me with her efforts to constantly improve the lives of those around her (while on her own educational journey). To me this further demonstrates her commitment as a force for positive change and I would be remiss not to nominate her."

NORTHERN VIRGINIA COMMUNITY COLLEGE - MEDICAL

Mary Pat O'Brien

Best Teacher, Health Sciences and Nursing

"Always positive and knowledgeable regarding subject matter"

NORTHWEST MISSOURI STATE UNIVERSITY

Margaret Drew

Most Helpful to Students, Education

Patricia Thompson

Best Teacher, Education

NORTHWEST VISTA COLLEGE

Don Lucas

Best Researcher/Scholar, Psychology

"Don Lucas makes research fun to do. His students enjoy working with him on all projects."

NOVA SOUTHEASTERN UNIVERSITY

Barbara Dastoor

Best Researcher/Scholar, Management

Barbara Dastoor has ample experience in studying and managing human behavior in many different cultures. She has lived and worked in several different countries. She spent 15 years working in Human Resources, serving as a supervisor of education for the City of Kansas City, Missouri, and program director of training and development for the Texas Coalition for Juvenile Justice. She has taught business for 20 years. It is this background, she says, that enables her to bring a cross-disciplinary and cross-cultural view of human behavior in human resource management, organizational behavior, and leadership to the classroom.

Her classes begin with an invitation. "I invite students to create a conscious, intentional learning community during the course so that it carries over to doctoral students completing their dissertations," said Dastoor. "At the same time, I familiarize student learners with the most rigorous academic standards and create expectations that they will adhere to them." For Dastoor, the rewards come when she sees her students go beyond what they thought they could accomplish and develop as professionals, writers, researchers,

Jeff Fountain

Most Helpful to Students, Management

Jeffrey Fountain wants his students to see the possibilities. "Not everyone is going to get a job right away with a professional sports team," said Fountain. "But there are many opportunities in sports management at the collegiate and amateur levels. I try to help students define their goals and chart a path for reaching them."

Fountain has practice at setting and reaching goals. He earned his Ph.D. at the age of 27 and joined the Huizenga School soon after. His interest in sports management began as a player when he was a boy and grew as he began working in sports-related jobs. He currently spends his summers as head director of a youth sports camp in Maine, where he directs the activities of 375 campers and 130 staffers from around the world.

In the classroom, Fountain challenges students with real-world activities, such as re-working a college athletic budget to comply with Title 9 requirements and incorporating stock and bond lessons with the NCAA March Madness tournament. In addition, Fountain oversees the practicum/field experience courses, in which his students have the opportunity to intern with local professional sports organizations.

Charles Golden

Best Researcher/Scholar, Psychology

"Dr. Golden's research on neuropsychological issues for forensic clients is timely and important to assist other psychologists and attorneys in evaluating defendants who have committed various serious crimes."

Dr. Charles J. Golden is a 1975 graduate in clinical psychology from the University of Hawaii. He is currently a Professor of Psychology at Nova Southeastern University in Fort Lauderdale where he is director of the Neuropsychology Assessment Center. He is nationally and internationally known for his clinical research in the field of neuropsychological and psychological assessment, He has given over 1000 invited presentations/workshops, and published over 400 articles, book chapters, and books. He is a Fellow in the American Psychological Association (APA) and a past president of the National Academy of Neuropsychology. He holds a Diplomate in Clinical Psychology, Clinical Neuropsychology, and Psychological Assessment. In recognition of his clinical and research accomplishments, he has received the Distinguished Neuropsychologist Award from the National Association of Neuropsychologists in 2003. He is known for his work in the assessment of individuals charged with serious aggressive crimes as well as for the development and evaluation of tools for neuropsychological assessment.

OAKLAND UNIVERSITY

Doug Baltz

Best Teacher, Education
Most Helpful to Students, Education
Most Helpful to Students, Education
Best Teacher, Education

OCEAN COUNTY COLLEGE

Barbara Napolitano

Best Overall Faculty Member, Business

OHIO STATE UNIVERSITY

Anil Arya

Best Researcher/Scholar, Accounting

Professor Arya's research has developed insights about accounting practices and highlighted the unique role played by accounting numbers in decentralized organizations where information is critical. His articles have studied issues such as earnings management, real options and control problems, information system design, team effectiveness, and historical cost reporting. His work has appeared in many academic journals, including Journal of Accounting and Economics, Journal of Accounting Research, The Accounting Review, Review of Accounting Studies, Journal of Economic Theory, The Rand Journal of Economics, Management Science, and Issues in Accounting Education. Professor Arya teaches in the MBA, MAcc, undergraduate Honors, and the Ph.D. Accounting programs. In the classroom, Arya attempts to develop an understanding of accounting as more than a collection of rules by highlighting economic forces that might give rise to accounting as we know it. This leads to an interdisciplinary approach, and one that capitalizes on synergies between research and teaching activities.

Bruce Bellner

Best Teacher, Economics

Eun Bin Chung

Best Researcher/Scholar, Political Science
Best Teacher, Political Science
Most Helpful to Students, Political Science

Douglas Crews

Most Helpful to Students, Anthropology

"Very encouraging and supportive of student's aspirations. Easy to discuss anthropological topics with him and is always ready to enter into a conversation."

Dr. Crews is currently working with data collected from American Samoans, African Americans of central Ohio, and Yanomami and Cofan Indians of the Brazilian and Ecuadorian Amazon. He currently is co-investigator of a project exploring maternal health, and pregnancy outcomes among the Buthia of Sikkim State, India. Dr. Crews has a joint faculty appointment in the School of Public Health and is a member of the Graduate Interdisciplinary Specialization in Aging Coordinating Committee. Dr. Crews' graduate students are actively conducting research in Brazil, Ecuador, India, Cayo Santiago, and the United States, in addition to conducting molecular studies of the HLA system and candidate genes for diabetes, obesity, and blood pressure.

William Dancey

Best Teacher, Anthropology

"Very interested in teaching. Actually reads through student papers instead of just giving straight As, thereby reducing his workload (like some other professors in the department). Even in undergraduate classes, he is able to maintain a sophisticated level of discussion and teaches students how to critique the works they are reading. I learned to write a research paper only after I took his class."

An archaeologist who is involved in the investigation of culture change during the Woodland Period of central Ohio, including the demise of the Hopewell Phenomenon, through regional and site specific analysis. Currently under analysis are collections from an Adena burial mound (Galbreath Mound), a Middle Woodland settlement (Murphy Site), a Late Woodland settlement (Water Plant), intensive surveys of select regions, and test excavations of numerous settlement localities.

Dr. Dancey's research interests include testing of methods of recovering and analyzing regional scale archaeological data. He is also interested in researching internal settlement structure along with analytic approaches to measuring functional and social properties of settlements. Other interests include lithic production system analysis of chert tools and debitage in sedentary and mobile settlements in central Ohio.

Cynthia Long

Best Teacher, Nursing

"Cindy Long has the highest standards for her students in the Health Assessment course and in clinical courses. Students learn so much from her in the first year of nursing that they use throughout the program and their nursing careers. Many students recall "what and how Cindy taught" them for years. She is absolutely one of their most memorable teachers. As a faculty member working with Cindy, I am continually amazed with and appreciate her organization to lead these very large courses. With Cindy's precise and organized approach to assessment, the students are able to give better care to their patients."

Cynthia Long's clinical interest is in medical surgical nursing. Past experiences include oncology nursing, community health nurse practitioner and parish nursing. Areas of interest include care of the patient with diabetes, wound care, end of life care as well as new strategies in nursing education.

Certifications include: licensed Registered Nurse in Ohio; AHA CPR Instructor.

Ms. Long has been a practicing nurse since 1977. She is currently a Clinical Instructor in the nursing fundamentals and health assessment course at the College of Nursing, serving as course head during Winter quarter, and instructing students on the clinical units in medical-surgical areas.

Ms. Long serves as the Chairperson of the Health and Wellness Committee at Epiphany Lutheran Church in Pickerington, Ohio. She also serves as a CPR/BLS instructor for undergraduate nursing students and faculty seeking renewal.

Mary Alice Momeyer

Best Overall Faculty Member, Nursing
Best Researcher/Scholar, Nursing
Best Teacher, Nursing

Mary Alice Momeyer has been with the College of Nursing since 2001 teaching in both the undergraduate and graduate programs. She has been responsible for didactic and clinical teaching related to care of older adults and pharmacology for advance practice nurses. She completed post-masters certification programs at the Ohio State University and Northern Kentucky University to become dually certified as both an Adult and Geriatric Nurse Practitioner. Mary Alice earned the Doctor of Nursing Practice (DNP) degree from Chatham University in Pittsburgh, Pa. In addition to her academic responsibilities at the college, Mary Alice practices as a Geriatric Nurse Practitioner working in assisted living facilities providing primary care.

Mark Moritz

Best Overall Faculty Member, Anthropology
Best Researcher/Scholar, Anthropology

"Very dedicated to the profession. Thinks very objectively. Very forthright and honest in his critiques. Stays above petty politics so that he can dedicate himself to researching the questions he is interested in. Also expects this of his students. Sets high standards.

Has high standards. Is very committed to research. At the same time, he is also very committed to teaching and experimenting with teaching. Overall, a very sincere, honest, and hardworking professor who deserves to be at an R1 institute."

My research focuses on the transformation of African pastoral systems. I examine how pastoralists adapt to changing ecological, political and institutional conditions that affect their lives and livelihoods. I have been conducting research with pastoralists in the Far North Region of Cameroon since 1993. The long-term research has resulted in strong collaborations with local researchers, which has allowed me to develop a new interdisciplinary with colleagues at the Ohio State University and the University of Maroua in Cameroon.

Dr. Moritz's cv.pdf

[pdf] - Some links on this page are to .pdf files, which require the use of Adobe Acrobat Reader software to open them. If you do not have Reader, you may use the following link to Adobe to download it for free at: Adobe Acrobat Reader

Stephanie Moulton

Best Overall Faculty Member, Public Administration
Best Researcher/Scholar, Public Administration
Best Teacher, Public Administration
Most Helpful to Students, Public Administration
Best Researcher/Scholar, Public Affairs, Political & Policy Sciences
Best Teacher, Public Affairs, Political & Policy Sciences

Diane Sheets

Most Helpful to Students, Nursing

Eric Spires

Best Overall Faculty Member, Accounting

Professor Spires has undergraduate degrees from Eastern Kentucky University in Accounting and Music Theory. He was employed on the audit staff of Touche Ross & Co. for approximately four years.

Professor Spires's research interests involve decision making, both in accounting and auditing contexts and more generally. His articles have appeared in The Accounting Review, the Journal of Business, Finance and Accounting, Multivariate Behavioral Research, Organizational Behavior and Human Decision Processes, and other journals.

Larry Tomassini

Most Helpful to Students, Accounting

Professor Tomassini previously served as Department Chair for Accounting & MIS (1993–2000) and Academic Director of the Master of Accounting Program (2000–2002). Before coming to Ohio State, he was Head of the Department of Accountancy and the Ernst & Young Distinguished Professor at the University of Illinois and the Peat Marwick Mitchell Centennial Professor in Accounting at the University of Texas.

Rick Young

Best Teacher, Accounting

Professor Young served on the faculties at the University of Texas at Austin and the University of Iowa before returning to Ohio State. He has been on the editorial board of The Accounting Review and Journal of Management Accounting Research. His work is in the area of accounting, information and control and has published his research in the Journal of Accounting Research, The Accounting Review, Review of Accounting Studies, Contemporary Accounting Research, Journal of Economic Theory, Management Science, and Journal of Management Accounting Research. He is currently the Editor of Journal of Management Accounting Research and will continue through 2010. Professor Young has acted as Director of the Honors Accounting Program and the AMIS Ph.D. Program and has also served on the Fisher College Personnel Committee.

OHIO UNIVERSITY

Haley Duschinski

Best Teacher, Anthropology

Roger Gilders

Best Overall Faculty Member, Health Sciences and Nursing

Stephen Howard

Best Researcher/Scholar, International Studies
Most Helpful to Students, International Studies

"I am voting for Dr. Howard because, in the last 9 or so years I have worked with him, he has tirelessly worked to promote global consciousness–mostly by using African examples–in US students. He was also instrumental in the recruitment of graduate students from Africa with the goal of–often in collaboration with their home institutions–providing them with international experience and academic knowledge. He is, moreover, a strong believer in knowledge transferability and an impassioned supporter of African development by African people."

Juli Miller

Best Researcher/Scholar, Health Science
Best Researcher/Scholar, Health Sciences and Nursing

Deb Murray

Best Teacher, Health Science
Best Teacher, Health Sciences and Nursing

Sharon Rana

Best Overall Faculty Member, Health Science
Most Helpful to Students, Health Science
Most Helpful to Students, Health Sciences and Nursing

Associate Professor - Exercise Physiology, School of Applied Health Sciences and Wellness

Nancy Tatarek

Most Helpful to Students, Anthropology

OHIO UNIVERSITY SOUTHERN

Beth Delaney

Most Helpful to Students, Nursing

Molly Johnson

Most Helpful to Students, Nursing

OKLAHOMA STATE UNIVERSITY: OKLAHOMA CITY

Jeff Brewer

Best Researcher/Scholar, Business
Best Teacher, Business
Best Overall Faculty Member, Business

Amber Hefner

Best Teacher, Business
Best Researcher/Scholar, Business

Lisa McConnell

Best Overall Faculty Member, Business
Most Helpful to Students, Business
Most Helpful to Students, Business

OLD DOMINION UNIVERSITY

Aaron Arndt

Best Teacher, Marketing

Aaron D. Arndt (Ph.D. - University of Oklahoma) has been an assistant professor at Old Dominion University since August, 2008. He has published in the Journal of Retailing, Journal of Personal Selling and Sales Management, International Journal of Logistics Management and Business Horizons. His primary research interests include sales management, interpersonal selling and negotiations–particularly across cultures. Since this line of research spans multiple levels of analysis, he also studies multi-level theory and methodology. His primary teaching interests include personal selling, sales management, negotiations, retailing, research methods, and international marketing.

Michelle Carpenter

Most Helpful to Students, Marketing

David Cook

Best Teacher, Business

"Patiently explains material and offers to help."

Education University of Kentucky, 1997 Major: Production and Operations Management Degree: Ph. D. Bowling Green State University, 1992 Major: Production and Operations Management Degree: M.B.A. Bowling Green State University, 1991 Major: Production and Operations Management Degree: B.S.B.A Licensures and Certifications A+ Certified Sponsoring Organization: CompTIA Date Obtained: 2001-01-01 Network+ Certified Sponsoring Organization: CompTIA Date Obtained: 2001-01-01 Expertise Decision Modeling

Elizabeth Esinhart

Most Helpful to Students, Political Science

Education Duke University School of Law, 1979 Degree: J.D. Mount Holyoke College, 1976 Degree: B.A. Licensures and Certifications Attorney License Sponsoring Organization: Virginia State Bar Date Obtained: 1981-10-08 Attorney License Sponsoring Organization: Minnesota State Bar Date Obtained: 1979-10-05

John Ford

Best Teacher, Marketing

"His understanding and enthusiasm makes learning interesting."

Professor of Marketing and International Business and Eminent Scholar Education University of Georgia, 1985 Major: Business Administration - Marketing Degree: Ph. D. The University of Georgia, 1983 Major: Marketing Degree: M.B.A. Yale University, 1971 Major: English Degree: B.A. Research Interests International Advertising, Role Portrayals in Advertising, Not-for-profit Marketing Strategy, and Cross-Cultural Research Issues. Expertise Marketing Advertising International Business

Mahesh Gopinath

Best Researcher/Scholar, Marketing
Most Helpful to Students, Marketing
Most Helpful to Students, Business

"Dr. Gopinath is willing to help all students be better researchers. He helps all students, not just the "top of class" ones."

Sylvia Hudgins

Best Overall Faculty Member, Business

"She is an asset; helpful, understanding, knowledgeable, and encouraging to all students to make difficult material understandable."

Sylvia Hudgins received her Doctorate in 1987 from Virginia Tech. She teaches classes in corporate finance and financial institutions. Much of her research focuses on the empirical analysis of commercial banks and thrifts. She examines questions concerning management, regulation, and legislation. The journals publishing her research include: Journal of Money, Credit, and Banking, Financial Management, Journal of Financial Economics, and Economic Inquiry. Professionally she serves on the Board of Directors of the Eastern Finance Association and on the Board of Editors for the Financial Management Association's Survey and Synthesis Series. In the community, she serves on the Board of Directors of the Old Dominion University Credit Union, the Canterbury Center for Campus Ministry, and Crystal-Lind Wellness Center. She is presently writing a textbook on commercial bank management with Peter Rose of Texas A&M University that will be published by McGraw-Hill in 2005.

Yuping Liu-Thompkins

Best Researcher/Scholar, Business

Yuping Liu-Thompkins, Ph.D. is Professor of Marketing and E. V. Williams Faculty Fellow at Old Dominion University. She received her Bachelor's Degree in Marketing from Renmin University of China in 1996 and her MBA and Ph.D. in Management (Marketing Concentration) from Rutgers University in 2002. Education Rutgers University, 2002 Degree: M.B.A. Rutgers University, 2002 Major: Management/Marketing Degree: Ph. D. Remin University of China, 1996 Major: Marketing Degree: B.A. Research Interests Loyalty Programs, Habit, and Internet Marketing Expertise Internet Marketing Internet Marketing/Social Media/Interactive Communications Customer Relationship Management Loyalty Programs/Customer Loyalty/Consumption Habit

Edward Markowski

Best Researcher/Scholar, Business

"He encourages students to analyze research and apply what they have learned to their own work."

Anusorn Singhapakdi

Best Overall Faculty Member, Marketing
Best Researcher/Scholar, Marketing

"He deals with all students in a variety of contexts and helps each one."

Glen Sussman

Best Teacher, Political Science
Best Researcher/Scholar, Public Affairs, Political & Policy Sciences

Professor Sussman received his Ph.D. in political science from Washington State University. He joined the faculty at Old Dominion University in 1992 where he teaches undergraduate courses in the field of American politics. As an Affiliate Faculty member in GPIS, he offers three seminars – Global Environmental Policy, Politics of Climate Change: Comparative Perspectives, and Comparative Political Behavior.

His research focuses on environmental politics and policy and the politics of climate change.

He is currently working with two colleagues on the 2nd edition of their co-authored book, American Politics and the Environment (Longman, 2002). SUNY Press offered an advance contract in support of the project.

He is also working on a survey research project that involves state legislators in coastal states and their political orientations about sea level rise.

Professor Sussman's scholarship and professional activities include 5 books, over 100 journal articles, book chapters, professional papers and research essays and approximately 100 lectures, interviews, panels, and speaking engagements.

He served two terms as chair of an eighteen member department, 2000–2006.

Steve Yetiv

Best Researcher/Scholar, Political Science

312

ORAL ROBERTS UNIVERSITY

Sonny Branham

Best Overall Faculty Member, Public Affairs, Political & Policy Sciences

"Students appreciate how caring he is."

Professor Sonny Branham, now in his 15th year at ORU, brings with him more than 25 years of industry and life experience both in journalism and the banking industry. He and his wife Judy distinguished themselves in journalism and small business ownership as the youngest publishers of a weekly newspaper in Kentucky, owners of The Cumberland County News in Burkesville, KY from 1974 through 1980. They doubled the circulation of the newspaper and tripled the advertising revenue before selling their award-winning publication to their former journalism college professors in 1980. During those five years, the newspaper won state and national acclaim for writing and photography, including Best Locally Written Column, Best Sports Picture, and Best Newspaper Overall in Circulation Division. Branham was selected for a term as President of the Kentucky Weekly Newspaper Association, the youngest person selected by his peers for the position He continues to be a member of the Tulsa Press Club and the National Press Club in Washington, DC. Branham's distinction in the community for his honesty, integrity, hard work and community activism made him a prized product for the financial service industry outlets in the area. He chose to go with a bank that offered him stock and the opportunity for professional development. He attained a Certificate from the Kentucky School of Banking in Lexington, KY in 1982. After moving up the corporate ladder from teller, to assistant cashier, to Vice-President and Chief Operations Officer, to Vice-President and Chief Financial Officer, Branham eventually left the bank in 1980 to begin as an adjunct instructor at ORU in the Fall Semester of 2000. (During the 1980's and 1990's he taught social studies as an adjunct instructor at Lindsey Wilson College, a United Methodist Church affiliated school.) From his years of journalism, small business ownership, financial service industry and academic experience, Professor Branham brings his leadership and vision for community activism, civic responsibility and social concern. In his classes in government and politics, Prof. Branham stresses these attributes for his students to possess as they make their transition to young professionals in the workplace. Prof. Branham has served in leadership roles both in the Tulsa community and at ORU, including his long-running tenure as a director the United Nations Association of Eastern Oklahoma, a member of the Tulsa Global Alliance, and as a private political consultant. He also serves on the board of a number of advocacy groups and non-governmental organizations (NGOs), providing extensive networking and potential career opportunities for his students. Prof. Branham serves on numerous university committees including the Christian Worldview Committee, Green Campus Committee and the Quest Whole Person Scholarship Committee. He is also the academic advisor for the award-winning Oklahoma Intercollegiate Legislature delegation, the Model Arab League Delegation and the long-time advisor to the Model United Nations delegation. He also places students in different political campaign opportunities and helps secure internships for students interested in politics and government. He describes himself as "a half-way decent civics teacher who loves the students," and not a scholar or researcher. However, he has produced two textbooks used for the American Government and Politics (GOV 101 class). Custom publications designed particularly for ORU include "I Pledge Allegiance to the Flag: Readings in American Government" published by McGraw-Hill and "And to the Republic for Which it Stands," a comprehensive textbook from Pearson Publishing Company. Prof. Branham is acknowledged for his strong faith and passion for faith integration, focusing on developing future political leaders to be Christ-like in character. He challenges students to live their faith and hear God's voice in their chosen professions, especially in the popular Christian Faith and Government class. Professor Branham, who earned both bachelor and master degrees from Western Kentucky University in Bowling Green, completed the ORU Teaching Excellence Program. He was rewarded with promotion to the rank of assistant professor. In the Fall Semester of 2012, Branham was named a Sabbatical Fellow at the American Studies Program (ASP) in Washington, DC. ASP is a distinguished component of the Council for Christian Colleges and Universities (CCCU) Best Semester program. He continues a strong working relationship with leadership at CCCU due to the intimate connection he established with the group while on ORU sabbatical leave. - See more at: http://web.oru.edu/academics/faculty_profile.php?id=109&k=#sthash.T3nI2AHb.dpuf

Winston Frost

Best Researcher/Scholar, Public Affairs, Political & Policy Sciences

I was a miracle baby. Born with a collapsed lung, I faced the possibility of severe brain damage due to lack of oxygen. However, due to the prayers of many and God's grace, I survived unscathed and un-handicapped. However, it is only in recent years that I truly have understood just how faithful, kind, patient and loving God really is. I accepted Christ as my Savior on April 13, 1966. In many ways, I had the model upbringing and was not involved in drugs or alcohol abuse. I was a good student and an athlete. This allowed me to attend the Air Force Academy on a scholarship. My time at the academy prepared me for the challenges I would face in the future.

My college days were a time of spiritual complacency. My walk was in many ways a crawl, but still God was faithful. God simply wasn't the central focus of my life at that time. This changed after college, as I was called to attend a Christian law school and thought I was on my way to fulfilling God's ultimate purpose for my life. I had visions of preserving America's Biblical heritage and stopping the moral decline of our nation. I was going to be part of a generation of lawyers who would make a difference in the courts of our nation. That in turn led to becoming involved in Christian legal education. I spend the next 15 years of my career being a Christian attorney. I had the opportunity of leading clients to Christ, standing up for my faith in the public square, and, after being replaced as editor of the bar journal for quoting John 3:16, I was chosen to be a full time law professor and later elevated to the position of dean of the law school.

It was here that my life took a dramatic turn and my faith was truly tested. I voluntarily resigned from the law school, and my marriage resulted in a divorce. However, God did not abandon me. God allowed me to move on and foiled the efforts of my enemies to ruin me. I returned to practicing law, started my own law school, got into a Doctoral program, served as the development director for a Christian school, and went to the Northern Cheyenne Reservation for two years on a mission trip. By the summer of 2008, God had given me the opportunity to return home to be with my ailing father. This past year my father passed away after his bout with cancer.

If I can impact others in the same positive way that he impacted me and can use my experience to teach others to trust God, grow in His word and live by faith I will truly be fulfilling God's will for my life. In short, the fact that I am still healthy and alive, still moving forward and now have the opportunity to lead others to an education and a closer relationship with God is a testimony to the seeds my father planted and to God's unending grace.

ORANGE COAST COLLEGE

Arabian Morgan

Best Overall Faculty Member, Business

ORANGE COUNTY COMMUNITY COLLEGE (SUNY)

Robert Cacciatore

Best Overall Faculty Member, Criminal Justice

"Gives 100% for student achievement"

OREGON STATE UNIVERSITY

Erik Larson

Best Researcher/Scholar, Business

OWENS COMMUNITY COLLEGE: TOLEDO

Barb Dinardo

Best Teacher, Business

OZARKS TECHNICAL COMMUNITY COLLEGE

J. C. Walker

Best Overall Faculty Member, Business

PALOMAR COLLEGE

Patrick O'Brien

Most Helpful to Students, Counseling
Most Helpful to Students, Education

"I hear wonderful comments from his students. They appreciate his assistance and learn a lot from this wonderful man."

PELLISSIPPI STATE COMMUNITY COLLEGE

Barbara Jenkins

Best Overall Faculty Member, Education
Best Researcher/Scholar, Education
Best Teacher, Education
Most Helpful to Students, Education
Best Researcher/Scholar, Education

Terenia Schumann

Best Teacher, Education
Best Overall Faculty Member, Education

Catherine Shafer

Most Helpful to Students, Education

PENSACOLA STATE COLLEGE

John Atkins

Best Overall Faculty Member, Business
Best Researcher/Scholar, Business

Carla Rich

Best Overall Faculty Member, Business

Jane Spruill

Most Helpful to Students, Education

PEPPERDINE UNIVERSITY

Carrie Wall

Best Overall Faculty Member, Education
Best Teacher, Education

"Dr. Wall is diligent and passionate about her work with Teacher Education. She is a favorite teacher but also an interested and very helpful academic advisor. She leads the Department with integrity and professionalism.

Teaching is Dr. Wall's passion! She models excellent preparation, wise implementation, and compassionate follow-through as a teacher. Her students, who are aspiring teachers, are blessed to have her as a mentor, guide, and role-model!"

PEPPERDINE UNIVERSITY'S GRAZIADIO SCHOOL OF BUSINESS AND MANAGEMENT

Alfred Hagan

Best Researcher/Scholar, Business

"Competent, fair and a good teacher."

Bob McQuaid

Best Overall Faculty Member, Business
Best Teacher, Business

"Competent, fair and works with students. Willing to do and has done any job. He has volunteered for most non-paid faculty posts and does an excellent job in each of them."

Dr. McQuaid is noted for applying operations research techniques to real-world situations. In 12 years with Abbott Laboratories and General Dynamics, he made significant contributions in manufacturing engineering and planning, project management, and manufacturing supervision. Dr. McQuaid is especially interested in the development and application of practical analytical tools to solve complex operations problems.

Edward Rockey

Most Helpful to Students, Business

"Always available for students"

Dr. Rockey presents programs on creative problem solving, communication, stress management, and leadership for corporations such as Procter & Gamble and Prudential Insurance and for smaller companies. He wrote, narrated, produced, and marketed the workbook/cassette album Successful Time Management and authored Communicating in Organizations (Winthrop Management Series). He has lectured in several countries including at an international conference in New Zealand which published his work on coaching students to write and present original, current, actual case studies which class members attempt to solve. Dr. Rockey maintains high interest in reporting research on how executives and entrepreneurs apply visual thinking, and he is presenting "Icons of the Managerial Mind" at the 2011 conference of the Creative Problem Solving Institute.

PIEDMONT COLLEGE

Madge Kibler

Most Helpful to Students, Education

Patsy Mapp

Best Teacher, Education

Franklin Shumake

Best Researcher/Scholar, Education

Joseph Wisenbaker

Best Teacher, Education

PIERCE COLLEGE (ALL)

Douglas Edison

Most Helpful to Students, Business

"Doug is funny, outgoing and everybody's "Father". He is a straight shooter with a huge heart and lots of real world wisdom."

Officer Candidate School, Newport, RI - Lieutenant, U.S.N.R. International Program for Port Planning and Management - University of New Orleans, LA (Valedictorian) Certification in and a license to use the Myers-Briggs Type Indicator, CPP, Inc. Quality Circle Institute, Red Bluff, CA - Certified Facilitator for Total Quality Management Outside Edge, Inc., Seattle, WA - Certified to lead group dynamics, group development, conflict management, and facilitation

Tom Phelps

Best Teacher, Business
Best Researcher/Scholar, Business

"Tom is friendly, knowledgable, and entertaining. He has written the degree for our Business Transfer program and has mentored dozens of faculty members over the past couple of decades."

PIMA COMMUNITY COLLEGE

Nancy Christie

Best Teacher, Psychology

"Excellent teaching skills."

Amy Cramer

Best Teacher, Business

Rita Flattely

Best Overall Faculty Member, Psychology

Rita Flattley has been teaching psychology at PCC, first as an adjunct and then as a full time faculty member, since 1982. She holds an A.A. in Social Services from PCC as well as a B.A. in Psychology and M.Ed. in Counseling and Guidance from the University of Arizona, and has been taking continued graduate work in Educational Psychology with an emphasis in instructional media.

Rita has been influenced by the ideas of cognitive psychologist Howard Gardner and his theory of multiple intelligences, and therefore likes to give varied types of classroom activities and assessments so that students can express their intelligences in their own way.

Ron Jorgensen

Best Teacher, Psychology
Most Helpful to Students, Psychology

Jeffrey Neubauer

Best Researcher/Scholar, Psychology

Don Roberts

Most Helpful to Students, Economics

Kirk Spiker

Best Teacher, Economics

PITTSBURG STATE UNIVERSITY

David Hurford

Best Researcher/Scholar, Psychology

PRINCE GEORGE'S COMMUNITY COLLEGE

Kimberly Veney

Most Helpful to Students, Health Sciences and Nursing
Best Researcher/Scholar, Health Sciences and Nursing

QUINNIPIAC UNIVERSITY

Rowena Ortiz-Walters

Best Overall Faculty Member, Management

"For her leadership and overall caring for students and faculty"

Rowena Ortiz-Walters is Department Chair and Professor of Management at Quinnipiac University. She received her Ph.D. from the University of Connecticut and has published in The Journal of Organizational Behavior, Journal of Vocational Behavior and Journal of Developmental Entrepreneurship. Her research interests include examining mentoring relationships as a career developmental tool for women and racial minorities, issues of diversity in the workplace and the entrepreneurial ventures of racial minorities and women. She is a founding member of the B-WISE (Business Women in Search of Excellence) initiative at QU and a member of Connecting Women, a SOB Advisory Board committee, and has served as advisory board member for a study of gender diversity for the Harvard Medical School. More recently, she is co-founder of the Center for Women and Business.

QUINSIGAMOND COMMUNITY COLLEGE

Dagne Yesihak

Best Teacher, Criminal Justice

RADFORD UNIVERSITY

Ian Clelland

Best Overall Faculty Member, Management

Kenna Colley

Best Overall Faculty Member, Special Education

Kenna Colley is an Associate Professor in Special Education Programs. Her work in the field of special education focuses on the inclusion of students with disabilities in general education classrooms, teacher collaboration, positive behavior supports, assessment, and response to intervention. She teaches courses in assessment, positive behavior supports, introduction to and characteristics of students with high-incidence disabilities, instructional strategies for inclusive classrooms, and current trends and issues in special education. She supervises early field experience students and student teachers in rural and suburban settings.

Dr. Colley assisted in creating and piloting the co-placement of elementary and special education interns in fully inclusive general education classrooms. She is a co-director of Project MERGE, a project funded by the U.S. Department of Education's Office of Special Education Programs. Project MERGE focuses on preparing special education and general education teachers who can work together in K–12 classrooms to meet the needs of all children — with and without disabilities.

Dr. Colley is principal investigator of the Training and Technical Assistance Center (T/TAC) housed at Radford University. Funded by the Virginia Department of Education, RU's T/TAC serves 34 school divisions.

Lesile Daniel

Best Teacher, Special Education

Dr. Leslie S. Daniel is an associate professor in the Special Education: General & Adapted Curriculum five-year program. She is the program area leader for special education and also coordinates the Certificate of Autism Studies that she developed for Radford University. Dr. Daniel teaches at both the undergraduate and graduate level.

Courses taught include EDSP 400/500 Introduction to Autism Spectrum Disorders (ASD), EDSP 401/501 Approaches for Supporting and Teaching Individuals with ASD, EDSP 402/502 Expanding Social Competence for Students with Autism Spectrum Disorders, EDSP 462 Proactive Classroom Management and Positive Behavior Supports, and EDSP 670 Proactive Classroom Management and Advanced Positive Behavior Supports.

Dr. Daniel's expertise is enhanced by her years teaching students with disabilities in both general and special education settings in preschool and in elementary and middle schools. She was coordinator providing consultations, professional development, and extended support to 34 school divisions in southwest Virginia through the Virginia Department of Education's Training and Technical Assistance Centers (T/TAC) at Virginia Tech and Radford University. Prior to teaching, she worked in group homes supporting adults with disabilities to live full lives in their communities.

Gary Fetter

Most Helpful to Students, Management

Stephen Owen

Best Overall Faculty Member, Criminal Justice

RAMAPO COLLEGE - NEW JERSEY

Asha Mehta

Most Helpful to Students, Nursing
Most Helpful to Students, Health Sciences and Nursing

Elaine Patterson

Best Overall Faculty Member, Nursing
Best Researcher/Scholar, Nursing
Best Teacher, Nursing
Best Researcher/Scholar, Health Sciences and Nursing
Best Teacher, Health Sciences and Nursing
Best Overall Faculty Member, Health Sciences and Nursing

RARITAN VALLEY COMMUNITY COLLEGE

Michael Fagan

Most Helpful to Students, Finance

Karen Gutshall-Seidman

Most Helpful to Students, Human Services

I believe that everyone can learn and it is our job as educators to unlock the capacity of each student, allowing them to make the most of the educational opportunity before them. Each student presents his or her own set of unique strengths, aspirations and challenges and as educators we have the ability to build them up or squash them down. I'd like to think of myself as a builder!

Kimberly Schirner

Best Teacher, Education

"Kimberly is an authentic educator who models what our profession is all about for her students in effective and productive ways!"

Education Currently pursuing an Ed.D, Educational Leadership, Nova Southeastern University, Orlando, FL M.Ed., Teaching and Learning, Gratz College, Melrose Park, PA B.S., Early Childhood Education, University of Maryland College Park Campus, College Park, MD A.A., Communications, Centenary College, Hackettstown, NJ Educational Philosophy Change and continuity are the most recurrent themes in my own educational experience as a teacher and as a student. A great deal has changed for instance, since the classics of Greek political philosophy were written more than 2000 years ago, yet Aristotle's Politics never seem to become dated. Some issues, in other words, seem to endure between various ages, civilizations and peoples, and can often provide a startling perspective for anyone who imagines that no one else may have come to similar crossroads. I have been teaching Biographical Sketch here at RVCC since the Fall of 1982, having previously taught at Triton College in Illinois, as I was pursuing graduate studies at the University of Chicago. My dissertation was devoted to the US-Soviet Cold war relationship through five post WW II presidential administrations, as I sought to identify American views of Soviet motives during that lengthy confrontation. Before I began teaching full time, I had a varied and highly instructive career as a forklift operator, material handler and warehouse foreman, as well as additional time as a short-order chef. Very good opportunities to meet and mingle with a wide cross section of humanity, and an experience I value more and more as the years pass. I am interested first and foremost in US foreign policy and international relations, but I have also continued to work in ancient and modern political theory, American government and politics, US Constitutional history and legal theory. I also maintain an active interest in historical research, and have taught history at RVCC since I've been here. Courses Taught Introduction to Political Science (POLI 101) American Government and Politics (POLI 121) International Relations (POLI 231) U.S. History: Beginnings to 1877 (HIST 201) U.S. History: 1877 to Present (HIST 202) World Civilization II (HIST 102) The Bill of Rights Comparative Government Global Patterns of Racism (HIST 204) last modified 3/12/2013 by IIJS . . .

RENSSELAER POLYTECHNIC INSTITUTE

Qiang Wu

Best Researcher/Scholar, Management

Qiang Wu's current research interests include accounting conservatism, debt contracting, tax avoidance, earnings management, corporate governance, and behavior accounting. His research works have appeared in The Accounting Review, Journal of Financial Economics, Contemporary Accounting Research, Accounting Horizons, Journal of The American Taxation Association, Journal of Accounting, Auditing and Finance, Journal of Business Finance and Accounting, Financial Management, Journal of Financial Research, and several other academic journals. More details about his research work could be found on SSRN at http://ssrn.com/author=976747 or Google Scholar at http://scholar.google.com/citations?user=hQlsdB8AAAAJ&hl=en

RESURRECTION UNIVERSITY

Nancy Reese

Best Overall Faculty Member, Health Sciences and Nursing

RHODE ISLAND COLLEGE

Lynn Blanchette

Most Helpful to Students, Nursing

Joanne Costello

Best Overall Faculty Member, Nursing

Claire Creamer

Best Teacher, Nursing

Academic Background
    B.S. in Nursing Rhode Island College M.S. in Nursing Regis College, Weston, MA Certifications CPNP from PNCB Courses Taught
    NURS 224 Health Assessment NURS 346 Nursing of Children & Families Areas of Interest Pediatrics/Emergency Medicine

Karen Hetzel

Best Overall Faculty Member, Health Sciences and Nursing

Yolande Lockett

Most Helpful to Students, Health Sciences and Nursing

B.S. Rhode Island College M.S.N. Indiana University Ph.D. University of Connecticut Courses Taught
 NURS 316 Physical Assessment NURS 375 Transition to Professional Practice NURS 506 Advanced Health Assessment Areas of Interest
 Primary Health Care of Children and Adolescents Maintains practice as a certified pediatric nurse practitioner Recent Scholarly Activity
Attended national conference for pediatric nurse practitioners Provided professional development courses for school nurse teachers and disability nurses in the state

Patricia Quigley

Best Researcher/Scholar, Nursing

Sylvia Ross

Best Teacher, Health Sciences and Nursing

Jeanne Schwager

Best Researcher/Scholar, Health Sciences and Nursing

ROGER WILLIAMS UNIVERSITY

Garret Berman

Best Teacher, Psychology

Kelly Brooks

Best Overall Faculty Member, Psychology

Frank Dicataldo

Best Researcher/Scholar, Psychology

Pamela Elizabeth

Best Researcher/Scholar, Psychology

Grace Medeiros

Best Overall Faculty Member, Psychology
Best Teacher, Psychology
Most Helpful to Students, Psychology
Best Teacher, Medicine
Most Helpful to Students, Medicine
Best Overall Faculty Member, Medicine

"Medical Doctor and wonderful professor."

Judith Platania

Best Researcher/Scholar, Psychology

Jessica Skolnikoff

Best Researcher/Scholar, Medicine

Dr. Jessica Skolnikoff is a cultural anthropologist who investigates youth dispositions toward physical activity across the United States. Her work explores how children develop attitudes and habits related to exercise and physical activity. Present data collection consists of interviewing families and students of middle-school age focusing on views and practices in their culture. Skolnikoff's cross-cultural research highlights the relevance of social and cultural values that affect long-term beliefs and behaviors about physical activity.

ROOSEVELT UNIVERSITY

Steven Meyers

Best Researcher/Scholar, Psychology

I have developed innovative strategies for students to become agents of social justice and create change beyond helping individuals. My students explore how the course material connects with social issues and legislation, and then contact their state representatives and senators to advocate.

Melissa Sisco

Best Overall Faculty Member, Psychology
Best Teacher, Psychology
Most Helpful to Students, Psychology
Best Researcher/Scholar, Medicine
Best Teacher, Medicine
Most Helpful to Students, Medicine
Best Overall Faculty Member, Medicine

SAN ANTONIO COLLEGE

Christy Woodward Kaupert

Best Teacher, Political Science
Best Teacher, Public Affairs, Political & Policy Sciences

I am the Academic advisor for all Political Science majors at SAC, am a team advisor for the Honors Academy and also serve as the Department's Internship Coordinator.

I am on part-time assignment in the Office of Research and Institutional Effectiveness and am responsible for fulfilling requests regarding the demographic profile of our college, its students and faculty.

SAN DIEGO STATE UNIVERSITY

Carmen Bianchi

Best Teacher, Business

Michael Cunningham

Best Overall Faculty Member, Management

Dr. Michael R. Cunningham was dean of the College of Business Administration at San Diego State University from June 20th, 2011 through June 30th, 2013. An award winning teacher, he's taught management, entrepreneurship, and international business strategy at California Polytechnic State University, San Diego State University, and New York University.

Prior to his academic career, Dr. Cunningham, a New York native, was a successful entrepreneur. He founded Cunningham Graphics International (CGII) in 1989, and took the company public on NASDAQ in 1998. By the time the company was sold to Automatic Data Processing (ADP) in 2000, it had operations in seven countries and employed more than 1500 people.

In 2007, Dr. Cunningham returned to the corporate world by taking the helm at Diversified Global Graphics Group (DG3), where he restructured and led a management leveraged buyout alongside a private equity firm. He remained with DG3 as president and Chief Executive Officer until 2010.

Michael Cunningham's global business expertise perfectly matches San Diego State University's highly ranked international business programs and the College of Business Administration's mission, "Leadership for the Global Marketplace."

Michael Cunningham earned his master's degree in graphic communications, management and technology (1996), and his Ph.D. in administration, leadership and technology (2005) at New York University. He received his bachelor's degree in marketing and business management from the University of Massachusetts Amherst.

A proud father of three adult children, Dr. Cunningham lives in Rancho Santa Fe.

David DeBoskey

Most Helpful to Students, Accounting

David G. DeBoskey is an Associate Professor in the Charles W. Lamden School of Accountancy and served as the KPMG Faculty Fellow from 2011–2012. Since joining the College of Business Administration faculty in 2006, he has taught upper-division undergraduate accounting courses for the Charles W. Lamden School of Accountancy and specialized graduate MBA courses in the Sports and Executive MBA programs at SDSU.

Alex DeNoble

Best Teacher, Management
Best Overall Faculty Member, Business

Alex F. DeNoble is a Professor in the Management Department in the College of Business at San Diego State University (SDSU). He is also Executive Director of SDSU's Lavin Entrepreneurship Center. His primary areas of expertise include entrepreneurship and corporate innovation, technology commercialization and strategic management. He has conducted research in these areas and has taught related classes in the University's undergraduate, graduate and executive MBA programs. He has published articles in such journals as the Journal of Business Venturing, the Journal of High Technology Management Research, the Journal of Technology Transfer, International Marketing Review, and Entrepreneurship: Theory and Practice.

His other professional activities encompass both executive training and strategic consulting. Recent assignments have included business plan development consulting for new and existing entrepreneurial firms, market research and analysis for technology-based companies and entrepreneurship training for Taiwanese, German, Russian, Japanese, Mexican, Middle Eastern and U.S. executives. Over the past several years, he has conducted training programs or consulted with such companies as Siemens Corporation, QUALCOMM, Delta Electronics (Taiwan), the U.S. Russia Center for Entrepreneurship, Banco Nacional de Commercio Exterior (the National Export Bank of Mexico), NEC Electronics USA, Shell Technology Ventures, and Orincon Technologies (now a part of Lockheed Martin). Dr. DeNoble is the recipient of several awards including the 2008 Monty Award from SDSU for Outstanding Faculty Contributions, 2004 Educator of the Year award from San Diego's T-Sector Magazine; the 2001 Gloria and Edwin Appel Award from the Price-Babson Fellows program for excellence in entrepreneurship education, and the 2000 Ernst & Young Entrepreneur of the Year Award.

Damon Fleming

Best Overall Faculty Member, Accounting
Best Researcher/Scholar, Accounting
Best Teacher, Accounting

Damon M. Fleming is an Associate Professor and Ernst & Young Faculty Fellow in the Charles W. Lamden School of Accountancy. Professor Fleming currently teaches courses on financial reporting, financial statement analysis, financial accounting research, and professional judgment and decision making in the Master of Science in Accountancy (MSA) program.

Professor Fleming's research uses theories from psychology, behavioral decision research, and economics to investigate accounting issues in the areas of auditing, ethics, financial reporting, and taxation as well as cross-cultural issues in these areas. He has published over twenty papers in accounting journals including Accounting Horizons, Behavioral Research in Accounting, The International Journal of Accounting, Issues in Accounting Education, Journal of Business Ethics, Journal of Accountancy, and Strategic Finance. In addition, he serves on the editorial board for Issues in Accounting Education and Review of Accounting and Finance as well as an ad hoc reviewer for several high quality scholarly journals. He also regularly presents research at national and international research conferences.

Professor Fleming received his Ph.D. from Virginia Tech and his B.S. and M.S. degrees in accounting from San Diego State University. He is also a CFA® charterholder and Certified Management Accountant (CMA). Prior to entering academia, Professor Fleming was a principal at a venture capital firm in southern California. He is a member of the American Accounting Association, CFA Institute, California Society of CPAs, and Institute of Management Accountants.

SAN FRANCISCO STATE UNIVERSITY

Vivian Chavez

Best Teacher, Health Education

Michele Eliason

Most Helpful to Students, Health Education

Mickey has a background in nursing and psychology and has had clinical and research experience in hospitals, clinics, community-based treatment agencies, and a women's prison. Her research interests are in the areas of substance abuse and treatment, human sexuality studies, LGBT & queer studies, sexual health, and LGBTQ health. She is widely published in these areas. Her latest book, Improving Substance Abuse Treatment: An Introduction to the Evidence-Based Practice Movement, was published by Sage in 2007. Her forthcoming co-authored volume on LGBTQ health will be published by Lippincott Publishers.

Theresa Hammond

Best Overall Faculty Member, Accounting
Best Teacher, Accounting

William Hefter

Most Helpful to Students, Accounting

Su-Jane Hsieh

Best Researcher/Scholar, Accounting

Dr. Su-Jane Hsieh, professor of accounting at San Francisco State University.

Yikuan Lee

Best Overall Faculty Member, Public Affairs, Political & Policy Sciences

Professor Yikuan Lee is an associate professor in the International Business department at San Francisco State University.

Before joining SF State, she held a visiting assistant professor position in the Marketing & Supply Chain department at Michigan State University.

Her research interests include commercialization of innovative products, network effects, new product development, and strategic marketing management in high technology arenas. Much of her work focuses on how firms integrate marketing and technology competencies.

She received the Best Dissertation Award and the Best Paper Award at the 1999 Product Development and Management Association (PDMA) International Conference. She also won the 2000 Edl and Edith Darger Dissertation Prize in Management in recognition of outstanding academic achievement.

Emma Sanchez

Best Researcher/Scholar, Health Education

Yuli Su

Best Teacher, Business

Gerardo Ungson

Best Researcher/Scholar, Business

Professor Gerardo R. Ungson is the Y.F. Chang Endowed Chair, and a professor of international business at San Francisco State University. His research covers strategies in digital-based environments; global strategic alliances, strategic implementation, organizational design and change; executive compensation; entry market strategies in China; and global strategies of U.S., European, and Asian firms. His works have been published in 40 papers in academic journals and 5 books. He is an active lecturer in executive management programs in the United States, the Netherlands, Vietnam, and Southeast Asia. He previously taught at the University of Oregon, the Amos Tuck School of Business, Dartmouth College; the College of Business, University of California, Berkeley; the College of Business, Pennsylvania State University; International University of Japan, Niigata, Japan; and Nijenrode, The Netherlands School of Business. He received his B.S. in management engineering at Ateneo de Manila University, Philippines in 1969, an M.B.A., Management, and a Ph.D., Business Administration, at Pennsylvania State University in 1978. Contact Information Email: bungson@sfsu.edu Website: Office Hours: T 1–4; F 11–12 Phone: 405-3749 Department: International Business Location: SCI 304 Advising Duties: None Intellectual Contributions Coordination and Control: Evolutionary Perspective (2005) Changing Paradigms and Challenges for International Business (2002) Timing of Entry in International Markets: An Empirical Study of U.S. Fortune 500 Firms in China (2000) The Determinants of Timing in Sequential Entry Decisions: An Exploratory Analysis (2000) The Determinants of Timing in Sequential Entry Decisions: An Exploratory Analysis (2000).

Yim-Yu Wong

Most Helpful to Students, Business

Susan Zieff

Best Overall Faculty Member, Kinesiology
Best Researcher/Scholar, Kinesiology
Best Teacher, Kinesiology

Dr. Susan Zieff specializes in the socio-cultural study of physical activity with a specific interest in environmental and policy change through active living. She is the Director of the Active Living Across the Lifespan Research Group within PACE Lab and she is a member of the national CDC-funded Physical Activity Policy Research Network (PAPRN), Shape Up SF (local active living collaborative) and the SF Physical Activity Council. Recent projects include: evaluation of the health and physical activity outcomes of participants of Sunday Streets SF (a local ciclovia [open streets] initiative); assessment of the economic impact of Sunday Streets; active living collaboratives; state school-based physical activity and physical education policies; policy-maker and neighborhood perspectives on physical activity; and state obesity plans. She uses mixed methodology in her investigations. Her current projects include a collaboration with Bogota, Colombia comparing open streets initiatives and a study of the local Play Streets project, an off-shoot of Sunday Streets.

SAN JACINTO COLLEGE - NORTH

Patricia Ferrell

Most Helpful to Students, Nursing

Kerri Hines

Best Overall Faculty Member, Nursing
Best Overall Faculty Member, Health Sciences and Nursing

SAN JOAQUIN DELTA COLLEGE

Rosalind Gottfried

Best Researcher/Scholar, Anthropology

"She has demonstrated inquisitiveness, commitment, knowledge, excellence and dedication for her discipline. Her students are quick to mention she promotes exchange of ideas and enriches them greatly."

Rosalind received her M.A. and Ph.D., in sociology, from Brandeis University in Massachusetts and her B.A. in sociology, with a psychology minor from Rutgers University in New Jersey. She also has an M.A. in counseling from Webster University, New Mexico. She has been teaching for more than thirty years, mostly at community colleges. She has been at SJDC since 1996. In addition to teaching, she has worked as a counselor, CASA, pet therapist (with her standard poodle). She is the proud mother of two daughters she raised as a single parent. She teaches Introduction of Sociology, Social Problems, and has also taught Gender, Social Psychology, Family, and Race and Ethnicity.

Elizabeth Maloney

Best Overall Faculty Member, Psychology
Best Teacher, Psychology
Most Helpful to Students, Psychology

Norman Solomowitz

Best Teacher, Psychology

"Mr. Solomowitz is very helpful, informative, interacts with students and role plays different scenarios. He demonstrates situations/events on a personal level. He is very clear in explaining the material."

SAN JOSE STATE UNIVERSITY

Daryl Canham

Most Helpful to Students, Nursing

Diane Stuenkel

Best Overall Faculty Member, Nursing

Diane Stuenkel RN, EdD is a professor in The Valley Foundation School of Nursing at SJSU. She has taught all aspects of Care of the Adult/Medical-Surgical nursing since joining the faculty in 1995. Her clinical background includes 15 years in critical care, as well as urgent care, cardiology, cardiac rehabilitation, and phone triage/advice. She was the Patient Education-Staff Development Manager for 5 years for a Bay Area medical group. Dr. Stuenkel served for 9 years as the Curriculum Coordinator and the Professional Development Coordinator for the SON. She currently serves as Assistant Director.

Dr. Stuenkel is active in several professional organizations including Sigma Theta Tau International Honor Society of Nursing. She is a charter member of the San Francisco Bay Area End of Life Nursing Education Consortium. Her research interests focus on predicting student nurse success on the NCLEX-RN and the work environment of staff nurses.

Dr. Stuenkel earned her BSN from Valparaiso University, her MS from the SJSU School of Nursing, and her doctorate from the University of San Francisco.

SANTA ANA COLLEGE

Rosemarie Hirsch

Best Overall Faculty Member, Nursing

"Rosemarie is the current Assistant Director for the Nursing Program as well as teaches the 4th semester nursing students. She is truly doing an outstanding job tackling both roles while also serving as an integral part of preparing the Nursing Department for our ACEN accreditation for Spring 2015. Rose is also a key leader in research and development for guiding the nursing department in a complete revision of curriculum. She is currently the only faculty at SAC that is credentialed as a "Nurse Educator" by the National League for Nursing (NLN). I fully support this nominee and feel she is most deserving of the Faculty of the year award!!"

SANTA MONICA COLLEGE

Elaine Roque

Best Teacher, Kinesiology

Professor Elaine Roque played professional indoor volleyball in the U.S and Europe and professional beach volleyball for 19 seasons. A former collegiate All-American at U.C.L.A. and Utah State, Professor Roque has published articles in national magazines and written and edited books for the Amateur Athletic Foundation. She also has a CD ROM titled "Core Training for Volleyball" that is sold worldwide and used for coaching training programs.

She has coached both men's and women's collegiate teams at the community college, NAIA and NCAA Division I levels as well as beach volleyball. She has worked with the U.S.A. junior national team and was a member of U.S.A. Volleyball's Coaches' Cadre. Professor Roque currently teaches volleyball, yoga, fitness classes, weight training and various majors' classes. She holds a B.A. from Utah State University and an M.A. from U.C.S.B. She is a member of the National Strength and Conditioning Association, AAPHERD, Iyengar Yoga National Association of the US, and the American Volleyball Coaches' Association.

SETON HALL UNIVERSITY

Donna Mesler

Best Overall Faculty Member, Nursing

Donna Mesler is a Pediatric Nurse Practitioner who loves teaching undergraduate nursing students. Her passion is cultural competence in nursing students and she has been the lead faculty on study abroad trips to the Philippines and Costa Rica with her students. She has also researched and published on this topic. She is the Director of the Family Nursing Department in the College of Nursing at Seton Hall University and has been recognized as a Leader and Mentor. She is currently the President of the Gamma Nu Chapter of Sigma Theta Tau International.

SETON HILL UNIVERSITY

Daniel Gray

Best Overall Faculty Member, Education

SHIPPENSBURG UNIVERSITY OF PENNSYLVANIA

Ming-Shuin Pan

Best Teacher, Business

"eager to do in-depth research"

SINCLAIR COMMUNITY COLLEGE

Mullins

Most Helpful to Students, Nursing

"Connie is so kind in all of her student interactions. She is exceptional in guiding the students to be the best they can be."

Mary Cox

Best Overall Faculty Member, Nursing

"Mary is very involved in the nursing department - she is an active member of the curriculum committee and has always been the "sunshine" to faculty who have needed uplifting."

Linda Johnson

Best Teacher, Nursing

"Linda is dedicated to educating future nurses! She teaches from her experiences in her practice. Her classroom is student centered and is current in evidence based practice."

SLIPPERY ROCK UNIVERSITY

Nancy Shipe

Most Helpful to Students, Medicine

"Momma Shipe is loved by students. She tells them to pull up their big boy pants and holds them accountable to high standards. She leads by example!"

SOUTH TEXAS COLLEGE (ALL CAMPUSES)

Cynthia Salinas

Best Overall Faculty Member, Nursing

"She has been working in the Vocational Nursing Program in the Starr County campus for 5 years, commuting from McAllen every day. She has worked well and impacted many lives in Rio Grande City because of her untiring efforts with students in the VN program."

Judith Sevilla-Dela Cruz

Best Researcher/Scholar, Nursing

Graduated with my BSN 25 years ago, experienced in different areas in nursing field, got my MSN-Education 2008 and teaching nursing in South Texas College since 2003. Married for 20 years and with 2 awesome high school children.

Jayson Valerio

Most Helpful to Students, Nursing

"He is just an exceptional instructor and program chair. Accommodating and very kind hearted."

SOUTHEAST MISSOURI STATE UNIVERSITY

Valerie Blackmon

Most Helpful to Students, Management

Peter Kerr

Best Teacher, Business

Heather McMillan

Best Researcher/Scholar, Management

Heather McMillan is an Assistant Professor of Management in the Department of Management and Marketing in the Donald L. Harrison College of Business at Southeast Missouri State University. She has completed her Ph.D. in Management from the University of Tennessee, Knoxville. She also received a Master of Business Administration in Health Care Human Resources and a Bachelor of Business Administration in Human Resources from East Tennessee State University. Additionally, she is certified as a Professional in Human Resources by the Society for Human Resource Management.

She has 8 years progressive human resource management experience in hospital organizations, culminating as HR Manager for a 50-bed inpatient rehabilitation hospital. She has taught a variety of management classes at the University of Tennessee and East Tennessee State University including Evaluating Training Design, Business Communications, Operations Management and Management Information Systems.

Dr. McMillan has a number of publications in academic journals, as well as multiple refereed conference proceedings, including the Southern Management Association and the Academy for Human Resource Development. She has also presented research at the National Council of Family Relations Conference and the Families and Work Research Conference sponsored by Brigham Young University. She has been a reviewer, session chair and discussant at the national meetings of the Academy of Management, Academy of Human Resource Development and Southern Management Association. Finally, she served as Conference Program Coordinator for the Academy of Human Resource Development's 2007 and 2008 conferences.

Dr. McMillan is a member of the Academy of Management, the Academy of Human Resource Development, the Southern Management Association, the Society for Human Resource Management, Phi Kappa Phi and Beta Gamma Sigma.

Kang Hoon Park

Best Researcher/Scholar, Business
Best Overall Faculty Member, Business

Willie Redmond

Most Helpful to Students, Business

Dana Schwieger

Best Overall Faculty Member, Management
Best Teacher, Management

SOUTHEASTERN LOUISIANA UNIVERSITY

Lana Auzenne

Best Teacher, Nursing

Lana Lee Auzenne Nursing Faculty; Registered Nurse
   Southeastern Louisiana University Baton Rouge, Louisiana
Southern University and Agricultural and Mechanical College at Baton Rouge
   Employment History:
   Southeastern Louisiana University (2007–Present)
   Education:
   Southern University and Agricultural and Mechanical College at Baton Rouge

Ashley Bowers

Best Teacher, Kinesiology & Health Studies

Eileen Creel

Best Overall Faculty Member, Nursing
Most Helpful to Students, Nursing

Karen Hill

Best Teacher, Nursing

Mitzie Meyers

Best Teacher, Nursing
Most Helpful to Students, Nursing

Kristie Riddle

Best Overall Faculty Member, Nursing

Bovorn Sirikul

Most Helpful to Students, Education

Penny Thomas

Best Overall Faculty Member, Health Sciences and Nursing

SOUTHERN ILLINOIS UNIVERSITY - CARBONDALE

David G Gilbert

Best Researcher/Scholar, Psychology

David Gilbert received his B.S. from the University of Washington in 1970 and earned his Ph.D. from Florida State University in 1978. He completed his internship at Linwood V.A. Hospital in Augusta, GA. The recipient of numerous research grants, including seven from the National Institutes of Health, Dr. Gilbert has attained international recognition for ground-breaking research in the study of psychological and biological basis of substance use/abuse (nicotine, antidepressants, alcohol, marijuana (THC), and caffeine). His research is broad and integrative, including environmental, genetic, and biological factors contributing to individual differences in cognition, affect, motivation, coping, and substance abuse. His laboratory utilizes state-of-the-art technology, including computerized eye-tracking, fMRI, EEG, ERP brain imaging, and emotional stimulus-presentation systems (including three 128 channel EEG systems). He is the author of Smoking: Individual Differences, Psychopathology and Emotion, edited Personality, Social Skills, and Psychopathology and has published over 75 articles in these areas.

SOUTHERN METHODIST UNIVERSITY

Brian Fennig

Best Teacher, Wellness

Brian Fennig came to Southern Methodist University in the spring of 2001 and joined the Wellness Department as a lecturer, where he teaches both Personal Responsibility and Wellness and Individual Fitness. He earned his Bachelor of Kinesiology and Health Science in 1989 and his Master of Education in 1990 from Stephen F. Austin State University. He is a three-time HOPE honoree and has been recognized as an outstanding faculty member by the National Residence Hall Honorary. He has participated in many Mustang Corral sessions as a faculty mentor, involvement speaker, and sneak preview provider. He currently serves as a Faculty Affiliate and annually leads incoming freshmen in their common reading discussion.

Prior to coming to Southern Methodist University, Brian taught high school biology, worked as a fitness specialist at Presbyterian Hospital of Dallas, and earned his second Master's Degree in Liberal Arts at Southern Methodist University in 1997. With an emphasis in myth, pop music, and modern music technology, he completed his Ph.D. in Humanities at the University of Texas at Dallas in December of 2013.

In his spare time, Brian writes and performs with his band, Fennig, as a drummer and singer. He has been active in the local poetry community and has several small press publications. Brian is an avid soccer fan and has worked for and/or attended the FIFA World Cup finals in The United States, Japan/Korea, Germany, South Africa and Brazil.

SPARTANBURG METHODIST COLLEGE

Mary Jane Farmer

Most Helpful to Students, Psychology

SPRING ARBOR UNIVERSITY

Terry Darling

Best Teacher, Psychology
Most Helpful to Students, Psychology
Best Teacher, Medicine
Most Helpful to Students, Medicine
Best Overall Faculty Member, Medicine

Jan Yeaman

Best Overall Faculty Member, Psychology
Best Researcher/Scholar, Psychology
Best Researcher/Scholar, Medicine

Ph.D. in Health Education (Health Psychology) from University of Maryland, College of Health and Human Performance, 1994 M.A. in Counseling Psychology from Biola University, Rosemead School of Psychology, 1979 Honorable B.A. in Clinical Psychology from Laurentian University, 1975.

ST. CLAIR COUNTY COMMUNITY COLLEGE

Brent Forsgren

Most Helpful to Students, Political Science

ST. CLOUD STATE UNIVERSITY

Jane Bagley

Most Helpful to Students, Nursing

"Jane interacts with first semester nursing students and starts them in their path to professional nursing. She is an excellent role model and sets high expectations."

Patricia Bresser

Best Overall Faculty Member, Nursing
Best Overall Faculty Member, Health Sciences and Nursing
Most Helpful to Students, Health Sciences and Nursing

Theresa Heck

Best Overall Faculty Member, Health
Best Researcher/Scholar, Health
Best Teacher, Health

"She has contributed significantly to the re-organization, and curriculum development of our program."

Sue Herm

Best Teacher, Nursing
Best Teacher, Health Sciences and Nursing

Joyce Simones

Best Researcher/Scholar, Nursing
Best Researcher/Scholar, Health Sciences and Nursing

ST. JOHN FISHER COLLEGE

Jennifer Mathews

Most Helpful to Students, Pharmacy

Dr. Mathews received her Ph.D. in Pharmacology from the University of Rochester School of Medicine and Dentistry. She began her tenure at St. John Fisher College in 2007. Dr. Mathews teaches in the Department of Pharmaceutical Sciences and her courses include: Systems Pharmacology I–V, Diversity, and Substances of Abuse.

Dr. Mathews' research is focused on the scholarship of teaching and learning. Research projects include novel in class exercises, community outreach and models of assessment. Of particular interest are topics related to opioid pharmacology and cultural competency. Dr. Mathews has mentored pharmacy and undergraduate Science Scholar students.

In 2010, Dr. Mathews was recognized by St. John Fisher College (Diversity Innovations Award) and the American Association of Colleges of Pharmacy (Innovations in Teaching Award, Honorable Mention) for her implementation of Deaf Strong Hospital at the Wegmans School of Pharmacy. This role-reversal program was designed to teach the first-year students about techniques for overcoming communication barriers as well as some of the specific challenges in communicating with deaf or hard-of-hearing patients. Dr. Mathews served as the faculty mentor for Salia Farrokh (2009) and Rosemary Garbowski (2011), recipients of the AACP/Wal-Mart Scholarship, which encourages pharmacy students to consider a career in academia. In 2007, only one year after joining the faculty at SJFC, Dr. Mathews was voted Teacher of the Year by the student body of the Wegmans School of Pharmacy.

Dr. Mathews has also been recognized by the scientific community, receiving travel awards from the International Narcotics Research Conference and the National Institute of Drug Abuse. While working on her M.A. degree at the University of Northern Colorado, she received the Dean's Citation for Outstanding Thesis and a Grant in Aid of Research from the Sigma Xi Honor Society. Dr. Mathews is active in several scientific and educational organizations including: The American Society of Health-Systems Pharmacists The American Association of Colleges of Pharmacy The College on the Problems of Drug Dependence.

ST. LOUIS COMMUNITY COLLEGE AT FOREST PARK

Michael Downey

Best Teacher, Culinary Arts

ST. LOUIS COMMUNITY COLLEGE AT MERAMEC

Michael Fuller

Best Researcher/Scholar, Anthropology

ST. PETERSBURG COLLEGE (ALL CAMPUSES)

Julie Adamich

Best Researcher/Scholar, Accounting

Biography

I have earned a Bachelor of Arts in Accounting from the University of South Florida, a Masters of Business Administration with a concentration in Accounting from Florida Institute of Technology and a Ph.D. in Curriculum and Instruction - Higher Education Administration from the University of South Florida. As you can see I "love to learn". I believe that continuing your education throughout life is both "stimulating" and "fun" and can be achieved with "persistence" and "patience". While earning my degrees over the past 25 (or so) years, I have spent approximately eight years on auditing, consulting and tax engagements for an international and a large local accounting firm and approximately five years as a cost control and pricing administrator in the defense industry. I can relate to the challenges of budgeting time for work, family, and school. Thus, I find on-line and blended instruction to provide the flexibility needed to reach these goals. Since 1986, I began teaching financial, managerial, intermediate, cost accounting, tax and auditing to both undergraduate and graduate students. I enjoy sharing my profession and encourage each of you to consider accounting as a profession.
* Member of the American Institute of Certified Public Accountants.

Joni Esser

Best Overall Faculty Member, Nursing
Best Teacher, Nursing

I am originally from Cleveland, Ohio and moved to Florida in 2004. I graduated from St. Alexis Hospital School of Nursing with my RN diploma and have worked in many areas of nursing, including medical surgical, intensive care, emergency and pediatrics. After attaining my Master of Science in Nursing degree in 1997, I became a Certified Pediatric Nurse Practitioner. Prior to moving to St. Petersburg, and accepting a position in the trauma program at All Children's Hospital, I was employed in a private pediatric practice. I have taught at St Petersburg College since the fall of 2005. I received my Doctorate in Nursing at Case Western Reserve University in the Spring of 2008, and a Certified Nurse Educator in 2012. My special nursing interests are trauma/emergency nursing and pediatric nursing. Nursing has been an integral part of my life for many years and I enjoy sharing my knowledge with students. Outside of the classroom, I enjoy working on my house and yard. I also run and bike on the Pinellas Trail. I live very close to the Gulf of Mexico, and go there often to kayak and walk on the beach. I have a wonderful partner, John and three grown children. I am blessed with five grandchildren – Lucy, Harper, Finley, Claire and Everett.

Joanne Goot

Most Helpful to Students, Nursing

Beverly Hindenlang

Best Researcher/Scholar, Health Sciences and Nursing

Leslie Honig

Most Helpful to Students, Health Sciences and Nursing

In 1975 she received a BA in social services from SUNY at Binghamton. After graduation, she decided to pursue a career in nursing and in 1978 she received her MSN and FNP certification from Pace University/New York Medical College. Over the next 25 years, she practiced pediatric and neonatal/obstetric nursing. In 2001, she began teaching as an adjunct in the SPC College of Nursing. The following year, she began teaching full time and has since taught at all levels in the CON. Her favorite is Level III where they combine med/surg with pediatric nursing!! As an educator, she see her role as mentor, coach, motivator, and partner in education.

Brent Rudolph

Best Teacher, Health Sciences and Nursing

Heidi Stein

Best Researcher/Scholar, Nursing

My educational background is a Master of Science in Nursing Education from University of South Florida. My goal as a teacher is to be flexible and be responsive to both the educational needs of the students and the service needs of the hospitals who are essential partners to the effective implementation of the nursing program/curriculum. My Philosophy of Nursing Education in the university or community college setting deals primarily with the adult learner. Life experiences, self motivation, self direction, and values have a direct impact on the teaching and learning experiences. The foundation for nursing education should be based on problem-solving situations whose solutions have been previously validated. These situations involve technical, intellectual, and interpersonal competencies. Person/Client A person is a unique, social, rational, spiritual, and biological individual entitled to respect and individual consideration. The person strives to maintain homeostasis through interaction with the community and move toward self actualization. Nursing as defined by the International Council of Nurses (ICN) encompasses autonomous and collaborative care of individuals of all ages, families, groups, and communities, sick or well and in all settings. Nursing includes the promotion of health, prevention of illness, and the care of the ill, disabled and dying people. Advocacy, promotion of safe environment, research, participation in shaping health policy and in patient and health systems management, and education are also key nursing roles (ICN, 2004). As a professional nurse, I apply my knowledge of nursing which involves assessment, diagnosis, planning, implementation and evaluation of patient outcomes. The environment is a dynamic complex of systems and subsystems which impact the individual. The forces of the current environment are transforming healthcare and its delivery. Trends to be considered include: wellness, management of chronic conditions, consumer empowerment, shift of care in the community, aging population, cultural diversity, fiscal accountability, personal and professional accountability, collaboration, expanding scientific knowledge and technology. Health is a state of being experienced in a unique way by each individual. When ones basic needs are met, and homeostasis is maintained, a state of wellness is achieved. The processes of teaching and learning are active and reciprocal. The learner attempts to achieve desired competency in application of knowledge while the teachers role is one of facilitator in this change of behavior. These changes are manifested in development of cognitive, psychomotor and affective abilities in the learner. Nursing education is an evolving body of knowledge built upon the biological, physical, and human sciences. Nursing education is responsive to the rapidly changing healthcare systems and reflective of new partnerships between students, teachers and clinicians. Nursing education provides collaborative opportunities with clinical sites to facilitate learning the roles of provider of care, manager of care, and becoming a member of the discipline of nursing. Nursing education focuses on evidenced based, clinically competent care in an ethical manner to all clients regardless of health problems, personal attributes, third party payors, and public demands. (NLN, 2003)

STARK STATE COLLEGE

Bryan Gerber

Best Teacher, Medicine

"Dynamic and passionate professor dedicated to student success."

STETSON UNIVERSITY

Nick Maddox

Best Teacher, Management

Nicholas Maddox has taught at Stetson University since 1985. His teaching focus incorporates new leadership and management paradigm concepts and practices for the enhancement of organizations. His classes feature intensive experiential learning methods.

Maddox has been an active trainer and consultant throughout his career often focusing on helping individuals and organizations become more skilled in the management of the human resource. He is a certified hypnotherapist and intends to build a performance consulting business in the near future. Maddox coordinates the Management Program in the School of Business Administration. He has been an active futurist for the past 30 years.

STETSON UNIVERSITY COLLEGE OF LAW

Cynthia Hawkins DeBose

Best Teacher, Law

STEVENSON UNIVERSITY

Anna Kayes

Best Teacher, Business

SUFFOLK COUNTY COMMUNITY COLLEGE

Lorraine Sanso

Most Helpful to Students, Health Sciences and Nursing

SUNY CANTON

Patricia Shinn

Best Overall Faculty Member, Nursing

SUNY OSWEGO

Laura Brown

Best Overall Faculty Member, Psychology

Laura Brown is a lifespan developmentalist interested in family relationships across the lifespan, especially those between grandparents, parents and children. Other research areas include adoptive families, service-learning in Gerontology and family kinkeeping.

SUNY PLATTSBURGH

Byrne Degrandpre

Best Researcher/Scholar, Education
Best Teacher, Education
Most Helpful to Students, Education

SUSQUEHANNA UNIVERSITY

David Bussard

Best Researcher/Scholar, Business
Most Helpful to Students, Business
Best Overall Faculty Member, Business

David Bussard, Associate Professor of Business; B.A. 1965, Bucknell University; M.B.A. 1969, University of Michigan; Ph.D. 1991, University of Pennsylvania Wharton School of Business. (1978, 1991)

TARRANT COUNTY COLLEGE (ALL)

James Ariail

Most Helpful to Students, Business

Regina Cannon

Best Overall Faculty Member, Business
Best Overall Faculty Member, Business
Best Teacher, Business

Talia Dancer

Most Helpful to Students, Business
Most Helpful to Students, Business

Jesse Garcia

Best Teacher, Business

Vincent Giardino

Best Teacher, Government
Most Helpful to Students, Public Affairs, Political & Policy Sciences

Karen Haun

Best Researcher/Scholar, Business

Thomas Kemp

Best Overall Faculty Member, Economics

"Dr. Kemp is constantly updating his economic knowledge base and has proved a trust worthy source of information."

Sally Proffitt

Best Researcher/Scholar, Business

Jaye Simpson

Best Teacher, Business

Holmes Suneye

Best Teacher, Economics

Professor in the Economics department at Tarrant County College (all), Fort Worth, TX

Tetsuya Umebayashi

Most Helpful to Students, Nursing

Associate Professor of Nursing at Tarrant County College

TENNESSEE STATE UNIVERSITY

Chun-Da Chen

Best Researcher/Scholar, Business

"outstanding publication record in International Finance"

Own Johnson

Best Overall Faculty Member, Health Sciences and Nursing

"Owen Johnson was very helpful in assisting me during my period of applying for faculty tenure."

Achintya Ray

Best Researcher/Scholar, Business

"Internationally recognized, expert for health care economics"

Dr. Achintya Ray is an Associate Professor of Economics and an Associate Editor of the Journal of Developing Areas (Project Muse, Johns Hopkins University). Dr. Ray earned his M.A. and Ph.D. degrees in Economics from Vanderbilt University. His research interests are in Industrial Organization, Health Economics and Development Economics. Dr. Ray has published his research in many internationally reputed peer reviewed journals like the Manchester School, Journal of Economics, Economics of Innovation and New Technology, Economics Bulletin etc. Dr. Ray has provided his services as a peer reviewer to many leading international journals including American Journal of Preventive Medicine, American Journal of Public Health, Journal of Epidemiology and Community Health, Manchester School, Applied Economics, Applied Financial Economics, British Medical Journal, Economics and Human Biology, Social Science and Medicine, and Bulletin of the World Health Organization. He is a noted columnist whose op-eds have been published by leading dailies around the world. He has presented invited talks at many internationally reputed institutions including the World Institute of Development Economics Research of the United Nations University, The George Washington University, Wharton School of the University of Pennsylvania, University of Waterloo, Whittier College, and Indian Institute of Management. Dr. Ray has served as expert examiner of doctoral dissertations of many reputed international universities.

TENNESSEE TECH UNIVERSITY

LaNise Rosemond

Best Overall Faculty Member, Physical Education
Most Helpful to Students, Physical Education

"Lanise goes out of her way to help each and every student."

Jeremy Wendt

Best Researcher/Scholar, Education

TEXAS A&M UNIVERSITY AT COLLEGE STATION

Melinda Lou Grant

Best Overall Faculty Member, Physical Education
Best Teacher, Physical Education
Most Helpful to Students, Physical Education
Best Teacher, Education
Most Helpful to Students, Education
Best Overall Faculty Member, Education

Short Bio 1997–present Associate Chair, Physical Education Activity Program, Texas A & M University 1994–present Lecturer, Kinesiology, Texas A & M University Summers 1996–2000 Director, Camp Aggieland 1992–1994 Physical Education Instructor, Douglas Elementary School, Tyler Independent School District, Tyler, TX 1985–1994 Instructor, Kinesiology Department, Tyler Junior College 1976–1980 Graduate Assistant and Instructor, School of Physical Education, West Virginia University.

Susan Lowey

Best Researcher/Scholar, Education

TEXAS A&M UNIVERSITY AT CORPUS CHRISTI

Lisa Comparini

Best Overall Faculty Member, Psychology

Mary Fernandez

Best Researcher/Scholar, Counseling
Best Teacher, Counseling
Most Helpful to Students, Counseling
Most Helpful to Students, Education
Best Teacher, Education

Gina Glanc

Best Researcher/Scholar, Psychology

Research Interests: Dr. Glanc received her Ph.D. from Case Western Reserve University in Cleveland, Ohio in 2008. She is currently conducting research in human memory. More specifically, she is investigating the effects of orthographic neighborhood size and retrieval practice on recognition and recollection processes. She is also involved in collaboration with the Department of Neuroscience at Rice University to study memory changes in aging adults. Teaching Interests: Physiological Psychology, Sensation and Perception, Measurement and Statistics, General Psychology

Amy Holihan

Best Teacher, Psychology

Dr. Houlihan's research interests involve the application of social psychology to physical health. In particular, she does work on health risk behavior among adolescents and young adults (e.g., risky sexual behavior, heavy drinking). Much of this research involves examining social and personal characteristics that predict risky behavior, such as stressful situations and low self-control. Recent projects include a study on the attitudinal effects of the initiation of sexual behavior and a study of predictors of risky sexual behavior among Hispanic/Latino college students. Dr. Houlihan is also interested in psychosocial predictors of HPV vaccination and HPV-related attitudes.

Miguel Moreno

Most Helpful to Students, Psychology

Dan Pearce

Best Overall Faculty Member, Education
Best Researcher/Scholar, Education
Best Teacher, Education
Most Helpful to Students, Education

Daniel Pearce, Ph.D., is Professor of Education in the Department of Educational Leadership, Curriculum and Instruction at Texas A&M University-Corpus Christi. Before coming to TAMUCC in 1993, Dr. Pearce served on the faculties of Northeastern Illinois University, Montana State University-Billings, and Western Illinois University. His doctoral degree is from Michigan State University, and he received his master's degree from Wayne State University and his bachelor's degree from Albion College.

Ric Ricard

Best Overall Faculty Member, Counseling
Best Overall Faculty Member, Education
Best Researcher/Scholar, Education

http://education.tamucc.edu/bio/files/ricard_vita.pdf
    PERSONAL INFORMATION DOB: January 30, 1963 Address: 340 Jackson Place, Corpus Christi, TX 78411 H: (361) 852-8525 W: (361) 825-3725 E-mail: Richard.ricard@tamucc.edu Married: Melissa Ricard (Administrative Law Judge); Two children (Alexander 21; Benjamin 18) EDUCATION Harvard University, Cambridge, Psychology, A.M., 1986; Ph.D., 1991. Columbia University, New York, Exchange Scholar, 1988–1990. Institute of Psychology, Moscow, USSR Visiting Scholar, Fall 1988. University of California, San Diego, B.A. Psychology, 1985. Recent Awards and Acknowledgements Association of Perioperative Registered Nurses (AORN) 2005 Journal Writers Award TAMUCC College of Education Research Award (2003) Who's Who of American Teachers (2002) Educational Awards and Fellowships Ricard, R.J. (1985–1990). Harvard Merit Fellowship, Department of Psychology, Harvard University Ricard, R.J. (1984). Minority Access to Research Careers (MARC) Fellowship, University of California, San Diego Ricard, R.J. (1984). Minority Biology Research Fellowship, University of California, San Diego

Steven Seidel

Most Helpful to Students, Psychology

TEXAS A&M UNIVERSITY - SAN ANTONIO

Mishaleen Allen

Best Overall Faculty Member, Education
Best Teacher, Education

"Mishaleen Allen is an awesome professor and faculty member! I would like to nominate her as the best overall faculty member."

Dr. Mishaleen Allen joined the TAMU-SA family as adjunct faculty in 2008 and appointed to the rank of Assistant Professor at Texas A&M University-San Antonio Fall 2009 serving as Department Chair of Curriculum & Kinesiology and college administration from 2009–2011. She teaches undergraduate and graduate courses for general and special education educator preparation as well as the Special Education program's Educational Diagnostician certification and specialized Masters of Education in Special Education with Instructional Specialist, Educational Diagnostician, and Autism/Emotionally Disturbed focus options.

An active faculty participant in the College of Education and Human Development's Model for Success Initiative (MSI) since conception, Dr. Allen has had the privilege of working with administration to facilitate the innovative pre-service teacher preparation program within San Antonio Independent School District (SAISD) as Liaison (2010–2011) as well as Cohort Faculty/Supervisor (Cohort 2, 2011–2014; Cohort 5, 2014–2015). Dr. Allen provides community service and consultative support to a variety of local organizations including, but not limited to, City of San Antonio Head Start, SA Reads, American Sunrise Learning Center, and San Antonio Busy Bodies where she serves as board president while mentoring current and past students in various fields.

Dr. Allen received her higher education training at Texas Woman's University (Denton, TX) earning a Ph.D. in Special Education (1997) with additional emphasis in Family Studies, Administration, and Program Evaluation. She also earned an M.Ed. in Assessment/Supervision (1995), M.S. in Family Studies (1993), and B.S. in Vocational Home Economics Education (1990). Professionally, Dr. Allen served 16 years in public schools as a general and special educational teacher, educational diagnostician, and campus administrator prior to moving full time into higher education and educational consultation. Areas of research and specialization include assessment, increasing rigor in pre-service teacher programs through video assessments/self-reflection/PLNs, program development/evaluation, technology integration for UDL, and experiential learning.

Shelley Blackburn-Harris

Best Overall Faculty Member, Education
Best Teacher, Education
Most Helpful to Students, Education
Best Overall Faculty Member, Education
Most Helpful to Students, Education

Dr. Shelley Harris is an Associate Professor in Curriculum and Instruction at Texas A&M San Antonio. She teaches undergraduate and graduate courses in learning theory, instructional strategies and methodology. As a professional in the education community, her teaching experience includes working as an elementary, middle and secondary education teacher. In addition to teaching and research, she offers a Jaguar Camp each summer for graduate students hosted at American Sunrise Learning Center. Dr. Harris's areas of research and specialization include multimodal teaching, experiential learning, affective teaching, and effective instructional models.

Nancy Compean-Garcia

Best Overall Faculty Member, Education

Nancy Compean-Garcia is an Assistant Professor in Bilingual/ESL education at Texas A&M University-San Antonio. Dr. Garcia has served as the Senior Director for the Teacher Preparation and Certification Center. She has worked in the areas of teaching, research, service and mentoring undergraduate and graduate students in higher education since 1999. In addition, Dr. Garcia has been instrumental in teaching in the Bilingual/ESL Education Program and developing the Dual Language Education curriculum for EC-6 degree plans at TAMU-SA. Dr. Compean-Garcia has continued to participate in international, state, and national conferences that promote the development of the bi-literacy, bi-cultural and social capital. In 2002, Dr. Garcia taught an Effective Teaching Strategies course for university faculty in Jing-Hua University, China and in Frankfurt, Germany. She has published several articles in journals such as the Journal of Border Educational Research, the Texas Child Care Quarterly Journal, and Early Years-Texas Association for the Education of Young Children Journal. She is currently a board member and Chair-Elect of The Hispanic Women's Network of Texas-San Antonio Chapter.

Suzanne Mudge

Best Overall Faculty Member, Education
Best Overall Faculty Member, Education

TEXAS CHRISTIAN UNIVERSITY

Ellen Broom

Most Helpful to Students, Psychology

Ellen Broom earned a Ph.D. from the University of North Texas. Her areas of research interests include behavioral interventions for children and youth with autism and behavioral disorders, empathy levels of incarcerated youth and psychometric testing with children. Broom also is a writer in academics and genres of public interests. She speaks at events addressing a wide range of topics in psychology, education, parenting, breast cancer and mindfulness.

TEXAS SOUTHERN UNIVERSITY

Ingrid Haynes-Mays

Best Overall Faculty Member, Education

Ingrid Haynes-Mays is currently an Associate Professor in the Department of Curriculum and Instruction for Texas Southern University. Dr. Haynes-Mays received her Ph.D. in Curriculum and Instruction with emphasis in TESOL from the University of Mississippi, her Masters of Education in Reading from Texas Southern University and her Bachelor of Science in Elementary Education from Texas Southern University. Dr. Haynes-Mays' research interests include areas related to literacy and language development. She has presented and published numerous articles on the above topics. She has also co-authored the book entitled "A Recipe for Hands-On Activities for teaching Phonemic Awareness in the Primary Grades" – a wonderful book that provides teachers and parents with activities for improving phonemic awareness and phonics.

Holim Song

Best Researcher/Scholar, Education

Holim Song is an assistant professor of Instructional Technology in the College of Education at Texas Southern University, where he has taught since 2005. Song's primary research focus is in faculty's technology use in the classroom, instructional design methods integrating instructional technology, and instructional strategies for web-based instruction. He recently published, "Handbook of Research on Instructional Systems and Technology" (1st edition, New York: Hershey, 2008). Song has also written many articles published in journals such as International Journal of Information and Communication Technology Education, and International Journal of Web-Based Learning and Teaching Technologies. Song has provided Blackboard training and technology use for the College of Education Faculty professional development. Song has also served as a College of Education web master and Blackboard administrator. He earned his Ed.D. and M.A. at the University of Houston.

Chu-Sheng Tai

Best Overall Faculty Member, Finance

"Has the interest of students at heart. Fair to all students."

Chu-Sheng Tai, PhD, is an Associate Professor of Finance at the Jesse H. Jones School of Business at Texas Southern University. He earned his PhD from The Ohio State University in 1999. Before joining Texas Southern University in fall 2004, he taught at Pittsburg State University (1999–2000) and Texas A&M University - Kingsville (2000–2004). Dr. Tai's teaching experience has covered a broad range of areas in finance, including investment, international finance and corporate finance courses. He has taught undergraduate and graduate classes. Dr. Tai's main research areas are investments and international finance, and his research has been published in numerous journals, including Journal of Banking and Finance, Journal of Futures Markets, The Quarterly Review of Economics and Finance, Journal of International Financial Markets, Institutions, and Money, Journal of Multinational Financial Management, The Emerging Markets Review, International Review of Financial Analysis, Research in International Business and Finance, and Journal of Academy of Finance.

TEXAS STATE UNIVERSITY - SAN MARCOS

Patricia Parent

Best Teacher, Political Science

Dr. Patricia Parent is a Senior Lecturer in the Department of Political Science. She received her B.A. in education with a social science composite and her M.A. in political science from Southwest Texas State University. Additionally she earned her Ph.D. in American politics from The University of Texas. Dr. Parent has been part of the political science department since 1984 where she has since served the department in a variety of capacities, including undergraduate internship coordinator from 1989–present, graduate coordinator from 2006–present as well as serving on the Lecturer and Women's Course Committee, among others. She teaches a variety of junior and senior level courses, with her specialty being campaigns and elections, Congress and state legislatures, women in politics, and public policy. For the university, Dr. Parent has served as an advisor for undeclared majors in the College of General Studies from 1987–2002, participant in the Women's Studies Council in 1985,1986 and 2002. She also serves as the faculty advisor for the College Republicans and is a member of the Business and Professional Association of South Austin, Manchaca among other memberships.

Ram Shanmugam

Best Researcher/Scholar, Health Science
Best Researcher/Scholar, Health Sciences and Nursing

"Ram Shanmugam"

Ram Shanmugam, Ph.D., is an elected Fellow of the International Statistical Institute. His areas of research, teaching & consulting interest are operational research in health administration, clinical trials, informatics, six-sigma methodologies, environmental models, financial forecasts, and multivariate data collection, analysis & interpretation. Dr. Shanmugam has been invited to give seminars and guest lectures in international meetings. He serves as an associate editor in several journals.

TEXAS TECH UNIVERSITY

Raymond Edwards

Most Helpful to Students, Physical Education

Mark Isidro

Best Researcher/Scholar, Physical Education

David Lektzian

Best Researcher/Scholar, Political Science

THE CATHOLIC UNIVERSITY OF AMERICA

Andrew Abela

Most Helpful to Students, Business

Dr. Abela is the Dean of the School of Business & Economics and Associate Professor of Marketing at the Catholic University of America, in Washington, DC. His research on the integrity of the marketing process, including marketing ethics, Catholic Social Doctrine, and internal communication, has been published in several academic journals, including the Journal of Marketing, the Journal of the Academy of Marketing Science, the Journal of Business Ethics, and the Journal of Markets & Morality, and in two books. He is the co-editor of A Catechism for Business, from Catholic University Press, and winner of the 2009 Novak Award, a $10,000 prize given by the Acton Institute for "significant contributions to the study of the relationship between religion and economic liberty."

Dr. Abela also provides consulting and training in internal communications; recent clients of his include Microsoft Corporation, JPMorganChase, and the Corporate Executive Board. Prior to his academic career, he spent several years in industry as brand manager at Procter & Gamble, management consultant with McKinsey & Company, and Managing Director of the Marketing Leadership Council of the Corporate Executive Board. He holds a B.Sc. from the University of Toronto, an MBA from the Institute for Management Development (IMD) in Switzerland, and a Ph.D. in Marketing and Ethics from the Darden Business School at the University of Virginia. He and his wife, Kathleen, live in Great Falls, Virginia with their six children.

Martha Cruz-Zuniga

Best Researcher/Scholar, Business

Dr. Martha Cruz Zuniga is an assistant clinical professor of economics and is also the Director of Economics programs. She specializes in development economics, international economics and monetary economics with particular focus on migration, remittances, microfinance, poverty, education, and international transmission of shocks. She is especially interested in issues affecting developing countries.

Dr. Cruz Zuniga first joined The Catholic University of America faculty in 2006 as an Assistant Professor and taught Economic courses on both the undergraduate and graduate levels. From 2009 to 2011, Dr. Cruz-Zuniga held the position of General Manager at The Royal Flowers Group and Director of Finance and Administration at One World Telecom in Miami, FL. Dr. Cruz Zuniga returned to The Catholic University of America in 2012. She is currently involved in seeking grants, serving as Academic Advisor for thesis of master students at the Integral Economic Development Management program as well as teaching Economic courses to undergraduate and graduate students. Dr. Cruz Zuniga has also held appointments at Western Michigan University and has previously worked in private companies in Ecuador.

One theme in her current research explores the effects of dollarization in developing economies. In addition, she also examines the effects of international monetary shocks in economic growth. Another theme of her current research is how integral economic development can be reached through development initiatives that include microfinance and workers' training. Her earlier work has been published in economic journals and includes research on remittances and international transmission.

Dr. Cruz Zuniga has presented her work in various economic conferences including the Southern Economics Conference, the Midwest Economics Conference, and the Federal Reserve of St. Louis-Missouri Economics Conference. She has also presented at the Commission on the Status of Women, a division of the United Nations in New York.

Dr. Cruz Zuniga obtained her Ph.D. in Economics from Western Michigan University. She also holds a M.A. in Economics from Western Michigan University and a B.A. in Economics from The Pontifical Catholic University of Ecuador.

Mary Njai

Best Researcher/Scholar, Accounting

THE COLLEGE AT BROCKPORT (SUNY BROCKPORT)

Pilapa Esara

Most Helpful to Students, Anthropology

Lynne Gardner

Most Helpful to Students, Health Science

Jennifer Ramsay

Best Overall Faculty Member, Anthropology
Best Researcher/Scholar, Anthropology

I have worked on several multi-disciplinary projects. oFr example, in Israel I worked with the Combined Caesarea Expeditions, the Excavations at Khirbet Qana and the site of Shivta. nI Italy I have worked on material from the excavations at Horace's Villa, the Kaucana project and the Nordic Excavations at Nemi, Julius Caesar's Villa.

Jennifer Ratcliff

Best Researcher/Scholar, Psychology

"Dr. Ratcliff consistently involves undergraduates in her program of research-from conception of a study through publication. She often takes students to present at regional and national conferences and is also very involved with the college Study Abroad program and has made at least 3 trips to Europe, New Zealand and is planning her next trip with students to Africa in collaboration with the English department. Dr. Ratcliff's research as well as her teaching and creation of new course offerings are focused on critical prejudice and a quickly building volume of research and community service related to concerns of the LGBT community. She has also received awards for her advisement-awards for which students nominated her. She is a well-liked professor by faculty and students alike and embodies the personal and professional qualities we strive to help students realize in both our undergraduate and master's program."

Dr. Ratcliff's primary research program focuses on understanding the processes by which individuals develop and maintain both positive and negative attitudes toward marginalized outgroups. To this end, her research explores two antecedents for outgroup liking and prejudice: (a) cultural displays exhibited by the group in question, (e.g., pride displays), and (b) sociocultural norms (e.g., expectations regarding how men and women should behave in a given society). Her work in this area additionally elucidates the distinct implications of outgroup liking and prejudice for behavior toward marginalized groups. This work has been applied to understanding how to bring about positive relations between marginalized groups and majority groups in America (e.g., gay men and heterosexual individuals; Caucasian Americans and African Americans), as well as in Israel (e.g., Arab Israelis and Jewish Israelis).

In her second line of research, Dr. Ratcliff explores the influence of perceptual factors on social judgment. More precisely, she examines the impact of perceptual constraints—such as the relative salience of one actor over another during an ongoing interaction—on the way people interpret and ultimately evaluate videotaped criminal confessions.

Tiffany Rawlings

Best Teacher, Anthropology

THOMAS JEFFERSON UNIVERSITY

Frances Gilman

Best Researcher/Scholar, Health Sciences and Nursing

Fran joined the Department of Radiologic Sciences at Jefferson College of Health Professions in 1993 as an adjunct faculty member and became the Program Director in 1998. She was promoted to Assistant Professor in 2004 and was appointed Chair of the Department in 2003.

Fran has authored and implemented numerous curricula providing a variety of opportunities for students in Radiologic Sciences. These include Magnetic Resonance Imaging, Computed Tomography, Radiation Therapy, Medical Dosimetry, Women's Imaging and PET/CT. She currently teaches in the Radiography, Computed Tomography and Magnetic Resonance Imaging programs.

Fran earned her Certificate in Radiologic Technology from Northeastern University and Boston University Medical Center in 1974. While working as a technologist at Boston City Hospital, she completed her Baccalaureate degree in Health Administration at Northeastern University in 1981. Fran worked in clinical practice at BU Medical Center for ten years, initially in diagnostic radiography, and later specializing in cardiovascular-interventional imaging and computed tomography. She left clinical practice in 1982 to pursue opportunities in sales and marketing in the medical imaging industry.

Fran moved to the Philadelphia area in 1984 where she continued her career in sales, marketing and technical support. She earned her Master's Degree in Health Administration from St. Joseph's University in 1988. In 1989, Fran began her career in education as a part-time faculty member at the Community College of Philadelphia. While teaching at CCP she was also an adjunct faculty member at Thomas Jefferson University before accepting her current position in 1998. She attained advanced certifications in computed tomography, magnetic resonance imaging and cardio-vascular interventional imaging.

Fran is actively involved in radiography professional societies nationally and locally. From 1995–1999, she held positions as President and Board Chair for the Philadelphia Society of Radiologic Technologists and was an affiliate delegate in the House of Delegates at the American Society of Radiologic Technologists national convention. She is a graduate of the American Society of Radiologic Technologists Leadership Academies for Educators and Technologists and has served on a number of committees for the American Society of Radiologic Technologist and the Association of Educators in Radiologic Sciences. She most recently was selected and participated in the Association of Schools of Allied Health Professions Leadership workshop.

David Jack

Best Researcher/Scholar, Nursing
Best Teacher, Nursing
Most Helpful to Students, Nursing
Best Teacher, Health Sciences and Nursing
Most Helpful to Students, Health Sciences and Nursing
Best Overall Faculty Member, Health Sciences and Nursing

David Jack, PhD, MSN, RN, CPN, CNE 130 South Ninth Street Edison Building, Suite 1228 Philadelphia, PA 19107

(215) 955-5349

Education

PhD, Widener University MSN, Villanova University, Philadelphia, PA BSN, LaSalle University, Philadelphia, PA Diploma, Helene Fuld School of Nursing Certification

Certified Pediatric Nurse Certified Nurse Educator University Appointment

Assistant Professor Research & Clinical Interests

Nursing Education/Pediatrics Medical/Surgical, Obstetrical Nursing Pediatric violence, Aggression in young children Professional Memberships

NAPNAP (National Association of Pediatric Nurse Practitioners) Presentations

Acceptance of Poster Presentation: Test Blueprint: Mapping Cognitive Levels with the NCLEX

Categories for Examination Construction in an Associate Degree Nursing Program presented at the NLN Summit Meeting in September 2005

Acceptance of Poster Presentation: Sticks and Stones: Pilot Study Data of the Modified Aggression Scale presented at How Does Your Garden Grow, Penn State, Interdisciplinary conference on children and behaviors April 2007

Acceptance of Poster Presentation: The Effects of a Violence Prevention Program on Self-reported Acts of Aggression on Kindergarten-aged Children, Crozer-Keystone Health System Research Symposium held at Villanova University on October 2009

Oral Presentation of research data: The Effects of a Violence Prevention Program on Self-reported Acts of Aggression on Kindergarten-aged Children, Christiana Care Health System 4th Annual Nursing Research Conference November, 2009

TOWSON UNIVERSITY

Sharon Buchbinder

Best Researcher/Scholar, Health Science
Best Researcher/Scholar, Health Sciences and Nursing

Dr. Sharon Buchbinder received her undergraduate BA in Psychology from the University of Connecticut in 1973, and an MA in Psychology from the University of Hartford in 1976.

She worked in healthcare as an Intravenous Therapist from 1973 to 1980, where she was greeted by patients with "Not again!" each time she entered their rooms. Unhappy with working only eighty hours a week, she returned to school for an AAS in Nursing from Maria College in Albany NY. While attending school part-time, she was employed by the State of New York Department of Mental Hygiene and worked on a nurse exit survey examining reasons for turnover among NY state psychiatric nurses.

Bitten by the research and policy bug, she went to work for the National Commission on Nursing in Chicago IL and the American Hospital Association's Division of Nursing, also in Chicago. She then worked for the American Medical Association (AMA) (Chicago) from 1984–1990. While at the AMA she held a variety of positions in Science and Technology, Medical Education and Marketing, Membership and Strategic Planning. While employed full-time at the AMA, colleagues stalked her at the photocopier and told her life would be incomplete without a doctorate. Gullible and gleefully unaware of the work ahead, she pursued her PhD in Public Health Sciences, majoring in Health Resources Management, with a collateral in Economics. Dr. Buchbinder's dissertation was "Physician Job Satisfaction and Prediction of Likelihood of Practice Change," the defense of which completed her PhD requirements in 1992. Her family rejoiced, safe in the knowledge that she was not interested in pursuing a law degree.

Dr. Buchbinder moved to Baltimore, and took a postdoctoral fellowship in children's mental health services where she looked at the impact of international medical graduates on the provision of service for the severely mentally ill. Subsequent to the postdoc, she took a position in the Johns Hopkins' School of Medicine (JHU SOM) as Research Coordinator of the Evaluation of the Hawaii Healthy Start Program, a child abuse and neglect prevention program that has been in existence for 15 years. The job required several trips a year to Hawaii, and while it was a tough job, someone had to do it. During her time at JHU SOM, she successfully pursued grant funding for "Primary Care Physician Job Satisfaction and Turnover" and was promoted to the faculty at JHU as a Research Associate.

Lured to Towson University by the prospects of big bucks and new audiences of undergraduate and graduate students for her old jokes, Dr. Buchbinder became an Assistant Professor in 1996. In 2000, she grew into an Associate Professor and Coordinator of the undergraduate Health Care Management Program. She was promoted to Full Professor in March of 2005 and appointed as Chairman of the Department of Health Science in June, 2005. She also conducts health care management research and provides relevant and effective service to her department, college, university, and discipline, and is Immediate Past Chair of the Board of the Association of University Programs in Health Administration (AUPHA).

Dr. Buchbinder and her husband, Dr. Dale Buchbinder, have a son who GRADUATED on January 7, 2007 (YAAAAY!!) with a BS in English from Towson University. When not attempting to make students and colleagues laugh, she can be found exercising on an elliptical while reading, golfing, deep sea fishing, playing with her cats and dog, writing healthcare management textbooks, romantic short stories, and novels with a weird blend of horror, mystery, and romance. For a glimpse at Dr. Buchbinder's alter ego, go to www.sharonbuchbinder.com.

Linda Caplis

Best Overall Faculty Member, Health Science
Best Teacher, Health Science
Best Teacher, Health Sciences and Nursing
Best Overall Faculty Member, Health Sciences and Nursing

Islam Elshahat

Best Teacher, Accounting
Most Helpful to Students, Accounting
Best Teacher, Business
Most Helpful to Students, Business

Martin Freedman

Best Researcher/Scholar, Accounting
Best Researcher/Scholar, Business

Jack Fruchtman

Best Overall Faculty Member, Public Affairs, Political & Policy Sciences

"Best ever."

Jack Fruchtman Jr., Professor of Political Science and Director of the Program in Law and American Civilization, has taught at Maryland's Towson University since 1978. He has published six book-length studies on many eighteenth-century figures and constitutional issues. He has edited or annotated another five. His most recent books are The Political Philosophy of Thomas Paine (2009) and The Supreme Court: Rulings on American Government and Society (second edition 2014). He also serves as Series Co-Editor of The Enlightenment World, published by Pickering & Chatto. He is currently working on a history of the United States Constitution.

Professor Fruchtman teaches courses on constitutional law and politics with an emphasis on the casebook method. Students are expected to read and understand the cases handed down by the U.S. Supreme Court as they prepare to discuss and debate them during each class session. He teaches these courses somewhat like law school classes but with the understanding that undergraduates and law students are not similarly situated. Students should expect that by the end of the first few weeks of class, he will know their names and will not hesitate to call on them to respond to questions.

The courses are typically those that investigate the following: the origins of the Constitution; the separation of powers; the science of federalism, the Commerce Clause; civil rights; civil liberties; the equal protection of the laws, including segregation, desegregation, resegregation, and affirmative action; the rights of the accused; the death penalty and the right to privacy.

As the director of the program in Law and American Civilization, he advises students, oversees the curriculum, and conducts senior thesis research, writing, and presentation. He is also charged with guiding the LWAC majors who engage in internships.

Carrie McFadden

Most Helpful to Students, Health Science
Most Helpful to Students, Health Sciences and Nursing

Katherine Rabon

Most Helpful to Students, Nursing

"Kathy always has time for her colleagues and her students. She thinks of innovative ways to engage her students and keep them interested and learning. She mentors her adjuncts and her colleagues readily and without reserve."

BSN 1986 University of South Carolina, Columbia, SC MS-Nursing 2009 Towson University, Towson, MD Post-Baccalaureate Certificate in Nursing Education 2009 Towson University, Towson, MD Experience: 29 years in mainly acute care adult health (medical-surgical, cardiac, and management); focus the past 9 years in nursing education-undergraduate with a specialty in simulation.

Andrew Schiff

Best Overall Faculty Member, Accounting
Best Overall Faculty Member, Business

TRIDENT TECHNICAL COLLEGE

Jane Benton

Most Helpful to Students, Health Sciences and Nursing
Best Overall Faculty Member, Health Sciences and Nursing

Lori Fischer

Best Overall Faculty Member, Health Science

"Over the past year I have watched Lori work endless hours on the accreditation renewal for the PTA Program without complaint. The whole time she continued to serve the needs of her students with a smile. She also managed to assist me with questions pertaining to college procedures. I feel that we are very fortunate to have such a dedicated faculty member and colleague."

I gradated from Medical University of South Carolina in 1986 with my Bachelor of Science in Physical Therapy. I worked at Roper Hospital as a staff PT, Team Leader and Center Coordinator of Clinical Education. In 1993, I began my career in academia at Trident Technical College as the Academic Coordinator of Clinical Education for the PTA program until 2002 when I became the Program Coordinator. In 1996, I once again graduated from MUSC with a Master of Science in Health Professions Education. I have also worked PRN in home health, outpatient PT as well as in a skilled nursing facility. I was the co-author and project PT of the PHILE II grant from 2002 until 2005. I served as the secretary of SCAPTA from 2005 until 2007 and have been a member of SCAPTA and the APTA since 1986. I have served on Trident Technical College's Faculty Council since 2001.

Krista Gentry

Best Overall Faculty Member, Allied Health
Best Researcher/Scholar, Allied Health
Best Teacher, Allied Health
Best Researcher/Scholar, Health Sciences and Nursing

Melicent Middlebrook

Most Helpful to Students, Accounting

Suzie Walters

Best Teacher, Health Sciences and Nursing

Deborah White

Most Helpful to Students, Allied Health

TROY UNIVERSITY

Mary Ann Hooten

Best Overall Faculty Member, Psychology

"Mary Ann Hooten has several years of experience as a teacher, spends office time with students, helps them with research and is very productive as the best teacher."

TULSA COMMUNITY COLLEGE

Phoebe Baker

Best Overall Faculty Member, Psychology
Best Researcher/Scholar, Psychology
Best Teacher, Medicine
Best Overall Faculty Member, Medicine

Cathy Furlong

Most Helpful to Students, Psychology
Best Researcher/Scholar, Medicine

Eric Lange

Best Researcher/Scholar, Education

Sarah Plunkett

Best Overall Faculty Member, Nursing

"She is always on the cutting edge of current Best Practices. She is instrumental in designing and helping the faculty write our new curriculum, utilizing the QSEN competencies and other standards for ACEN accreditation. As an instructor, she teaches the use of techniques to best help students retain the material and think their way into becoming excellent nurses. She is the best nursing instructor I have ever seen!"

Krena White

Best Teacher, Psychology
Most Helpful to Students, Medicine

TUSCULUM COLLEGE

DiAnn Casteel

Best Teacher, Education
Most Helpful to Students, Education
Best Overall Faculty Member, Education

DiAnn B. Casteel, associate professor of education, has taught in Tusculum's graduate education program for eight years and will begin teaching in the Residential College educational classes in the fall 2004 semester. She has 30 years of experience as a teacher and administrator in the Greene County School System. She has taught in grades K–12, in Chapter 1 classes, and as an evening and summer instructor for high school and adult students in a program designed to allow students to earn a high school diploma through individualized study. She has also served as assessment coordinator/career counselor for the Single Parent/Displaced Homemaker Project at the Greeneville-Greene County Center for Technology and was coordinator for Project CHOICE at the school. She has also served as special education coordinator, assistant principal, and principal of elementary schools. Dr. Casteel has also taught as an adjunct faculty member at Virginia Intermont College, Walters State Community College, and for Tusculum in the early 1990s. Dr. Casteel earned her doctorate in educational leadership and policy analysis at East Tennessee State University and also received an M.A. as a reading specialist from ETSU. She is also a Certified Novell Administrator. She has written book review articles for professional journal, "The School Administrator," and presented a paper to the Mid-South Educational Research Association. She has also served a grant reviewer for the U.S. Department of Education and the Tennessee Department of Education. Active in the community, Dr. Casteel has served and provided leadership in such organizations as 4-H, the Girl Scouts, and USS Greeneville Inc, the organization that helped work for a naming of a U.S. Navy submarine for Greeneville and provides support for the sailors on the boat. She was a member of Rotary International's first Women's Group Study Exchange Team to visit India. As a result of this trip, she began to contribute books and materials that led to the formation of a Library/English Medium School in Chirala, India, and also obtained sponsors for more than 20 cases of materials and books for use in the library/school. She has also been involved in many activities involving horses, one of her personal interests. She also enjoys camping, reading, swimming, and cooking.

UNIVERSITY AT BUFFALO (SUNY BUFFALO)

Nathanael D Carbrey

Most Helpful to Students, Business

Michael Duffey

Best Overall Faculty Member, Medicine

Debbie Grossman

Most Helpful to Students, Marketing

Professor Grossman worked in the field of marketing for several years before beginning her career in teaching. She continues to work as a marketing consultant while teaching full time. Her experience includes advertising, market research, product development/management and brand management. Her current teaching responsibilities include both undergraduate and graduate level marketing courses.

Arun Jain

Best Overall Faculty Member, Marketing
Best Researcher/Scholar, Marketing
Best Teacher, Marketing

Dr. Jain has been instrumental in "internationalizing" the school's curriculum, and has extensive international experience. In 1979, he was selected as the Senior Fulbright Hayes Scholar to the European Institute of Business Administration (INSEAD), France, to which he returned in 1987. Dr. Jain has played a key role in helping the former Soviet Union and other Eastern European countries move towards privatization and market-based economies. He has traveled, given lectures or taught executive education programs in Japan, China, Taiwan, Korea, Thailand, Singapore, Malaysia, Indonesia, India, Italy, France, Belgium, Germany, England, Scotland, Hungary, the former Soviet Union, Latvia and Mexico. Dr. Jain was selected as "Professor of the Year" by MBA students in 1989 and was honored by SUNY Chancellor D. Bruce Johnstone in 1990 with the "Chancellor's Medal" for excellence in teaching. He is author or co-author of more than 60 articles published in international journals. He serves as a consultant to major international corporations.

Kathleen Nesper

Best Teacher, Business

Prior to joining the University at Buffalo's faculty, Professor Nesper was a practitioner in the fields of public accounting and corporate accounting for over 17 years. In addition to her audit experience at two international public accounting firms, her corporate accounting experience includes: forecasting and operations analysis, budget and strategic business plan development and analysis, cost analysis (including Activity-Based Costing), overseas/joint venture analysis, product line profitability analysis, investment and capital budgeting analysis, and future product program analysis. Professor Nesper's current teaching responsibilities include the areas of cost and managerial accounting at both the undergraduate and graduate levels. She serves as faculty advisor for the School of Management's Accounting Association.

Richard Rabin

Most Helpful to Students, Medicine

Robert H Sanborn

Best Overall Faculty Member, Business

Robert H. Sanborn is an associate professor in the Accounting and Law Department. He has a doctoral degree from the University of Georgia, an MBA from Boston University and a bachelor's degree from Johns Hopkins University. He also attended the University at Buffalo for 30 credits of accounting coursework.

Prior to joining UB, Sanborn served on the faculty of the Robins School of Business at the University of Richmond since 1992 as an assistant professor and associate professor. He held visiting associate professor roles at Massey University in New Zealand in 1995 and the University at Buffalo in 2001–02. From 1988–92, he was an assistant professor in the McIntire School of Commerce at the University of Virginia. He also worked from 1973–77 as a staff accountant at P.G. Bixby & Co. (now Deloitte & Touche).

Inho Suk

Best Researcher/Scholar, Business

Claude Welch

Best Overall Faculty Member, Political Science
Best Researcher/Scholar, Political Science
Best Teacher, Political Science
Most Helpful to Students, Political Science

His academic specializations include African politics, the roles of armed forces in politics, and human rights. Welch has published widely on these subjects, with fourteen books and close to forty chapters and articles in academic journals. In 2006, he received the first-ever Lifetime Achievement Award given by TIAA-CREF and the SUNY Research Foundation. During his career at Buffalo, which started in 1964, Welch has chaired or served on ~100 dissertation committees. He is known for his high-quality teaching, having received awards from Political Science students, the Undergraduate Student Association, and the Chancellor's Award for Excellence in Teaching.

UNIVERSITY OF AKRON

James Sperling

Best Teacher, Political Science

Jim Sperling is professor of political science at the University of Akron. He has taught at Akron since 1988. Prior to that time he held appointments at Davidson College and the James Madison College, Michigan State University. He teaches World Politics and Government (the introductory course to comparative politics and international relations), International Politics and Institutions, and Comparative Security Policy. His publications have explored various facets of German foreign economic and security policy over the course of the postwar period as well as the problem of the new security agenda, global security multilateralism, and regional security governance in the contemporary international system. His publications include the coauthored NATO's Trajectory into the 21st Century (Palgrave 2012) and EU Security Governance (Manchester University Press 2007) and the coedited NATO after Sixty Years (Kent State University Press), National Security Cultures: Patterns of Global Governance (Routledge 2010), European Security Governance: The European Union in a Westphalian World (Routledge 2009), and Global Security Governance (Routledge 2009). He is editor of the Handbook on Governance and Security (Edward Elgar 2014) and is currently coauthoring with Mark Webber NATO: What's wrong with it? How to fix it? (Polity Press).

He is married to Joy Sperling, professor of art history at Denison University, president of the American Culture Association/Popular Culture Association, and author of Fragonard's Shoe: the sculpture of Jude Tallichet. His daughter, Victoria, graduated from Wellesley College in 2011 and is currently a PhD candidate at the Johns Hopkins University School of Medicine.

UNIVERSITY OF ALABAMA

Norman Baldwin

Best Researcher/Scholar, Public Affairs, Political & Policy Sciences
Best Teacher, Public Affairs, Political & Policy Sciences

Steve Borelli

Most Helpful to Students, Public Affairs, Political & Policy Sciences
Best Overall Faculty Member, Public Affairs, Political & Policy Sciences

Susan Gaskins

Best Researcher/Scholar, Nursing

Dr. Gaskins received a BSN from the Medical College of Virginia (Virginia Commonwealth University), an MPH from the University of Pittsburgh, and a PhD from the University of Alabama at Birmingham. Her background is Community/Public Health Nursing and she has worked in Public Health, Home Health, and Hospice Care. She teaches NUR 731: Philosophical, Theoretical and Conceptual Foundations for Advanced Practice Nursing and NUR 735; Population Health in the DNP program, and NUR 696: Doctoral Research Seminar in the EdD program. Her research focuses on psychosocial issues related to HIV/AIDS, especially stigma and disclosure. She recently completed a National Institutes of Health (NIH) funded study, Impact of AIDS Disclosure on Rural African American Men. Currently. she is senior investigator on a study funded by the Center for Disease Control and Prevention, Faith-based Anti-Stigma Initiative Towards Healing HIV/AIDS (Project FAITHH). This randomized clinical trial compares the effectiveness of an anti-stigma faith-based curriculum, a standard HIV/AIDS curriculum and the use of literature in decreasing AIDS related stigma and increasing HIV/AIDS knowledge in rural African American churches. Additionally she is co-investigator on a study, End of Life Needs of People Living with HIV/AIDS in Appalachia in Tennessee and Alabama funded by the National Institute of Nursing Research at NIH. This study is being conducted in collaboration with faculty at the University of Tennessee in Knoxville.

Dr. Gaskins is an AIDS Certified RN (ACRN) and is an active member of the Association of Nurses in AIDS Care (ANAC). Presently she serves on their Research and Annual Conference committees. Other professional memberships include Sigma Theta Tau International, the

American Nurses Association, and the Southern Nursing Research Society. Additionally she serves on the Board of Directors of West Alabama AIDS Outreach. She has been the recipient of the Capstone College of Nursing's Board of Visitors Outstanding Commitment to Teaching Award, and the Alabama League for Nursing Lamplighter Award. She is a fellow of the American Academy of Nursing and serves on their Expert Panel on Emerging and Infectious Diseases.

Frannie James

Best Teacher, Languages
Most Helpful to Students, Languages

Frannie has been involved in International Education at The University of Alabama since 1990. For much of that time, she has been at the ELI. However, she has also taught first-year composition to UA students. She currently teaches Introduction to Global Studies (CIP 200) and Culture and Human Experience (IHP 105) at UA. Each of these courses focuses on fostering intercultural awareness in degree-seeking UA students, many of whom are international students. Her current role at the ELI is coordinating Culturally Speaking. This is a class in which degree-seeking UA students meet with small groups of ELI students twice a week for discussions focused on issues of culture.

Paige Johnson

Best Overall Faculty Member, Nursing

Dr. Johnson has been teaching at the Capstone College of Nursing since 2005. She received her BSN from the University of Alabama Capstone College of Nursing in 1996 and her Master of Science in Nursing with an emphasis in case management in 2001. In 2013, Dr. Johnson earned her Ph.D. from the University of Alabama/University of Alabama at Birmingham in Health Education and Health Promotion.

Dr. Johnson began her nursing career in critical care and also has extensive experience in clinical research conducting trials in the areas of hypertension, diabetes, osteoarthritis, depression, bipolar disease, and post-operative pain management. She received certification as a certified clinical research coordinator in 2003. Most recently, Dr. Johnson's clinical focus has been in community and public health nursing.

Dr. Johnson teaches in the undergraduate program. Previously, she has taught Human Pathophysiology, Adult Health Nursing and in the RN Mobility program. Dr. Johnson presently teaches in Professional Nursing Practice: Community Health and Inquiry for Evidence-Based Practice in Nursing.

Fran Oneal

Best Overall Faculty Member, International Studies
Best Researcher/Scholar, International Studies

Roy Sherrod

Best Teacher, Nursing

Stephanie Wynn

Most Helpful to Students, Nursing

Dr. Wynn received a BSN from the University of Alabama Capstone College of Nursing. She also earned a MSN degree in Rural Case Management from the University of Alabama. Later, she earned a Post-Master's Certificate as a Psychiatric Nurse Practitioner and her DNP from the University of South Alabama. She is certified by the American Nurses Credentialing Center as a Psychiatric and Mental Health Nurse and a Family Psychiatric and Mental Health Nurse Practitioner. Professional memberships include American Nurses Association, National League for Nurses, and Sigma Theta Tau.

Dr. Wynn is the course leader for NUR 374 Professional Nursing Practice: Mental Health and NUR 102: Introduction to Nursing. She is responsible for classroom and clinical supervision of undergraduate nursing students. In addition to her faculty responsibilities, Dr. Wynn is the chair of the Student Life committee.

UNIVERSITY OF ALASKA FAIRBANKS

Ken Abramowicz

Most Helpful to Students, Accounting

Cameron Carlson

Most Helpful to Students, Business

Nicole Cundiff

Best Teacher, Business

Michael Davis

Best Researcher/Scholar, Accounting

Liz Ross

Best Teacher, Business

Liz Ross (legal name is Betty) is one of 64 Alaskan Natives with a doctorate and the only Alaskan Native with a doctorate in Business. Her family is from the rural village of Unalakleet, Alaska. She has been teaching college courses since 1999 and has been actively involved with students at the University of Alaska Fairbanks since 2004. As a former CEO of an Alaska Native Corporation and a volunteer mentor for rural villages, she understands the importance of education as a form of empowerment. Liz provides real-world experience into the classroom.

Liz is the faculty advisor for the Native Alaskan Business Leaders (NABL), the UAF Hillel Jewish Student Organization, and the MBA Program Director. Ross' first visit with students at UAF was in 2004 when she was a CEO speaker addressing NABL. NABL's underlying principles of culture, service, education, and leadership experience are areas that Liz utilizes in her teaching philosophy and mentoring philosophy.

UNIVERSITY OF ARIZONA

Sandy Klasa

Best Researcher/Scholar, Finance

Maha Nassar

Best Researcher/Scholar, Near Eastern Studies

Maha Nassar is a historian of the modern Middle East, with a focus on the twentieth-century Arab world. Her research explores intellectual constructs of ethnicity, culture and national identity in the Arab Middle East, analyzing how writers in the 1950s and 1960s utilized mass media, especially newspapers and literary journals, to challenge the dominant narratives of the nation-state. Her dissertation, "Affirmation and Resistance: Press, Poetry and the Formation of National Identity among Palestinian Citizens of Israel, 1948–1967," is the first systematic study of the Palestinian-Israeli popular press and its seminal role in creating an emerging sense of nationhood among Palestinian citizens of Israel. It also explores the social, cultural and political connections between Palestinians in Israel and the broader Arab world, despite their political and geographic isolation. She has conducted fieldwork in Egypt, Jordan, Lebanon and Israel/Palestine, and she is currently working on a book about the cultural history of Palestinians in Israel.

Spike Peterson

Best Teacher, Political Science

V. Spike Peterson is a Professor of International Relations in the School of Government and Public Policy, with courtesy appointments in the Department of Gender and Women's Studies, Institute for LGBT Studies, Center for Latin American Studies, and International Studies. During the spring of 2014, she was a Senior Research Fellow hosted by the Department of Geography at Durham University, England, and during 2008–2011 was an Associate Research Fellow of the London School of Economics. Her book publications include Global Gender Issues in the New Millennium (2010) and two earlier editions of Global Gender Issues (1999, 1993) with Anne Sisson Runyan; her own A Critical Rewriting of Global Political Economy: Integrating Reproductive, Productive and Virtual Economies (2003); and Gendered States: Feminist (Re)Visions of International Relations Theory (1992), which she contributed to and edited. She has published more than 75 journal articles and book chapters, most recently on informalizations of work in relation to structural inequalities and their corollary insecurities worldwide; global householding; gendering war and its economies; and queering marriage, citizenship and states/nations.

Peterson has been awarded a Rockefeller Bellagio Scholarly Residency (2008), a Udall Center Public Policy Fellowship (2007), a Fulbright Scholarship for research in the Czech Republic (1997, declined) and a MacArthur Foundation Research and Writing Grant (1996). She has held the position of Visiting Research Scholar at Durham University (2014), the London School of Economics (2007, 2008), University of Göteborg (2000), University of Bristol (1998) and the Australian National University (1995). She has guest lectured at numerous universities in the United States, Canada, Australia and Europe and been an invited speaker at international conferences in Europe, Asia and Latin America.

Peterson currently serves on the editorial boards of International Feminist Journal of Politics; International Theory; Globalizations; Journal of Women, Politics and Policy; New Political Science; Politics & Gender; and Perspectives: The Review of International Affairs. She received the SBS Dean's Award for Excellence in Upper Division Teaching (2014), the Magellan Circle Award for Teaching Excellence (2008) and the Provost's General Education Teaching Award (2001) at the University of Arizona, as well as the national Mentor Award of the Society for Women in International Political Economy (2000). She regularly teaches POL 150 (Politics of Difference: Race/Ethnicity, Class, Gender and Sexualities), POL 360 (Global Political Economy), undergraduate courses in Politics and Theory that are cross-listed with Gender and Women's Studies, and graduate seminars on contemporary social theory and global political economy.

Patricia Sparks

Best Teacher, Health Science

UNIVERSITY OF ARKANSAS FAYETTEVILLE

Margaret Reid

Best Overall Faculty Member, Political Science

Margaret F. Reid (Ph.D., University of Oklahoma, 1986) is also the chair of the political science department. Her research focuses on gendered workplaces, complexities involving the implementation of multi-actor policy partnerships, and sustainable community development (domestic and international). Her current projects center on strategic management issues involving nonprofit organizations and on cross-sectoral board of directors interlocks.

Reid is the creator and director of See Change (a 5–6 month nonprofit certificate for senior nonprofit managers) http://seechange.org. She is also engaged in various community development and nonprofit training projects. Reid is a member of the University of Arkansas's Women's Giving Circle.

A co-authored book with colleagues Kerr and Miller on Glass Walls and Glass Ceilings: Women's Representation in State and Municipal Bureaucracies. She has been author or co-author in research that has been published in among others Sex Roles, Journal of Women, Politics & Policy, Administration & Society, Public Administration Review, American Review of Public Administration, Review of Public Personnel Administration, Urban Affairs Review, State and Local Government Review, Journal of Management Information Systems, Information and Management, Journal of Asian and African Studies, International Journal of Tourism Research and in numerous edited works.

UNIVERSITY OF BALTIMORE

Ed Gibson

Best Researcher/Scholar, Public Affairs, Political & Policy Sciences
Best Overall Faculty Member, Public Affairs, Political & Policy Sciences

Being a latecomer to teaching heightens my appreciation for the exchange of ideas in the classroom. I spent the early part of my career as a programmer, supervisor and manager within the information technology industry, working for a company called EDS, better known for its famous founder, Ross Perot, who ran for president in 1992 and 1996.

Getting my M.B.A. from the University of Maryland enabled me to swap the language of bits and bytes for that of dollars and cents by joining a government contractor, Performance Engineering Corp., where I consulted with federal agencies, chiefly the federal courts, for more than a dozen years. Practicing the arcane art of cost-benefit analysis and related programmatic and financial techniques introduced me to bureaucratic decision-making processes. I gained a deep respect for the capabilities of public managers, who often labored under tremendous deficits of resources and excesses of guidance, yet managed to conduct the public's business. In particular, visiting federal courts, which fairly hummed with the steady disposition of cases and dispensing of justice, and interviewing judges and court executives impressed upon me the reality of high-performance public organizations.

Daily involvement with federal agencies forced me to confront a new set of issues stemming from missions undertaken in the public interest, in sharp contrast to the for-profit orientation of the IT industry, where I'd spent 15 years. Accordingly, I set about gaining a theoretical grasp of public administration by getting a Ph.D. from Virginia Tech—a 10-year endeavor! Joining the faculty of the University of Baltimore represents a fresh opportunity to tackle the challenges of public service, without the need to stoke the "bottom line" as a member of the "shadow government," with its attendant consequences of downsizing and contracting out.

I feel most fortunate to be part of the UB community. Aside from a business degree from the University of North Carolina, all of my education has been obtained while working. The challenges of the part-time student are second nature to me. While pleased to take my position in front of the class, I remember what it feels like from the other side and do my best to keep the learning experience mutual, engaging and light.

I live in Virginia's Shenandoah Valley with my wife of 25 years and our daughter, whenever she's home from college.

Marilyn Oblak

Best Researcher/Scholar, Management

Marilyn Oblak is a professor of Decision Science and Associate Dean of the Merrick School of Business. Dr. Oblak has served the University in several capacities including as Director of the Helen P. Denit Honors Program. She is currently serving on the Sylvan Beach Foundation Board.

UNIVERSITY OF CALIFORNIA BERKELEY

Ron Amundson

Best Researcher/Scholar, Health Sciences and Nursing

UNIVERSITY OF CALIFORNIA DAVIS

Dean Simonton

Best Overall Faculty Member, Psychology
Best Researcher/Scholar, Psychology
Best Teacher, Psychology
Most Helpful to Students, Psychology
Best Researcher/Scholar, Medicine
Best Teacher, Medicine
Most Helpful to Students, Medicine
Best Overall Faculty Member, Medicine

UNIVERSITY OF CALIFORNIA LOS ANGELES (UCLA)

Peter Aranella

Best Overall Faculty Member, Law

Peter Arenella teaches Criminal Law, Criminal Procedure, and seminars on moral agency and criminal law excuse theory. In 1999 he received the School of Law's Rutter Award for Excellence in Teaching. He is a nationally recognized criminal law and procedure scholar, writing about the relationship between criminal and moral responsibility by exploring competing conceptions of criminal culpability and moral agency at work in immaturity and mental disability defenses (e.g., insanity, diminished capacity, mental retardation). His moral agency work in philosophical and legal journals has generated considerable commentary and is frequently cited in ongoing debates about the justifications for criminal punishment. He has also written extensively on the privilege against self incrimination and grand jury practices.

Professor Arenella clerked for the Chief Justice of the Massachusetts Supreme Judicial Court and practiced criminal law as both a public defender and private counsel. He then taught at Rutgers, the University of Pennsylvania, and Boston University, where he won that university's prize for excellence in teaching.

State and federal courts, including the U.S. Supreme Court, have cited Professor Arenella's articles, and congressional committees have sought his advice and testimony. He gives annual lectures on Supreme Court decisions to federal and state appellate judges and speaks regularly on criminal law issues to civic groups, attorneys, judges, law faculty, and students across the country. He has worked as a legal consultant for ABC News, National Public Radio, and in Los Angeles KNX radio and KTLA Channel 5, and he achieved national prominence for his media commentary on the O.J. Simpson and other trials. He is a fanatical Boston Red Sox fan.

UNIVERSITY OF CALIFORNIA MERCED

Meaghan Altman

Best Teacher, Psychology

Jeff Gilger

Best Researcher/Scholar, Psychology

Alex Khislavsky

Best Teacher, Psychology
Most Helpful to Students, Psychology
Most Helpful to Students, Medicine
Best Overall Faculty Member, Medicine

Jim McDiarmid

Best Researcher/Scholar, Medicine
Best Teacher, Medicine

Michael Spivey

Best Overall Faculty Member, Psychology

Education: Ph.D., 1996 — University of Rochester M.A., 1995 — University of Rochester B.A., 1991 — University of California, Santa Cruz Awards: 2010 - William Procter Prize for Scientific Achievement, Sigma Xi

Research Interests: Professor Spivey's research interests include:

Interaction between language and vision Sentence processing Word recognition Visual attention Visual memory Eye movements Computational modeling Using an eyetracker and simultaneously recording the streaming x, y coordinates of the subject's computer mouse movements, he gets an online measure of some of the probabilistic representations (or "tentative interpretations") that get computed in real-time as the subject attempts to integrate various sources of visual and/or linguistic information.

One finding from this study is that spoken word recognition and syntactic parsing are immediately influenced by relevant visual context — thus compromising strict modular theories of language processing. This, and related, work is described in his book, The Continuity of Mind (2007).

Books: The Continuity of Mind

UNIVERSITY OF CALIFORNIA SAN DIEGO

Jonathon Hill

Most Helpful to Students, Business

Richard Kolodner

Best Researcher/Scholar, Medicine

The research performed in the Kolodner lab is focused on three areas of genetics. First, they use the yeast Saccharomyces cerevisiae to study the genetics of DNA mismatch repair and to identify the genes and pathways that prevent genome instability. Second, they are actively studying the human genetics of defects in mismatch repair genes and genes that prevent genome instability. Finally, they initiated several successful collaborative positional cloning projects that have resulted in the identification of the genetic basis of two human diseases, Cold Induced Auto-inflammatory Syndrome due to inherited mutations in the NALP3 gene and Congenital Tufting Enteropathy due to inherited mutants in the gene encoding EpCAM. The lab has identified a number of human mismatch repair genes and demonstrated that inherited mutations in two of these genes cause hereditary non-polyposis colon cancer, now called Lynch Syndrome, a common inherited human cancer susceptibility syndrome. They also demonstrated the genetic and epigenetic basis of mismatch repair defects in sporadic cancers. This work resulted in the development of now widely used diagnostic tests used in cancer treatment and diagnosis. More recently they have developed evidence that human genes that are homologs of the S. cerevisiae genome instability genes are also mutated in cancer and are candidate tumor suppressor genes.

Joel Sobel

Best Overall Faculty Member, Economics
Best Researcher/Scholar, Economics

Joel Watson

Best Teacher, Economics
Most Helpful to Students, Economics
Best Researcher/Scholar, Business
Best Teacher, Business
Best Overall Faculty Member, Business

Environmental and Resource Economics (Primary) Law & Economics Microeconomic Theory Research Interest: Economic Theory, in particular the analysis of contracts and institutions Research Statement: Joel Watson is an economic theorist who specializes in developing game theoretic models of contractual settings. His work addresses topics in several applied areas, including labor and macroeconomics, international economics, and the law. Watson is also the CEO of EconJobMarket.org, a non-profit company that serves as the central repository for advertisements, applications, and letters of recommendation for the academic economics job market. Watson has served as an editor and/or board member for the B.E. Journal of Theoretical Economics, Games and Economic Behavior, and the International Journal of Game Theory. His book Strategy: An Introduction to Game Theory is a popular text for advanced undergraduates and first-year graduate students.

UNIVERSITY OF CALIFORNIA SANTA BARBARA

Tom Dunne

Best Overall Faculty Member, Environmental Science & Mgmt.

Bren School of Environmental Science & Management

Steve Wiener

Best Researcher/Scholar, Political Science

UNIVERSITY OF CENTRAL ARKANSAS

Adam Bruenger

Best Researcher/Scholar, Kinesiology & Physical Ed.
Best Researcher/Scholar, Education

Shawn Charlton

Best Researcher/Scholar, Psychology & Counseling
Most Helpful to Students, Psychology & Counseling

Ellen Epping

Best Overall Faculty Member, Education

http://uca.edu/kped/facultystaff/ellen-epping-2/

Kimberly Eskola

Best Overall Faculty Member, Kinesiology & Physical Ed.
Best Teacher, Kinesiology & Physical Ed.
Most Helpful to Students, Education

Kim received a Bachelor's of Arts degree in Political Science from the University of Central Arkansas (UCA). After being employed with the American Heart Association as the Marketing Consultant for Schools, she pursued her Master of Science degree in Kinesiology at the University of Central Arkansas. She was the Health Director/Member Services Director for the Little Rock Athletic Club from 1995–2005. She is now a Clinical Instructor II/Coordinator in the Bachelor of Science in Kinesiology program at UCA. She is an ACSM Certified Exercise Physiologist and ACSM Health/Fitness Director. When she is not teaching and advising students in the Kinesiology program, she coaches and officiates youth and high school soccer. She holds a national D license to coach soccer and is a grade 7 soccer official. She has also completed four Ironman triathlons and over 40 marathons in 26 states.

William Lammers

Best Overall Faculty Member, Psychology & Counseling
Best Teacher, Psychology & Counseling

I earned my Bachelor's degree in Psychology from San Diego State University and my Masters and PhD in Experimental Psychology from Bowling Green State University. I am currently Professor of Psychology at the University of Central Arkansas. My teaching interests are quite varied but focus on Introductory Psychology, Statistics, and Research Design. Research interests focus on the scholarship of teaching, with a focus on the importance of student-instructor rapport. I and a colleague are currently working on a project to investigate rapport, goal commitment, and student learning in Statistics. I've received our University's Teaching Excellence Award and an Award for Innovative Excellence in Teaching, Learning, and Technology (International Conference on College Teaching and Learning).

Julie Meaux

Best Overall Faculty Member, Nursing

Betty Sessum

Most Helpful to Students, Kinesiology & Physical Ed.
Best Teacher, Education

UNIVERSITY OF CENTRAL FLORIDA

Paul Desmarais

Best Overall Faculty Member, Nursing
Best Teacher, Nursing

Born: September 30, 1947 Manchester N.H. Married: Cecile Gibeau Children: 4 Graduated: St. Anselm College 1973 BA degree in English. Ohio University - Zanesville 1978 - ASN. St. Anselm College 1984 - BSN. University of Massachusetts - Lowell 1994 MSN - Administration. University of Massachusetts - Lowell 2002 PhD Nursing/Health Promotion

April Fisher

Best Teacher, Public Administration

"She is extremely knowledgeable about her area."

Dr. Fisher has over 15 years of planning and zoning experience in both the public and private sector. She is an expert in zoning analysis, and land use reports and applications.

Debbie Hahs-Vaughn

Best Overall Faculty Member, Education

Debbie Hahs-Vaughn joined the Department of Educational and Human Sciences within the College of Education and Human Performance at the University of Central Florida in August 2003. She completed her Ph.D. in educational research (minors in applied statistics and educational foundations) at the University of Alabama in May 2003. Dr. Hahs-Vaughn earned a Bachelor's of Fine Arts in graphic design and Master's of Business Administration with an emphasis in marketing from Southwest Missouri State University (now known as Missouri State University) (1990 and 1995 respectively). Other past career experiences within higher education include serving as Manager of Graduate Student Services and Manager of Student Support Services (a federally funded TRIO program) both also at the University of Alabama. Additional career experiences that have helped to mold the research and teaching interests of Dr. Hahs-Vaughn include serving as Director of Educational Services for the Alabama Credit Union League, Marketing Director, and graphic designer.

Sigrid Ladores

Best Teacher, Nursing
Most Helpful to Students, Nursing
Best Teacher, Health Sciences and Nursing

Patricia Leli

Best Overall Faculty Member, Nursing
Most Helpful to Students, Nursing
Most Helpful to Students, Health Sciences and Nursing
Best Overall Faculty Member, Health Sciences and Nursing

Irene B. Pons

Most Helpful to Students, Law

At the age of two, Irene Beatriz Pons traveled from the small island of Puerto Rico with her mother and siblings on a big plane to the United States. As a first generation immigrant, life was not easy for her mother, but that did not impact her ability to fill her children with dreams, participate in their education, encourage their creativity and instill a sense of pride within them. In grade school, Irene told her family that someday she was going to be a lawyer so that she could take care of them and help others. Irene is a true testament to the American Dream: growing up on welfare while her Mother worked odd jobs babysitting and cleaning houses to make ends meet, yet Irene always knew that someday she would be fortunate enough to repay the sacrifices her mother made.

Growing up in Orlando, Irene worked her way through her University of Central Florida (UCF) bachelor's degree as a Disney cast member. Her most notable role was as Mary Poppins in the Magic Kingdom. Even then, her family stuck together, all finding jobs at the world renowned company. After graduation, Irene received her J.D. from Nova Southeastern University and quickly moved back to Orlando to become a legal professor at her undergraduate alma mater, UCF.

For nearly a decade, Irene has been a both a full-time Family Law attorney at her own firm and a full-time professor at UCF. Her classes include Family Law, Civil Litigation, Law & the Legal System, Legal Writing, Legal Research, Torts, and Entertainment Law. She also sits on various committees at the university and is the secretary of the Hispanic Bar Association's Orlando chapter, which has a growing membership of over 180 lawyers, judges and paralegals. Irene was also appointed by the Mexican consulate of Orlando to assist with immigrant Family Law issues.

Attorney Irene Pons Meyers has been happily married for two years. True to form, the couple was married at Disney's Beach Club surrounded by family and friends. What many people do not know about Irene, is that she was an art minor, has worked as a graphic designer and enjoys designing plates, key chains and mosaic tiles. When she's not teaching, creating art, working with clients or spending time with her family, Irene represents domestic violence victims for Harbor House and rescues Great Danes.

Irene's most rewarding cases involve adoptions by step-parents and grandparents, paternity rights and child support. "There are no winners in Family Law, but there is a sense of satisfaction when a case comes to a close giving people an opportunity to move on," says Irene. "I'm not there to tell you what you want to hear, but I am there to represent you zealously, be diligent, prompt, and do everything in my power to achieve a positive outcome."

Attorney Irene Pons Meyers' compassion and unprejudiced nature allows her to provide clients with the time and personalized attention that they deserve to successfully guide them through some of the most difficult times in their lives. She has her clients' best interests at heart, which is exemplified in her confident, composed, yet boldly aggressive representation strategy. She also has a great sense of humor and often jokes, "As much as I love being a Lawyer, I look forward to retirement as a professor." Her influential life experiences and dedication to her profession are what make Irene Pons Meyers one of the top up and coming lawyers to watch in Orlando, Florida.

Stephen Talbert

Best Researcher/Scholar, Nursing
Best Researcher/Scholar, Health Sciences and Nursing

UNIVERSITY OF CENTRAL OKLAHOMA

Kathy Brown

Best Teacher, Education

Ed Collins

Best Researcher/Scholar, Education

Degrees: Doctor of Philosophy: Educational Psychology, University of Oklahoma, 1976 Masters of Teaching: Counseling and Guidance, Moorehead State University & East Central Oklahoma State College, 1967 Bachelors of Education: Elementary Education/Special Education, Gonzaga University, 1966

General Information: Teacher training associated with Emotional Disturbance, Learning Disabilities, Tourette Syndrome, Autism, and Traumatic Brain Injury, Response to Intervention, and Positive Behavioral Interventions & Supports

Awards: UCO, College of Education, First Recipient of the Elizabeth H. Threat Diversity Initiative Award, 2006 Oklahoma Federation CEC, Special Educator of the Year Award, 2001 CEC Wayne Jeans Chapter, Special Education Award, 2000 Distinguished Educator Award, National Tourette Syndrome Association, 1999 Walraven Award, CEC Wayne Jeans Chapter 446, 1996.

Pat Couts

Most Helpful to Students, Education

Bryan Duke

Best Teacher, Education

Dr. Bryan L. Duke is the Assistant Dean & Director of Educator Preparation and Professor of Educational Sciences, Foundations & Research at the College of Education and Professional Studies at the University of Central Oklahoma. He has been in education for twenty-one years, serving his first eleven as a junior high and high school English teacher, as well as an administrator. In 2002, he was selected as Moore Public Schools' (Moore, OK) District Teacher of the Year. Currently, he teaches graduate courses in advanced educational psychology, advanced developmental psychology and managing secondary classrooms at the University. He served as president of the Oklahoma Association of Teacher Educators and as the Third Vice President of the Oklahoma Council of Teachers of English. Currently, he is the historian for the Oklahoma Association of Supervision and Curriculum Development. He has been a researcher with the Oklahoma A+ Schools research team for seven years. Additionally, he has written curriculum and participated in national institutes for teaching Shakespeare with the Folger Shakespeare Library (Washington, D.C.), Shakespeare and Company (Lenox, MA), and the Globe Theatre (London, England).

Kristi Frush

Best Overall Faculty Member, Education

Dr. Kristi Archuleta Frush joined the University of Central Oklahoma as an Assistant Professor in the Department of Adult Education and Safety Science in January of 2011. She has been teaching in higher education environments for over 12 years at both the undergraduate and graduate levels and her previous work experience has primarily been in the areas of marketing, management, instructional design, and program development.

Her university-wide service includes faculty governance, serving as a member of Faculty Senate from 2012–2014, where she co-wrote a resolution to help support Adjunct Faculty, was Chair of the Adjunct Affairs Committee, member of the Faculty Handbook Review Committee, and member of the E-Learning Advisory Committee. She has also served as a Co-Chair of the Prior Learning Assessment Action Team, a member of the Adjunct Faculty Action Team, is a Mentor for new faculty, and an online course evaluator for CeCe. She recently joined the UCO Latino Faculty and Staff Association and was also inducted as an inaugural UCO Service Learning Scholar.

College-wide, she is on the Global and Cultural Diversity Committee and previously was on the History Editorial Committee and Social Committee. At the Department level, Dr. Frush served as the Reach Higher Academic Program Coordinator for three and a half years, helping the program grow from 89 to 250 adult learners.

She is the Co-Chair of the Human Resource Development (HRD) Special Interest Group of the Commission of Professors of Adult Education (CPAE), serves on the membership committee of the Association of Adult and Continuing Education (AAACE), is on the steering committee of the Research to Practice Conference (R2P) and is a member of the Oklahoma Chapter of the Association for Talent Development (ATD), among other professional associations. Dr. Frush has given 40 presentations in 15 states and is a published academic and commercial author. She enjoys editing and reviewing submissions for publication and has a developing research agenda in the areas of diversity distance education, legal issues in higher education, self-directed and transformative learning.

Barbara Green

Most Helpful to Students, Education

Barbara Green, Ph.D (Oklahoma State University, 1992) has been a member of the Department of Advanced Professional and Special Services at the University of Central Oklahoma (UCO) since August, 1991. Prior to this, she served for two years as an instructor at Oklahoma State University, a junior high school special education teacher for two years, elementary education teacher for four years and taught music for one year at the elementary level. Dr. Green's doctorate was received in applied behavioral studies in 1992. Teaching assignments over the past twenty years include College Reading and Study Skills, Foundations of American Education/Field Experience, Teaching Individuals with Disabilities, Educational Strategies for the Young Child with Special Needs, Curriculum Development for Mild/Moderate Disabilities, Procedures for Mild/Moderate Learning Disabilities, Procedures for Mild/Moderate Mental Retardation and Secondary Special Education.

During my tenure at UCO, I have co-written and published one text supplement and two books for three required courses. I have had the honor of serving as a Ron McNair Scholar Mentor for five undergraduate students, all of whom presented research findings at various conferences. Additionally, I have supervised three graduate students' research studies, two were presented at national conferences and two of which were published in the ERIC Documents and the International Journal Diversity in Organizations. I have had the opportunity to give approximately 25 international, national and local presentations on the following topics: emotional and social intelligence, learning styles, parent involvement in the educational process and educational, emotional and social strategies to improve learning. Lastly, I am pleased to state that I assisted with obtaining and managing a teacher recruitment grant for five years. Eighty percent of those recruited graduated with various degrees in the field of teacher education.

Over the years I have been a member of the Learning Disabilities Association, Oklahoma Federation of the Council on Exceptional Children, and the African American National Studies Association. During my journey, I received the Hauptman Research Award, Elizabeth Threat Diversity Initiative Award, Ron McNair Mentor Award and the Delta Sigma Theta Educational Awareness Award.

"It is my hope that my professional involvement has positively impacted students and society. I would like to encourage students to continue to seek knowledge, wisdom and courage that would lead to fulfilling one's potential and making the world a better place to co-exist."

Donna Kearns

Best Overall Faculty Member, Education

Mark Maddy

Best Researcher/Scholar, Education

UNIVERSITY OF CINCINNATI

Sue Bourke

Most Helpful to Students, Criminal Justice

Professor Bourke received her B.S. degree from Eastern Kentucky University with a double major in Law Enforcement and Social Work, and an M.S. in Criminal Justice from UC. Prior to joining the faculty, she worked for the Kentucky Cabinet for Juvenile Justice as a juvenile counselor in a Day Treatment Program, was a Juvenile Court Probation Officer, and an Administrator for the Kenton County Juvenile Court. She began teaching as an adjunct instructor in the Criminal Justice Technology Program in 1986, and joined the faculty full-time in January 1996. Her area of expertise is corrections, particularly juvenile justice. Currently, she is the Director of Undergraduate Studies in the School of Criminal Justice, as well as Faculty Adviser to the Criminal Justice Field Placement Program and the Faculty Adviser to the Criminal Justice Society.

Elaine Hollensbe

Best Overall Faculty Member, Management

"General excellent student rapport"

Elaine Hollensbe is an Associate Professor of Management and Doctoral Program Coordinator for the Management Department. She completed her PhD in organizational behavior and human resource management at the University of Kansas. She is currently an Associate Editor of the Academy of Management Journal, and on the Editorial Board for the Journal of Organizational Behavior. Dr. Hollensbe has completed quantitative research on goal setting, compensation and self-efficacy, and qualitative research in the areas of identity, work-life balance, and emotion. Her current research is qualitative and focuses on organizational identity; the impact of mobile technology on work-home balance; role management; organizational identification in exterior employees and employees experiencing mergers; individual influences on sustainability; and intergenerational differences at work. Her research has been published in such journals as the Academy of Management Journal, Journal of Management, Academy of Management Review, Human Relations, and Human Resource Development Quarterly and has been recognized with national awards, including the Owens Scholarly Achievement Award, the Rosabeth Moss Kanter Award, and the Outstanding Publication in Organizational Behavior Award. Dr. Hollensbe teaches classes in the undergraduate, graduate, and doctoral programs and has received four teaching awards from the Lindner College of Business. She is a Fellow of the University's Academy of Fellows of Teaching and Learning, and is both a Research Fellow and a Teaching Fellow in the Carl H. Lindner College of Business. She also is involved in designing and facilitating executive programs and workshops on leadership, team building, and work-life balance. She has a BA in Journalism and Mass Communications from Iowa State

University, and an MBA from the University of Missouri-Kansas City. Prior to completing her PhD, she was an academic administrator and a communications consultant in Kansas City. - See more at: http://business.uc.edu/Elaine-Hollensbe.html#sthash.lQC3VVwc.dpuf

Yvette Pryse

Best Teacher, Nursing

Ruth Sieple

Best Overall Faculty Member, Business
Best Teacher, Business
Most Helpful to Students, Business

Mrs. Sieple has been a UC Instructor for 23 years. Prior to that her field experience was as a Product Line Manager for Square D Company, managing the production of electrical distribution equipment for both domestic and international markets. - See more at: http://business.uc.edu/departments/obais/faculty/ruth-seiple.html#sthash.M36Rtf3O.dpuf

UNIVERSITY OF COLORADO COLORADO SPRINGS

John Milliman

Best Teacher, Business
Best Teacher, Business

UNIVERSITY OF CONNECTICUT

Amy Anderson

Best Researcher/Scholar, Pharmacy

"She is a great scientist and devoted scholar."

Richard Kochanek

Best Overall Faculty Member, Accounting
Best Teacher, Accounting

Richard F. Kochanek is an emeritus professor in the Accounting Department at The University of Connecticut He received his B.B.A. and M.B.A. degrees from the University of Massachusetts and his Ph.D. degree from the University of Missouri. He joined the faculty at The University of Connecticut in 1972. For the past 15 years, Dick has taught the large mass sections of the introductory accounting course at UConn (about 1,000 students per year). He was head of the Accounting Department from 1997–2003. From 1988 to 1992, he served as Associate Dean for Academic Programs in the School of Business.

Over his 44 year career at UConn, Dick has published accounting textbooks and numerous articles on external financial reporting issues. Dick has received many teaching awards, both at the undergraduate and graduate level. These include outstanding accounting professor, outstanding professor in the School of Business, outstanding MBA Program professor, outstanding EMBA professor, the Alumni Association award for outstanding teaching at The University of Connecticut, one of the first four faculty designated as University Teaching Fellows at The University of Connecticut, Outstanding Educator of the Year by the Academic Affairs Committee of the University of Connecticut Undergraduate Student Government, the first "Faculty Award" given by Alpha Lamba Delta which is a National Honor Society for First Year Students at UConn, and was one of the first two faculty members inducted into the School of Business "Hall of Fame".

Dick always had a love for expression as an artist. Since 2003, he has had the great opportunity to sit in classes with faculty from the Art Department at UConn. Especially inspirational have been professors Kathryn Myers and Laurie Sloan. Dick feels very fortunate to have met and learned from such talented and dedicated professors and fellow students. He has gained a great respect and appreciation for the unique challenges and joys experienced by the artist. Dick believes that conceptualizing and creating works of original art has translated into a greater sensitivity and passion for his classroom teaching. He is a juried artist member of the Cape Cod Art Association, a member of The Printmakers of Cape Cod, and a member of the Falmouth Artist Guild located on Cape Cod. His work is regularly accepted and shown in juried shows.

At this point in his life, Dick's greatest message to share with his students is the realization of the enormous pool of talent that exists within each of us. To explore this inner talent, to grow as individuals each day, and to reach greater levels of expression: This is my fondest wish for all of us.

Olivier Morand

Best Teacher, Economics

Murphy Sewell

Best Overall Faculty Member, Marketing

UNIVERSITY OF CONNECTICUT AVERY POINT

Nancy Parent

Best Teacher, Anthropology

UNIVERSITY OF DENVER

Kevin Archer

Best Teacher, International Studies

Robert Uttarro

Most Helpful to Students, International Studies

UNIVERSITY OF FINDLAY

Melissa Cain

Best Overall Faculty Member, Education

"Dr. Melissa Cain is an outstanding professor at The University of Findlay. She is an expert in the area of reading and developed an online doctoral program that started at the University this year. We are proud of her accomplishments."

Dr. Cain taught in the primary grades for eight years. Her love of children's literature and her fascination with helping children learn to read lead her to pursue a master's degree and then a doctorate. She came to The University of Findlay in January 1992 when the education master's degree was just beginning. She developed and taught one of the core classes, EDUC 502: Collaboration: Education and Community. She taught many classes in the Elementary Education program, including Developmental Reading I and II, Language Arts Methods, and Children's Literature. She co-wrote the Early Childhood program when the Ohio licensure standards changed in 1997–1998 and wrote the Reading Core and Endorsement programs.

Dr. Cain has been the NCATE co-coordinator for the education unit since 1997 and the Dean of the College of Education since 2001. Under Dr. Cain's leadership, the College of Education developed its unit and candidate performance assessment system and successfully passed its April 2005 NCATE review with no conditions. Dr. Cain wrote a funded grant from the Ohio Department of Education to expand the College of Education's Assessment system. As part of that grant, the College was able to send ten program directors to the College LiveText User's Conference in Chicago in July 2005 and hire a consultant to assist in expanding the College's use of LiveText for generating candidate performance data.

Maria Gamba

Best Teacher, Business

Joseph Martelli

Best Overall Faculty Member, Business
Best Researcher/Scholar, Business
Most Helpful to Students, Business
Best Researcher/Scholar, Business
Best Teacher, Business
Most Helpful to Students, Business
Best Overall Faculty Member, Business

"Joe Martelli is an excellent researcher and is very well published"

Dr. Martelli is Associate Professor of Business at The University of Findlay where he teaches classes in both the undergraduate and graduate business degree programs.

Prior to Findlay, Martelli was a research scientist at the University of Michigan's Industrial Technology Institute in Ann Arbor. There he worked with corporations throughout the US in the design, implementation and evaluation of organizational systems aimed at giving American corporations, especially those in the automotive industry, competitive advantage in a global environment. Martelli was also a Corporate Human Resource Manager for Kellogg's in Battle Creek, MI. He worked on the design and implementation of HR and organizational systems in existing and new food processing and manufacturing facilities throughout Kellogg's worldwide organization. Martelli also has operations level HR experience as a Human Resource Manager at Kimberly-Clark's Pulp and Paper Mill in Memphis, Tennessee. Consulting clients have included Ford Motor Company, General Motors, General Electric, Motorola, Chrysler Corporation, as well as numerous small-and mid-sized corporations.

He has published articles in numerous professional and business journals, and has spoken at many international, national, state and local professional organizations and conferences. He is on the editorial board of the Journal of the North American Management Society (JNAMS) and serves as an editor for other professional journals as well. Martelli has a Doctorate in Human Resource Management and Industrial Technology from the University of Northern Iowa.

UNIVERSITY OF FLORIDA

Thomas Hurst

Best Overall Faculty Member, Law

Michael Warren

Best Teacher, Anthropology
Most Helpful to Students, Anthropology

UNIVERSITY OF GEORGIA

Bob Grafstein

Best Overall Faculty Member, Political Science
Best Researcher/Scholar, Political Science
Best Researcher/Scholar, Public Affairs, Political & Policy Sciences
Best Overall Faculty Member, Public Affairs, Political & Policy Sciences

With Fan Wen, a leader at China's National School of Administration, he coedited A Bridge Too Far? Commonalities and Differences between China and the U.S. (Rowman & Littlefield). He is also the author of Institutional Realism (Yale), Choice-Free Rationality (Michigan), and numerous articles in leading scholarly journals. He directs the Maymester China Study Abroad Program, which he created in 2005. He has also lectured extensively at Chinese universities. From 2002–2008 Grafstein served as Head of the School's Department of Political Science. He currently serves on the editorial board of the Journal of Theoretical Politics, and has served on the boards of the Journal of Politics and Social Science Quarterly. He received his A.B. degree from the University of Pennsylvania, and M.A. and Ph.D. from the University of Chicago.

Paul-Henri Gurian

Best Teacher, Political Science

Dr. Gurian earned his Ph.D. from the University of North Carolina at Chapel Hill in 1987. Since then he has been teaching and doing research in the Political Science Department at the University of Georgia. During that time, he developed both graduate and undergraduate courses on Campaign Politics. The graduate seminar now part of the regular curriculum, taught every other year by Dr. Gurian. The undergraduate course is now taught at least once each semester by various instructors. Dr. Gurian has won numerous teaching awards during his 29 years at U.G.A. Dr. Gurian's research focuses on presidential campaigns, mainly U.S. presidential primaries and general election campaigns, but also Korean presidential campaigns. Although his focus has been on campaign strategy and dynamics, he has also analyzed the impact of election rules, the news media and other aspects of campaigns. Dr. Gurian has been mentor (major professor) to seven Ph.D. students, all of whom now hold positions either as instructors at research universities or tenure track assistant professors (one is a research associate at Michigan State). Every four years, beginning in 1988, Dr. Gurian has organized and facilitated a weekly discussion group for S.P.I.A. faculty and graduate students focusing on U.S. presidential campaigns. In 2012, professors at Davidson College joined the discussion via video conferencing technology. For the 2016 campaign, he plans to expand the weekly video conference to include campaign experts at several other universities. Dr. Gurian has been asked to analyze U.S. presidential primary and election campaigns at several universities, political groups in Georgia, and recently at a university and an international conference in South Korea.

John Maltese

Most Helpful to Students, Political Science
Best Teacher, Public Affairs, Political & Policy Sciences
Most Helpful to Students, Public Affairs, Political & Policy Sciences

UNIVERSITY OF HAWAII AT HILO

Gene Johnson

Best Researcher/Scholar, Accounting

UNIVERSITY OF HOUSTON

Julia Babcock

Best Researcher/Scholar, Psychology

Shirley Ezell

Best Teacher, Health Sciences and Nursing

J. Leigh Leasure

Best Teacher, Psychology

Aditi Marwaha

Best Teacher, Pharmacology

Carla Sharp

Most Helpful to Students, Psychology

Dr. Sharp trained as a clinical psychologist (University of Stellenbosch, South Africa) from 1994–1997, after which she completed a Ph.D. in Developmental Psychopathology at Cambridge University, UK, 1997–2000. In 2001, she obtained full licensure as a clinical psychologist in the UK. From 2001–2004 she was appointed as a Research Post-doctoral Fellow in Developmental Psychopathology, Cambridge University. In 2004, she moved to the United States to take up an appointment as Assistant Professor in the Menninger Department of Psychiatry at Baylor College of Medicine. She obtained provisional licensure as Clinical Psychologist in Texas in 2008. In 2009, she was appointed as Associate Professor in the Department of Psychology at the University of Houston.

Her published work includes over 100 publications reflecting her interests in the social-cognitive basis of psychiatric problems and problems of behavioral health, and the application of this work in developing diagnostic tools and interventions. She has co-authored three books: An edited volume with Springer titled The Handbook of Borderline Personality Disorder in Children and Adolescents, an edited volume with Oxford University Press titled Social cognition and developmental psychopathology and a book with MIT Press titled Midbrain mutiny: Behavioral economics and neuroeconomics of gambling addiction as basic reward system disorder.

UNIVERSITY OF HOUSTON - CLEAR LAKE

Beth Hentges

Best Teacher, Psychology

Amanda Johnston

Most Helpful to Students, Psychology

Dr. Amanda Johnston is currently an Assistant Professor of Psychology at the University of Houston–Clear Lake. Her research focuses on understanding stereotypes about groups (primarily gender), social change, and motivation. Dr. Johnston approaches most of her research from a social role theory framework; this perspective emphasizes the critical nature of a group's position in the social structure for understanding both group differences and beliefs about groups. Her dissertation research explored how the beliefs that we have for why groups possess particular traits and behaviors contributes to the overall stability and legitimacy of the existing social structure. She is also interested in how the motivations underlying social change impact evaluations by others. Some of her most recent work explored how the goals women typically possess (e.g., caring for others) contribute to why women opt out of STEM (science, technology, engineering, and mathematics) fields.

Robert M. Jones

Most Helpful to Students, Education

30 years in higher education training preservice and inservice teachers. 7years experience with NASA in education, engineering and public affairs. 7 years experience in public schools in science and mathematics instruction and supervision.

David Malin

Best Researcher/Scholar, Psychology

Dr. Malin is Professor of Psychology and Neuroscience. A charter faculty member of UHCL, Dr. Malin has received the UHCL President's Distinguished Teaching and Distinguished Research Awards. His recent research primarily concerns laboratory study of nicotine dependence and age-related memory impairment. He leads a large student research group, known as the UHCL Rat Pack.

Mary Short

Best Overall Faculty Member, Psychology

Mary Short, Ph.D. is an Associate Professor of Clinical Psychology. She graduated from North Dakota State University in Fargo, ND with a B.S. in Psychology. She earned an MA degree from Mankato State University in Mankato, MN, and earned her Ph.D. in Clinical Psychology from the Western Michigan University in Kalamazoo, MI. She completed her internship at Boy's Town in Omaha, NE. She then completed a post-doctoral fellowship in Pediatric Psychology at the University of Texas Medical Branch in Galveston, TX

Jana Willis

Best Teacher, Education
Best Overall Faculty Member, Education

Dr. Willis obtained her Doctorate in Educational Psychology with Educational Technology Specialization from Texas A&M University. She earned a M.S. degree in Instructional Technology from UHCL. Dr. Willis has certification in secondary Computer Science and English. She has designed, developed, and taught a variety of online/hybrid instructional/educational technology courses. Prior to her work in higher education, Dr. Willis worked for 20+ years as a programmer in the business sector. She is an Associate Professor in in the Departments of Instructional Design and Technology and Curriculum & Instruction/Teacher Education at University of Houston-Clear Lake (UHCL). Her research interests include "serious games" in the K–12 classroom, technology integrated curriculum, teacher training, and self-efficacy. Dr. Willis has presented research findings at numerous conferences, published multiple book chapters, and several research articles in the International Journal of Early Childhood Environmental Education, the Childhood Education, The Journal for the Texas Association of Young Children Early Years, the Journal of Technology Integration in Teacher Education, and the Journal of Technology and Teacher Education. Dr. Willis has recently served as Co-Principal Investigator on two National Science Foundation (NSF) grants and one Defense Advanced Research Projects Agency (DARPA) grant.

Nancy Wright

Most Helpful to Students, Education

Nancy Wright is a Senior Lecturer at the University of Houston-Clear Lake where she teaches literacy and library science courses in the School of Education's teacher preparation program. She has years of experience in literacy instruction and the school library. Her teaching and research interests include quality enhancement through critical thinking concepts, and retention initiatives for first-generation college students.

UNIVERSITY OF ILLINOIS AT CHICAGO

Daniel Cervone

Best Researcher/Scholar, Psychology

"Dan is a serious scholar."

Daniel Cervone is Professor of Psychology at the University of Illinois at Chicago. He earned his Ph.D. in Psychology from Stanford University in 1985. In addition to his time at UIC, Dr. Cervone has been a visiting faculty member at the University of Washington and at the University of Rome "La Sapienza," and has been a fellow at the Center for Advanced Study in the Behavioral Sciences. He currently serves on the editorial boards of Psychological Review and the European Journal of Personality, and recently completed a term as Associate Editor at the Journal of Research in Personality.

UNIVERSITY OF ILLINOIS AT SPRINGFIELD

Vickie Cook

Best Researcher/Scholar, Education

In her early career in higher education, Vickie held a variety of positions at Kaskaskia Community College (1992–2005) including Dean of Continuing Education and University Alliance. Dr. Cook has served as a long time member of Illinois Council on Continuing Higher Education serving on the Executive Board, as well as having been an active member and board member for Illinois Council on Continuing Education and Training. She received the Charles V. Evans Research Award in 2004 & 2007 and Illinois Exemplary Leadership Award (ICCET) in 2005.

Nancy Forth

Best Teacher, Education

UNIVERSITY OF ILLINOIS AT URBANA-CHAMPAIGN

Wendy Heller

Most Helpful to Students, Psychology

Wendy Heller is Professor of Psychology in the Clinical/Community Division, former Director of Clinical Training and Associate Department Head in the Psychology Department, and a part-time Beckman Institute faculty member in the Cognitive Neuroscience Group. As of 2014, she was appointed Provost Fellow with a special focus on campus diversity. She holds a B.A. in Spanish and Psychology with Honors from the University of Pennsylvania and an M.A. and Ph.D. in Biopsychology from the University of Chicago. Her research investigates the neural mechanisms associated with emotion-cognition interactions and their implications for psychopathology. She is particularly interested in examining cognitive and emotional risk factors associated with the development or maintenance of anxiety and depression. She uses behavioral and neuroimaging methods such as neuropsychological task performance, functional magnetic resonance imaging (fMRI), electroencephalography (EEG), and event-related potentials (ERPs). She draws on psychological theories to model how fundamental emotion and personality constructs can be mapped onto brain systems to clarify the neural mechanisms of emotion and psychopathology. In turn, the neuropsychological and neuroimaging findings are used to inform psychological theories of emotion and psychopathology. Her work has been funded by the National Institute of Mental Health (NIMH).

David Irwin

Best Overall Faculty Member, Psychology

Daniel Newman

Best Researcher/Scholar, Psychology

Carol Packard

Best Overall Faculty Member, Business

Brent W Roberts

Best Teacher, Psychology

Brent W. Roberts is a Professor of Psychology in the Department of Psychology at the University of Illinois, in the Social-Personality-Organizational Division. Dr. Roberts received his Ph.D. from Berkeley in 1994 in Personality Psychology and worked at the University of Tulsa until 1999 when he joined the faculty at the University of Illinois, Urbana-Champaign. He received the J. S. Tanaka Dissertation Award for methodological and substantive contributions to the field of personality psychology in 1995. He also was awarded the Diener Mid-Career award in Personality Psychology from the Foundation for Personality and Social Psychology, The Theodore Millon Mid-Career award in Personality Psychology from the American Psychological Foundation, and was appointed as a Richard and Margaret Romano Professorial Scholar at the University of Illinois from 2008–2011. He recently received the 2012 Henry Murray Award from the Society for Personology and the American Psychological Association. He has served as the Associate Editor for the Journal of Research in Personality, as a member-at-large and Executive Officer for the Association for Research in Personality and serves on the Editorial Board of the Journal of Personality and Social Psychology, International Journal of Selection and Assessment, Personality and Social Psychology Review, and Perspectives on Psychological Science.

Norma Scagnoli

Best Overall Faculty Member, Business

"I nominate Dr. Norma Scagnoli to receive an award because without her, many of the other faculty would not be as good at teaching in today's technology-orientated learning/teaching

environment. Dr. Scagnoli continually keeps current with the latest e-learning advances, evaluates each for its value to learning and teaching, and supports other faculty in its diffusion across the college. Norma is incredibly generous with her time and expertise. Without her knowledge, skills and assistance, I could not be as good a teacher as I have been able to become."

UNIVERSITY OF IOWA

Sam Burer

Best Overall Faculty Member, Business

Sam Burer is Professor and Henry B. Tippie Research Fellow in the Department of Management Sciences at the University of Iowa. He received his Ph.D. in Algorithms, Combinatorics, and Optimization from the Georgia Institute of Technology, and his research and teaching interests include analytics, operations research, management sciences, and optimization.

His research has been supported by grants from the National Science Foundation, and he serves on the editorial board of Operations Research, SIAM Journal on Optimization, Mathematics of Operations Research, Management Science, and Optima. He also serves as a Council Member of the Mathematical Optimization Society, and as a Board of Directors Member of the INFORMS Computing Society.

For more information, please see Sam Burer's curriculum vitae (which includes links to PDF versions of his papers), Google Scholar profile, or official university web page.

Todd Houge

Best Teacher, Business

Ayca Kaya

Best Overall Faculty Member, Economics

Lon Moeller

Most Helpful to Students, Business

Lon Moeller is currently Associate Provost for Undergraduate Education and Dean of the University College. Prior to his appointment, Moeller served as Associate Dean for the Undergraduate Program in the Tippie College of Business. He is a full clinical professor in the Department of Management and Organizations and has also served as University Ombudsperson and as Co-Director for the Larned A. Waterman Iowa Nonprofit Resource Center.

Moeller earned his B.B.A., M.A., and J.D. from The University of Iowa. He worked in private law practice and served as System Legal Counsel for the University of Wisconsin System prior to becoming a faculty member in the Tippie College of Business. Moeller is a labor mediator and arbitrator and serves on several state and national labor arbitration panels. He is a frequent speaker on business ethics, conflict management, and negotiations. Moeller has co-written four books on the topics of conflict management, management, and negotiations and is currently writing and developing a digital business law textbook.

Thomas Rietz

Best Overall Faculty Member, Finance
Best Researcher/Scholar, Finance
Best Teacher, Finance

Gene Savin

Best Researcher/Scholar, Business

UNIVERSITY OF KANSAS

Allen Ford

Best Teacher, Accounting

James Heintz

Best Overall Faculty Member, Accounting

Ron Malcolm

Best Overall Faculty Member, Education
Best Teacher, Education

UNIVERSITY OF LA VERNE

Caroline Chizever

Best Overall Faculty Member, Business

Issam Ghazzawi

Best Teacher, Business

Rita Thakur

Best Overall Faculty Member, Business

"Her dedication to both the teaching profession and her students"

Professional Experiences: Practiced law in Gujarat High Court Consulting in Total Quality Management and Multiculturalism for business and educational institutions Courses Taught: Principles of Management Culture and Gender Issues in Management Organizational Theory Organizational Behavior Business Ethics Seminar in Management Research Interests: Culture and gender issues Total Quality Management

UNIVERSITY OF LOUISIANA AT LAFAYETTE

Mark Rees

Best Overall Faculty Member, Anthropology

UNIVERSITY OF LOUISVILLE

Joseph Gutmann

Best Teacher, Paralegal

"Prof. Gutmann has taught in the Paralegal Studies Program for over 25 years, and has always been a student favorite. He is dedicated to the program, the paralegal students, and the University as a whole!"

UNIVERSITY OF MAINE ORONO

Steven Butterfield

Best Overall Faculty Member, Education

Sandra Caron

Most Helpful to Students, Education

Wendy Coons

Best Teacher, Accounting

Christopher Nightingale

Best Researcher/Scholar, Kinesiology
Most Helpful to Students, Kinesiology
Best Researcher/Scholar, Education

Glenn Reif

Best Teacher, Education

UNIVERSITY OF MARYLAND - EASTERN SHORE

Emmanuel Onyeozili

Best Researcher/Scholar, Criminal Justice
Best Teacher, Criminal Justice
Most Helpful to Students, Criminal Justice
Most Helpful to Students, Education
Best Overall Faculty Member, Education

"Best overall teacher, Managing Editor of African Journal of Criminology and Criminal Justice; Founder of UMES Chapter of Alpha Phi Sigma"

UNIVERSITY OF MASSACHUSETTS

John Brigham

Best Researcher/Scholar, Political Science

Brenda Bushouse

Most Helpful to Students, Public Affairs, Political & Policy Sciences

My research interests include public policy with particular focus on the role of 501(c)3 nonprofit organizations in policy processes and an empirical focus on early childhood education and care.

Barbara Cruikshank

Best Teacher, Political Science
Best Overall Faculty Member, Political Science
Best Researcher/Scholar, Political Science
Most Helpful to Students, Political Science
Best Researcher/Scholar, Public Affairs, Political & Policy Sciences
Best Teacher, Public Affairs, Political & Policy Sciences
Most Helpful to Students, Public Affairs, Political & Policy Sciences
Best Overall Faculty Member, Public Affairs, Political & Policy Sciences

"*Barbara is supportive, inspiring, demanding, energetic and available to her students. She is always able to make materials accessible to students while allowing them to gain a complex, in-depth understanding of the texts.*
  Barbara is an incredible scholar, tireless and perfectionnist, active in many projects and about to finish a terrific book. She is also a wonderful, inspiring and supportive advisor, at every step of the way throughout one's PhD career she offers precise and helpful feedback while being very demanding and challenging (in the best sense of the term). Finally, teaching to undergraduates she is excellent at simplifying texts without doing violence to these, putting complex ideas to the level of the students, and allowing them in-depth analysis to train their critical skills."

Raymond La Raja

Best Teacher, Public Affairs, Political & Policy Sciences

Tatishe Nteta

Best Researcher/Scholar, Public Affairs, Political & Policy Sciences

Professor Tatishe M. Nteta is the Assistant Professor of Political Science at University of Massachusetts (Amherst, MA). His research interests lie at the intersection of the politics of race and ethnicity, public opinion, and political behavior. More specifically, my work examines the impact of changing demographics and shifts in the sociopolitical incorporation of racial minorities on American race relations, policy preferences, and participation.

M. J. Peterson

Best Overall Faculty Member, Public Affairs, Political & Policy Sciences

I am interested in understanding what conditions and factors facilitate or inhibit international cooperation among states, among states and non-state actors, or among nonstate actors. Some of my more recent papers are accessible on-line from the research page.

Jesse Rhodes

Most Helpful to Students, Political Science

Brian Schaffner

Best Overall Faculty Member, Political Science

UNIVERSITY OF MASSACHUSETTS - BOSTON

Elizabeth Bussiere

Most Helpful to Students, Political Science

Elizabeth Bussiere has been a fellow at the Bunting Institute at Radcliffe College and received a grant from the American Philosophical Society for a book on jury nullification and the decline of popular justice. Her publications include a book, (Dis)Entitling the Poor: The Warren Court, Welfare Rights, and the American Political Tradition (Penn State Press), which received an honorable mention from the American Political Science Association in 1998 for the Best Book on Women and Politics; an article on "The Failure of Constitutional Welfare Rights in the Warren Court," published in Political Science Quarterly; a revised version of that article, which appeared in Supreme Court Decision-Making: New Institutionalist Approaches (University of Chicago Press); and the lead article, "The 'New Property' Theory of Welfare Rights: Promises and Pitfalls," published by the Committee on the Good Society, a group consisting of the most prominent scholars in the social sciences, the humanities, and law.

Caroline Coscia

Best Teacher, Political Science

Professor Coscia has taught American government, public policy, and urban politics. She also teaches community portrait classes in the College of Public and Community Service.

Professor Coscia's experiences include serving as a Capital Hill intern, a legislative aide in the Massachusetts House of Representatives and two terms as an elected member of the Wakefield Planning Board.

Erin O'Brien

Best Researcher/Scholar, Political Science

Professor Erin E. O'Brien's research and teaching interests focuses on the politics of poverty and social welfare policy, voting access policymaking in the United States, and gender in political participation/representation. Her work employs a variety of methods and approaches to social science in order to examine the connections among social policy, political thought and action, inequality, and patterns of stratification associated with social groups.

Maureen Scully

Best Overall Faculty Member, Management

UNIVERSITY OF MASSACHUSETTS - DARTMOUTH

Steven White

Best Researcher/Scholar, Marketing

Dr. White is a professor of marketing & international business at the Charlton College of Business, University of Massachusetts Dartmouth. He earned his degrees from Cleveland State University (DBA, 1996; MBA, 1991) and Bowling Green State University (MA, 1987; BSBA, 1985) with coursework completed through intermediate-level Portuguese at the University of Massachusetts-Dartmouth (2004–2005). His doctorate is in business administration with a major in marketing and a minor in information systems. He joined the faculty of the Charlton College of Business in 1998.

His research interests include sustainable development, social media marketing, mobile marketing, international marketing, services marketing, global and social entrepreneurship, international business, global e-commerce, open source applications in global business and international business education. Academic journals in which his research has appeared include International Marketing Review, Journal of Business Research, Journal of Services Marketing, Thunderbird International Business Review, Journal of Marketing Management (U.K.), Journal of Marketing Education and International Business Review. He teaches undergraduate courses in marketing principles, social media marketing and international marketing and graduate courses in interactive marketing, marketing strategy and international business.

From January 2006 through May 2010, he served as chairperson of the Department of Management and Marketing in the Charlton College of Business and from June 2010 through June 2012, as director of the university's sustainability studies program during which time the university's graduate certificate in sustainable development was developed and launched. Since 2000, he has co-organized and guided 200+ students on alternative spring break task force missions to 24 cities in 15 countries (Argentina, Belgium, Brazil, Cape Verde, Chile, China, Costa Rica, England, France, Germany, Italy, Mexico, Peru, Portugal (including Azores) and Spain). Additionally, he has co-organized and guided trade missions in partnership with the Massachusetts Office of International Trade and Investment (MOITI).

Prior to entering academia Dr. White was vice president of an advertising agency, promoted sporting and special events, managed outdoor sporting goods stores/ski shops and served as an alpine ski instructor. Contact:

UNIVERSITY OF MASSACHUSETTS - LOWELL

Eve Buzawa

Best Researcher/Scholar, Criminal Justice

Dr. Buzawa's research interests and publications encompass a wide range of issues pertaining to policing, domestic violence, and violence against women.

Andrew Harris

Best Overall Faculty Member, Criminal Justice
Best Teacher, Criminal Justice

Andrew J. Harris is Associate Dean for Research & Graduate Programs in the College of Fine Arts, Humanities, and Social Sciences and Associate Professor in the School of Criminology and Justice Studies. An expert in public policy, he teaches in the areas of institutional and community corrections, substance abuse, sexual offending, crime and mental illness, and social policy.

Before joining the UMass Lowell faculty, Harris spent more than 16 years developing and managing innovative public sector policies and programs in both the criminal justice and human service arenas. Between 1990 and 1998, he held several management positions in New York City government, serving as a supervising analyst in the Mayor's Office of Management and Budget, as Assistant Commissioner in the New York City Department of Correction, and as Associate Executive Director for Correctional Health Services in the City's Health and Hospitals Corporation. Between 1998 and 2005, he served as Deputy Director of Health and Criminal Justice Programs at the University of Massachusetts Medical School.

Building on his prior professional work, Dr. Harris' research examines the implementation of public policies at the intersection of criminal justice and human service sectors. His current research agenda focuses on policy responses to sexual offending, and policies concerning justice-involved persons with serious mental illness. He is also Principal Investigator for a major national study, funded by the U.S. Department of Justice, designed to inform policies and practices concerning teen "sexting" and its related issues.

Harris has served as a consultant to state policy boards, public behavioral health agencies, community-based service providers, and state and municipal correctional systems. His work has appeared in a range of inter-disciplinary journals, including Criminal Justice and Behavior, Criminal Justice Policy Review, and Victims & Offenders. He also serves as editor of the Sex Offender Law Report, a nationally-circulated publication read by policymakers and legal and criminal justice professionals, and has been quoted and featured in a range of media outlets, including the New York Times and the Boston Globe.

A current listing of Dr. Harris' publications and presentations may be found on his faculty website.

Karen Devereaux Melillo

Best Researcher/Scholar, Nursing

Allan Roscoe

Most Helpful to Students, Criminal Justice

Allan D. Roscoe, B.S., M.A., Criminal Justice, University of Massachusetts, Lowell has been an instructor in the Criminal Justice Department since 1999. Professor Roscoe had a long and distinguished career as a Senior Special Agent with the Office of Special Investigations. His particular areas of expertise are Crimes of Violence, Terrorism/Counter-Terrorism, Homeland Security, National Security and Counter-Intelligence/Counter-Espionage.

Professor Roscoe had previously worked and lived in Asia and the Middle East and has worked extensively with various international police and governmental organizations. Professor Roscoe is the Undergraduate Coordinator for the Homeland Security Certificate Program as well as the Undergraduate Academic Advisor.

UNIVERSITY OF MASSACHUSETTS SCHOOL OF LAW

Kevin Connelly

Best Overall Faculty Member, Law

Professor Connelly began his legal career as an Assistant District Attorney in Boston. From 1986 to 1988, he served as chief of the Major Offenders' Bureau of the Suffolk District Attorney's Office. Thereafter, he worked in personal injury practice as Claims Counsel for the Commercial Union Insurance Company and engaged in the private practice of law, with an emphasis on criminal defense at trial and on appeal. From 1990 to 1992, Connelly served as an Assistant Attorney General in the Massachusetts Environmental Crimes Strike Force. From 1992 to 2005, he served as an Assistant District Attorney for the Bristol District in New Bedford, Massachusetts. He was chief of Homicide from 1998 to 2001 and chief of Appeals from 2001 until his retirement in 2005. He has extensive trial and appellate experience. Professor Connelly began teaching at the Southern New England School of Law in 2003, first as an adjunct professor and later as a visiting lecturer. He served as Faculty Advisor to SNESL's Family Law Moot Court team in Albany Law School's 2006 Gabrielli National Family Law Competition. In 2000, Connelly was elected Prosecutor of the Year by the Massachusetts District Attorneys' Association. From 2004 to 2007, he served on the Board of Regents of National College of District Attorneys, in Columbia, South Carolina. During that time, he also served as a member and co-Reporter of the Ad Hoc Committee on Revision of National District Attorneys' Association's National Prosecution Standards. His committee assignments were in the areas of pretrial discovery and trial conduct. Professor Connelly is a contributing author to the Massachusetts Superior Court Criminal Practice Jury Instructions, published in 1999 by the Massachusetts Continuing Legal Education Foundation. He is co-Author (with Hon. R. Marc Kantrowitz and Jennifer Bush, Esq.), of Closing Arguments: What Can and Cannot Be Said, 81 Mass. L. Rev. 95 (1996). In 2006, Professor Connelly was appointed as a member of the Massachusetts Supreme Judicial Court's Advisory Committee on Massachusetts Evidence Law. In September 2008, the Committee produced the Massachusetts Guide to Evidence.

Dwight Duncan

Best Overall Faculty Member, Law
Best Teacher, Law
Most Helpful to Students, Law
Best Teacher, Law
Most Helpful to Students, Law
Best Overall Faculty Member, Law

Rick Peltz

Best Researcher/Scholar, Law
Best Researcher/Scholar, Law

Peltz-Steele received his law degree from Duke University and a bachelor's in journalism and Spanish from Washington & Lee University. Peltz-Steele has won awards in teaching, research, and public service. He practiced commercial law in Baltimore and Washington, D.C., and taught law for more than thirteen years before coming to UMass Law. Peltz-Steele is author or co-author of qualitative and quantitative research articles in law and mass communication journals, as well as book chapters, a treatise in the law and mass communication field, a casebook in tort law, and a casebook in freedom of information law and policy. He is especially active in international media law and policy, having presented papers in Ireland, Malaysia, and South Africa, and having published in international and foreign journals, recently regarding privacy regulation in the European Union and indigenous identity in news reporting. He has served in various capacities for the American Bar Association, including the Tort Trial Insurance Practice Section committees on media and international law.

UNIVERSITY OF MEMPHIS

Robert Koch

Best Teacher, Nursing

"Dr. Koch uses unique teaching strategies that engage students and assists in knowledge retention. He is fair when reviewing and grading assignments, and gives feedback that students can use as they move forward in their program of study."

Robert W. Koch, DNSc, RN is an Associate Professor at Loewenberg School of Nursing, University of Memphis. He holds a baccalaureate and master's degree in nursing from the University of Tennessee Health Science Center and a doctorate in nursing administration and education leadership from Louisiana State University Medical Center in New Orleans. He holds memberships in Sigma Theta Tau International, the National League of Nursing, the American Organization of Nurse Executives, and the American Nurses Association.

Throughout Dr. Koch's professional career he has held several administrative positions such as nurse manager, director of nursing, assistant hospital administrator, and associate dean of the school of nursing. Koch holds positions on several organizational boards including The Regional Medical Center (The MED) L-TACH Board of Directors. He is the author of one textbook in nursing pharmacology and multiple journal articles. His most recent research activities include over 3 million dollars in federal funding to investigate the military/civilian interface during catastrophic wide scale disasters. As an accomplished presenter, Dr. Koch speaks regularly at national and international conferences and seminars. He teaches leadership and management at both the undergraduate and graduate levels.

Shirleatha Lee

Best Overall Faculty Member, Nursing

"Dr. Lee is very passionate about Nursing. She cares deeply for students and earns great respect from her peers. Without a doubt, she is a dreamer, thinker, and doer."

Dr. Lee is a Certified Nurse Educator and coordinates Medical Surgical Nursing I and also teaches Nursing Research. She challenges students to think critically in both the classroom and clinical setting while safely, effectively, and compassionately providing quality-nursing care.

Dr. Lee is a member of the University of Memphis IRB Committee, LSON Curriculum Committee, National League of Nursing, Sigma Theta Tau International, Southern Nursing Research Society, and Delta Sigma Theta Sorority.

Robin Poston

Best Overall Faculty Member, Business

"Dr. Poston is a very enthusiastic teacher, an accomplished researcher, and has many other accomplishments that have brought praise and respect to our MIS department."

Dr. Poston is a Systems Testing Research Fellow and Associate Director of the System Testing Excellence Program for the FedEx Institute of Technology at The University of Memphis, and she is an Associate Professor of Management Information Systems at the Fogelman College of Business & Economics at The University of Memphis. Dr. Poston is a recipient of the 2006–2007 University of Memphis Alumni Association Distinguished Teaching Award. She serves as an Associate Editor for the highly ranked Decision Sciences Journal, which publishes scholarly research that advances decision making. She also serves as an Associate Editor for another highly ranked journal, the European Journal of Information Systems, which publishes research about the theory and practice of information systems for a global audience. She also serves as the Organizing Chair for the International Research Workshop on Advances and Innovations in Software Testing. Formerly, she held the position of Vice-Chair for Sponsorship for the Special Interest Group on Human Computer Interaction.

Dr. Poston's research focuses on understanding how individuals use credibility information in decision support systems, web-based knowledge management applications, recommender and feedback systems, internet-based dissemination of information, and system testing management. Her research uses experimental laboratory techniques in a simulated on-line information environment and qualitative interview methods. She has published articles in publications such as Management Information Systems Quarterly, Decision Sciences Journal, Communications of the ACM, IEEE Computer, Information Systems Management, Journal of Organizational and End User Computing, International Journal of Electronic Business, Journal of Information Systems, International Journal of Accounting Information Systems, and in major international conference proceedings.

Dr. Poston has over 10 years of experience in the information systems field working for KPMG Consulting, Educational Computer Corporation, Meta Group Research, and Convergys, as well as consulting with several Fortune 500 companies and government agencies. Today, she works with local organizations, such as FedEx Corporation, First Horizons, St. Jude/ALSAC, Pinnacle Airlines, Rhodes College, and the University of Memphis IT Division to conduct projects and educational programs. Dr. Poston received her bachelor's degree in Computer Science from The University of Pennsylvania (1987), master's degree in Accounting from The University of Central Florida (1992), and Ph.D. in Management Information Systems from Michigan State University (2003).

Genae Strong

Best Researcher/Scholar, Nursing

"Dr. Strong has worked hard to produce multiple publications and has spoken at several conferences in the past few years. She has also obtained several certifications through her continuous education."

UNIVERSITY OF MIAMI

Marvin Jones

Best Researcher/Scholar, Law
Best Teacher, Law
Most Helpful to Students, Law

BIOGRAPHICAL SKETCH OF DONALD JONES
Donald Jones, a Baltimore native, is Professor of law at the University of Miami.

Professor Jones is one of the leading commentators in the United States on the civil and political rights of African-Americans and other minorities. He has published widely in the nation's best law journals. In 1997 Professor Jones was awarded the James Thomas prize by Yale University recognizing him as one the leading scholars in the country for that year writing in the area of "civil rights." Reaching out to audiences beyond the ivory tower, Professor Jones recently published an acclaimed book entitled, RACE, SEX, AND SUSPICION: THE MYTH OF THE BLACK MALE (GREENWOOD PRESS 2005).

A now familiar face in the local and national media, Professor Jones has appeared on PBS' Frontline, CNN's Burden of Proof; O'Reilly's The O' Reilly Factor and Michael Putney's The Week in Review, where he debated Ward Connerly. More recently, Professor Jones served as the legal expert for NBC 6 explaining to voters the evils of the Arizona Immigration Law. He is also a frequent contributor to the editorial pages of the THE MIAMI HERALD, THE SOUTH FLORIDA TIMES, AND THE MIAMI TIMES AND IN FOCUS MAGAZINE.

Professor Jones has taken a leadership role on many civil rights issues. In 2003 Professor Jones, in conjunction with efforts by Senator Fredericka Wilson, led an investigation into the impact of high stakes testing in inner city schools. In the same year, Professor Jones was hired as a Constitutional expert to draft the affirmative action plan for minority contracting in Dade County. In 2006 he was retained by the City of Miami to review their policy on deadly force. In 2007 Professor Jones was recognized by the National Bar Association as the "Outstanding Member" for that year. In 2009 Professor Jones was recognized as one of the most influential blacks in South Florida by Success Magazine.

Professor Jones has lectured nationally for many years. Professor Jones has been a keynote or featured speaker for the Law and Society Association, the National Bar Association, The Florida Association of State Judges, The National NAACP convention, The National Association of

Black Law Students, The Florida Conference of Black State Legislators and many others. Professor Jones continues to be a sought after speaker at many universities.

When not writing, teaching or appearing on TV, Professor Jones enjoys bike riding, theatre, chess, and listening to jazz.

Roger Kanet

Best Teacher, Political Science

"Amazing syllabi, great topics, warm and supportive personality, funny, passionate, and always fresh, Roger is the complete teaching package."

Leigh MacDonald

Most Helpful to Students, Law

Joseph Parent

Best Researcher/Scholar, Political Science

Sherry Porcelain

Best Teacher, International Studies

Sherri L. Porcelain, adjunct professor in the Department of International Studies and director of the disaster research program for global public health program at the University of Miami. Additionally, she is an adjunct faculty member in the Department of Epidemiology and Public Health at the University of Miami Miller School of Medicine since 1989. Professor Porcelain's teaching and research focuses on public health in world affairs and includes recent projects on disaster mitigation, program evaluation, injury prevention, and public health challenges in Latin America, Middle East, South East Asia and the United States. She serves as a consultant to local, national, and international organizations and brings the experiential and theoretical examination of human security challenges into the classroom setting.

Arthur Simon

Most Helpful to Students, Political Science

Arthur M. Simon holds business and law degrees from the University of Miami, a master's degree from the Harvard Kennedy School, and a Ph.D. in public administration from Florida State University.

His resume includes professional experience as an officer in the United States Marine Corps, an appellate attorney in Miami, a senior manager in the Florida Department of Banking and Finance, and a lobbyist for Associated Industries of Florida. In addition, he served 12 years in the Florida House of Representatives – representing the West Kendall area of Miami-Dade County.

Dr. Simon was signaled out and honored by the Association of Greek Letter Organizations as UM's most "outstanding faculty member" in the 2011–2012 academic year; and more recently, in 2014, he was recognized by the Office of Academic Enhancement "for extraordinary commitment to scholar excellence".

Joseph Uscinski

Best Researcher/Scholar, Political Science

"His work is interesting and policy relevant, timely and timeless, and has gained national attention."

Joseph E. Uscinski received his bachelor's degree from Plymouth State University, his Master's from University of New Hampshire, and his Doctorate from University of Arizona. His research has appeared in Journal of Politics, Political Research Quarterly, and Critical Review among other scholarly outlets. His first book, The People's News: Media, Politics, and the Demands of Capitalism (New York University Press, 2014) addresses how audience demands drive news content. His second book, American Conspiracy Theories (Oxford University Press, 2014) coauthored with Joseph Parent, examines why people believe in conspiracy theories.

Jonathan West

Best Overall Faculty Member, Political Science

"Profoundly decent, incredibly hard working, and beloved by just about everyone despite being in impossible situations sometimes. It's an honor just to work near the man."

Jonathan P. West is chair and professor of political science and director of the MPA program at the University of Miami. His research interests include human resource management, productivity, local government, and public service ethics. Professor West has published nine books and over 100 articles and book chapters. His most recent books are co-authored with Bowman, Public Service Ethics: Individual and Institutional Responsibilities (CQ Press, 2014), Berman, Bowman, and Van Wart, Human Resource Management: Paradoxes, Processes and Problems (4th ed., Sage, 2013) and with Bowman and Beck, Achieving Competencies in Public Service: The Professional Edge (2nd ed., M.E. Sharpe, 2010), with Bowman, eds. American Public Service: Radical Reform and the Merit System (Taylor & Francis, 2007), with Berman, eds., The Ethics Edge (ICMA, 2006) and with Sussman and Daynes, American Politics and the Environment (2002). For 16 years he has been the Managing Editor of Public Integrity, a journal co-sponsored by the American Society for Public Administration, the Council on Governmental Ethics Laws, the International City/County Management Association, the Council of State Governments and the Ethics Resource Center. He taught previously at the University of Houston and University of Arizona and served as a management analyst in the U. S. Surgeon General's Office, Department of the Army, Washington, D. C.

UNIVERSITY OF MICHIGAN

Sendil Ethiraj

Best Overall Faculty Member, Business Administration
Best Researcher/Scholar, Business Administration
Best Teacher, Business Administration

2002 THE WHARTON SCHOOL – UNIV. OF PENNSYLVANIA Ph.D. in Management Specialization: Strategy PhD DISSERTATION Four Essays on Innovation and Performance in Complex Systems EMPLOYMENT • Associate Professor of Strategy, Stephen M. Ross School of Business, September 2008 onwards. • Mike R. and Mary Kay Hallman Fellow, Stephen M. Ross School of Business, September 2003 onwards. • Assistant Professor of Strategy, Stephen M. Ross School of Business, September 2002–2008. AWARDS • PhD teaching excellence award at the Ross School of Business, 2005. • Sanford R. Robertson Chair in recognition of early career research and teaching excellence, 2005. • Mike and Mary Kay Hallman Fellowship at the Ross School of Business, 2003–present. • Finalist for Best Dissertation Prize of the Business Policy and Strategy Division of the Academy of Management, 2003. • Finalist for Best Dissertation Prize of the Technology and Innovation Management Division of the Academy of Management, 2003. • Awarded the Technology and Innovation Management Division Best Student Paper prize at the Academy of Management, 2002. • Winner of the SMS/Booz Allen & Hamilton best paper prize, Strategic Management Society Meetings, San Francisco, 2001. • Finalist for the McKinsey/SMS best conference paper prize, Strategic Management Society Meetings, San Francisco, 2001.

Judith Wismont

Best Teacher, Nursing

"Judy is a student advocate, leader and all round kind person who is an outstanding lecturer and fun faculty member."

Dr. Wismont has been involved with both the traditional and second career undergraduate nursing programs over the span of her 22 years with the U-M School of Nursing and she is currently responsible for the maternity nursing course in the traditional program. Dr. Wismont believes that while teaching facilitates learning, learning is an participatory process engaged in by students and, to that end, the components of her courses are structured in ways that best promote active learning opportunities. Didactically, she accomplishes this through scenario participation and small group work. Clinically, in union with unit based clinical mentors, Dr. Wismont facilitates student development in the areas of clinical competence as well as professional identity through the focused application of theory to clinical situations and ever increasing autonomy for patient care. Being a catalyst for growth and then seeing that growth actualized continues to bring Dr. Wismont joy and satisfaction.

UNIVERSITY OF MINNESOTA DULUTH

Jill Klingner

Best Teacher, Business
Most Helpful to Students, Business
Best Overall Faculty Member, Business

Jill Klingner is an Associate Professor of Health Care Management and Operations Management at the University of Minnesota Duluth; she joined UMD in Fall 2006. Dr. Klingner became a member of the Department of Economics in Fall 2013 when the Health Care Management Program was moved to the Department of Economics. She received her Ph.D. in Health Services Research, Policy and Administration from the University of Minnesota. Her research interest areas include rural health care, Medicare, health outcomes, access to supplemental insurance for disabled Medicare beneficiaries, quality of care, and quality measurement development and implementation. She also received her R.N. from the University of Minnesota. Dr. Klingner is a longtime member of the

University of Minnesota's IRB and has served as a mentor and advisor for health management students. She is a Registered Nurse with a wide range of clinical experience from pediatric intensive care to refugee work in Thailand. Prior to graduate school, Ms. Klingner worked at MedCenters/HealthPartners, a nationally recognized HMO, in quality and utilization review, analysis and development. Dr. Klingner teaches courses in Health Care Management and Operations Management.

Jennifer Schultz

Best Researcher/Scholar, Business

Dr. Schultz is an associate professor in the Department of Economics and Director of the Health Care Management Program. Her fields of research include health economics, pharmacoeconomics, and health policy. She is currently evaluating the effects of health insurance benefit costs on demand for full-time and part-time labor and retirement decisions; the effects of social capital on health in the U.S.; and taxation of unhealthy foods and purchasing behavior. Prior to joining the University of Minnesota, Dr. Schultz was a faculty member at Cornell University where she analyzed consumer decision-making in health care, use of health care information, and perceptions of quality differences across health care providers. Dr. Schultz's previous research has also included an evaluation of consumer driven health care (a health care purchasing arrangement by large employers), an investigation of the selection of health care provider groups by employees and families to determine their sensitivity to price and quality measures, and an empirical analysis of risk redefinition. As a research consultant for Ingenix, UnitedHealth Group, she analyzed health care utilization and costs for a variety of health conditions, such as rheumatoid arthritis, bipolar disorder, hyperlipidemia, cancer, migraine, and asthma. She received her Ph.D. from the University of Minnesota and her M.A. in Economics from Washington State University. She has published articles in the Journal of Health Economics, Health Services Research, Medical Care, American Journal of Managed Care, and Milbank Quarterly and has presented research at academic and professional conferences.

UNIVERSITY OF MISSOURI - COLUMBIA

Nancy Cheak-Zamora

Best Researcher/Scholar, Health Science

UNIVERSITY OF MONTANA - MISSOULA

Garry Kerr

Best Teacher, Anthropology

UNIVERSITY OF MOUNT UNION

Mandy Capel

Best Overall Faculty Member, Education
Best Overall Faculty Member, Education

Jennifer Martin

Best Researcher/Scholar, Education
Best Researcher/Scholar, Education

Jennifer L. Martin, Ph.D., is an assistant professor of education at the University of Mount Union. Prior to working in higher education, Dr. Martin worked in public education for 17 years, 15 of those as the department head of English at an alternative high school for at-risk students. She has served as a mentor to high school, undergraduate, and graduate students, as well as to new teachers in a variety of areas such as writing and publishing, career and leadership development, and advocacy. Dr. Martin is the editor of the two-volume series Women as Leaders in Education: Succeeding Despite Inequity, Discrimination, and Other Challenges (Praeger, 2011) which examines the intersections of class, race, gender, and sexuality for current and aspiring leaders from a variety of perspectives. She has conducted research, published numerous peer reviewed articles and book chapters on bullying and harassment, peer sexual harassment, educational equity, mentoring, issues of social justice, service-learning, and teaching at-risk students. In terms of current and future research, Dr. Martin is studying the cohort and online models in education degree programs. Her other research interests include school reform for social justice, women and leadership, and teacher leadership.

Pete Schneller

Best Teacher, Education

Teaching chose me. During my undergrad years, I was a camp counselor at Red Arrow Camp in Minocqua, Wisconsin. Actually, Chris Farley was a camper in my cabin for about two summers. Can you imagine going canoeing with a young Chris Farley? When I graduated from college I didn't have a teaching certificate; in fact, my first job was in sales. It appeared that I would be good at sales, but I was encouraged to oversell to my customers—to me, this was dishonest. I couldn't live with myself and do what I was told, so I quit and started working a general labor job. After a while of working with molten steel, I decided to pursue another career and went back to school to get a second degree in education. Now I don't feel dishonest in my job and only lie to my students when I play devil's advocate.

UNIVERSITY OF NEBRASKA - KEARNEY

Peter Longo

Best Overall Faculty Member, Political Science

B.A., Creighton University J.D., University of Nebraska - Lincoln Ph.D., University of Nebraska - Lincoln Hometown: Omaha-Bellevue, Nebraska Joined UNK Faculty: 1988

Constitutional Law, Public Policy

People who study constitutional law analyze how public policy develops through courts. Important questions driving inquiry in this subfield include:

How does the formal constitution play out in the policy process? What lessons can be learned from comparative constitutionalism and policy? Can "good constitutionalism" lead to a "good life"? On working with undergraduate students: "I am privileged to have the opportunity to work with undergraduate students. I like to encourage independent thought. Part of my role is to provide challenges and their task is to respond by answering the challenges and moving well beyond the initial challenge. Further, it is important for me to encourage students to be active and engaged citizens. I have seen many of our students make tremendously positive impacts on our campus and well beyond."

Favorite quasi-political quote: "It's better to burn out than it is to rust"—Neil Young

UNIVERSITY OF NEBRASKA - OMAHA

Randall Adkins

Best Overall Faculty Member, Political Science
Best Researcher/Scholar, Political Science
Best Teacher, Political Science
Best Researcher/Scholar, Public Affairs, Political & Policy Sciences
Best Teacher, Public Affairs, Political & Policy Sciences
Best Overall Faculty Member, Public Affairs, Political & Policy Sciences

Dr. Adkins joined the faculty at the University of Nebraska at Omaha in 2000 with previous teaching experience at Concord College and California University of Pennsylvania. He grew up in West Virginia and received his Bachelor's degree from Marshall University and Doctorate from Miami University in Oxford, Ohio. He primarily teaches courses on campaigns and elections, Congress, and the Presidency.

Jody Neathery-Castro

Most Helpful to Students, Political Science
Most Helpful to Students, Public Affairs, Political & Policy Sciences

UNIVERSITY OF NEVADA - RENO

MaryAnn Demchak

Best Overall Faculty Member, Special Education

Sheri Faircloth

Best Overall Faculty Member, Business

Ph.D., University of Texas - Arlington, 1997

Karl Geisler

Most Helpful to Students, Economics

Federico Guerrero

Best Teacher, Economics

Derek Kauneckis

Best Teacher, Public Affairs, Political & Policy Sciences
Most Helpful to Students, Public Affairs, Political & Policy Sciences

Derek Kauneckis' (Ph.D. Public Policy, Indiana University, Bloomington, 2005) research involves understanding local-level processes in public policy development, including local governance arrangements in environmental and science & technology policy. His research has included examining collaborative policy networks, the role of local governments in climate policy, and decision-making by water managers. Current projects include explaining the diversity of property rights arrangements to environmental resources, incorporating behavioral theories into policy design, and climate change adaptation policy. His expertise includes environmental decision-making, institutional analysis and development, and policy evaluation and analysis.

Jeanne Wendel

Best Overall Faculty Member, Economics
Best Researcher/Scholar, Economics

UNIVERSITY OF NEW HAMPSHIRE

Vicki Banyard

Best Overall Faculty Member, Psychology

Vicki Banyard, Ph.D. is a professor in the Department of Psychology with an affiliation with the Justice Studies Program. She is a research and evaluation consultant with Prevention Innovations. She received her Ph.D. in clinical psychology from the University of Michigan and has trained at both the Family Research Lab, University of New Hampshire and the Trauma Center in Boston. She conducts research on the long-term mental health consequences of interpersonal violence including resilience in survivors. She also conducts research on community approaches to prevention of interpersonal violence. She is part of a research team that has received funding from the NIJ, DOJ and CDC.

She is currently collaborating on a research project with Sewanee: The University of the South that evaluates the longest-running and original character writing program, The Laws of Life Essay Contest.

Katie Edwards

Best Researcher/Scholar, Psychology

Katie Edwards is an assistant professor of psychology and women's studies, and faculty affiliate of Prevention Innovations and the Carsey School of Public Policy. Dr. Edwards joined the UNH faculty in 2011 after completing her clinical internship at the Vanderbilt University-Department of Veteran's Affairs Consortium in Nashville Tennessee. She earned her Ph.D. in clinical psychology and graduate certificate in women's studies from Ohio University. Dr. Edwards teaches undergraduate and graduate-level classes on topics related to clinical and counseling psychology, women and gender studies, and interpersonal violence. Dr. Edwards' interdisciplinary program of research focuses broadly on better understanding the causes and consequences of interpersonal violence, primarily intimate partner violence (IPV) and sexual assault (SA) among adolescents and young adults. Specific areas of current research focuses on risk and protective factors for IPV and SA perpetration, victimization, and bystander intervention; disclosure, leaving, and recovery processes among survivors of IPV and SA; and individuals' reactions to participating in IPV and SA research. Dr. Edwards uses this research data to develop, implement, and evaluate IPV and SA prevention, intervention, and policy efforts.

Maureen Gillespie

Best Teacher, Psychology
Most Helpful to Students, Psychology

Becky Warner

Most Helpful to Students, Psychology

Peter Yarensky

Best Teacher, Psychology

UNIVERSITY OF NEW HAVEN

Martin O'Connor

Most Helpful to Students, Criminal Justice

"His dedication to students is legendary & spans decades."

Mr. O'Connor served on the New Haven Fire Department for thirty years and retired as chief of the department in 1998. He is a licensed attorney in Connecticut and an ordained deacon of the Archdiocese of Hartford. Mr. O'Connor also serves as the campus chaplain, faculty advisor to the President's Public Service Fellowship, and faculty member of the university's committee on community service.

Anshuman Prasad

Best Researcher/Scholar, Management

I draw upon inter-disciplinary perspectives, my scholarship deals with issues such as strategic management under conditions of globalization and postcoloniality, workplace diversity and multiculturalism, corporate legitimacy, resistance in organizations, and epistemological and methodological issues. My research activities during the past 5 years include the publication of a scholarly edited book with a major international university press, articles in high-impact factor peer-reviewed journals such as Human Relations and Organization, several refereed presentations at the prestigious Academy of Management Annual Meetings, a number of invited national and international presentations (including key-note addresses), and a series of Professional Enrichment Program (P.E.P.) Seminars at UNH that draw upon my ongoing research on globalization. Currently, I am working on a book that explores the multi-dimensional implications of ongoing global transformations, and co-editing the scholarly reference volume, Routledge Companion to Critical Management Studies.

UNIVERSITY OF NEW MEXICO

Melissa Binder

Best Teacher, Economics

"She is never distracted from the task of helping students reach a higher level of competency and understanding. She holds herself and her students to the highest standards of integrity."

Melissa Binder, Ph.D., has been a professor in the Department of Economics at the University of New Mexico since 1995 and the Director of ISR since the summer of 2013. Her research and policy interests include labor market and educational equity, and among her publications are analyses of the motherhood wage gap, gender equity in faculty pay, and the socio-economic achievement gap in higher education. As director of ISR, Dr. Binder participates in data and policy analysis for the New Mexico community in the areas of pay equity, family well-being, early childhood programs and K–12 initiatives.

Debra Brady

Best Overall Faculty Member, Nursing

"Dr. Brady recently retired from the College of Nursing and during her tenure she excelled as a nurse educator and researcher, publishing several articles. She held an administrative position as a team leader for the education team and she was excellent in this role: helpful to faculty when they had difficulties in a course assignment and always willing to meet with individuals and others to resolve any issues or conflicts. She provided a good role model for others following in her footsteps. This award would mean a great deal to her; recognizing the significant achievements she had during her time at the College of Nursing."

Matthew Lemberger-Truelove

Best Researcher/Scholar, Education

Robin Meize-Grochowski

Best Researcher/Scholar, Nursing

Dr. Meize-Grochowski, Regents' Professor of Nursing, is interested in symptom management in adults. Her most recent research focused on the use of meditation in older adults with postherpetic neuralgia, and she plans to extend this research to other persistent pain conditions. Dr. Meize-Grochowski also has an interest in spirituality, holistic nursing interventions (including healing touch), and quality of life in celiac disease and type 1 diabetes mellitus.

Originally from Farmington, NM, Dr. Meize-Grochowski received her BSN from Creighton University in Omaha, Nebraska, and her MSN and PhD in Nursing from The University of Texas at Austin. Since 1979, she has been involved in nursing education or higher education administration at UNM. She teaches graduate and undergraduate courses, including nursing research, honors, theory, gerontology, patient education, and teaching. Her latest funding (2012–2013) was from the NIH Pain Consortium, as Principal Investigator for a Center of Excellence in Pain Education.

Nancy Morton

Best Overall Faculty Member, Nursing
Most Helpful to Students, Nursing

Nancy Morton, MS, RN, CNE is currently Undergraduate Program Director at the College of Nursing, Health Science Center, University of New Mexico. She earned her Baccalaureate of Science in Nursing at California State University, Chico and her Master of Science in Nursing, Adult Health and Education tracks, at the University of Arizona. Nancy's clinical background is in adult acute and critical care. Since Nancy's employment at the College in 1994, she has taught many different undergraduate courses, predominantly adult acute and critical care, internships, and pharmacology. Nancy is also the Program Co-director of the Veteran's Administration Nursing Academy Partnership and the Co-chair of the New Mexico Nursing Education Consortium Curriculum committee.

Mark Parshall

Best Researcher/Scholar, Health Sciences and Nursing

UNIVERSITY OF NEW ORLEANS

Walter Lane

Best Teacher, Economics

"Excellent teacher, excellent colleague."

UNIVERSITY OF NORTH ALABAMA

Joyce McIntosh

Most Helpful to Students, Physical Ed

UNIVERSITY OF NORTH CAROLINA

Ralph Byrns

Best Teacher, Economics

"Great mentor and teacher - I learned so much from him about being a great teacher which shaped my own teaching and helped me succeed as one myself"

Prior to joining the Department of Economics at UNC-Chapel Hill as an Adjunct Professor in 2001, Ralph Byrns taught at Rice,

Clemson, Metropolitan State College at Denver, the University of Colorado at Denver, the University of Colorado at Boulder, Loyola University of Chicago, Greensboro College, and Duke. His research interests include the economics of personal relationships, the sociology of economics as a discipline, behavioral economics, and the logic of economic theoreticians vis-à-vis the art and science of economics.

Byrns' writings for economic education currently focus on Economicae: an Interactive Encyclopedia for Economics, and he's seeking a coauthor for his intermediate microeconomics text. His Economics text (with Gerald W. Stone, now in a 6th edition) has been used by more than one million students at over 1200 universities since 1981. He regularly teaches large sections of Introduction to Economics, upper division courses in History of Economic Doctrines and Financial Markets, and has been honored with numerous teaching awards. He also sponsors the Carolina Economics Club and enjoys debating economic issues and talking about possible career paths with students.

Professor Byrns holds a B.S. from Arizona State University (1965), and an M.A. (1972) and Ph.D. (1977) from Rice University. He and his wife, Patricia J. Byrns (M.D., Associate Dean, UNC Medical School) live in Chatham County. He enjoys reading about philosophy, science, and politics, plays golf wretchedly, and was the world champion backgammon player on the internet in 2002.

UNIVERSITY OF NORTH CAROLINA AT CHARLOTTE

Tina Heafner

Best Overall Faculty Member, Education
Best Researcher/Scholar, Education
Best Teacher, Education
Best Overall Faculty Member, Education
Best Teacher, Education

Tina L. Heafner, Ph.D. is Professor at the University of North Carolina at Charlotte. She is the lead social studies methods instructor for undergraduate and graduate programs in the Department of Middle, Secondary, K–12 Education. Tina serves as the Director of the M.Ed. in Secondary Education, the Minor in Secondary Education and the College of Education's Prospect for Success.

William Siegfried

Best Teacher, Psychology

UNIVERSITY OF NORTH CAROLINA WILMINGTON

Angela Housand

Most Helpful to Students, Education

Linda Mechling

Best Researcher/Scholar, Education

Mahnaz Moallem

Best Researcher/Scholar, Education

Mahnaz Moallem is a Professor of instructional technology and research and Program Coordinator of Instructional Technology Master's Program in the Watson School of Education at UNCW. She received her Ph.D. and her Master of Science degree in instructional systems design/instructional technology and her certification in program evaluation. She has NC Teaching Certification in the areas of instructional technology, curriculum and instruction, and elementary education. She has taught several undergraduate and graduate courses in the areas of instructional design/instructional technology and classroom evaluation and research. She has conducted both qualitative and quantitative research and received a national research award for the best qualitative study in the field of instructional technology. Dr. Moallem has long term experiences in using emerging technology in designing, developing, evaluating, and delivering undergraduate and graduate courses. She has been involved in several school-based research projects and served on many technology committees. Dr. Moallem is an active member of Association for Educational Communication and Technology and American Educational Research Association. She has presented and published in the area of instructional design and instructional technology extensively.

Denise Ousley-Exum

Best Teacher, Education

Jeanne Swafford

Best Overall Faculty Member, Education
Best Teacher, Education
Most Helpful to Students, Education
Best Overall Faculty Member, Education

UNIVERSITY OF NORTH DAKOTA

Robert Dosch

Best Overall Faculty Member, Accounting

Ric Ferraro

Most Helpful to Students, Psychology

Alison Looby

Best Overall Faculty Member, Psychology
Best Teacher, Psychology

Karyn Plumm

Most Helpful to Students, Psychology

Cheryl Terrance

Best Overall Faculty Member, Psychology
Best Researcher/Scholar, Psychology

Heather Terrell

Best Teacher, Psychology

Jeff Weatherly

Best Researcher/Scholar, Psychology

UNIVERSITY OF NORTH FLORIDA

William Ahrens

Best Overall Faculty Member, Nursing

Mr. Ahrens was born and raised in Springville, New York.
 He was commissioned an Ensign, Nurse Corps, United States Navy in 1970 and began active service at the Naval Hospital, Great Lakes, Illinois. During his Naval career he rose to the rank of Captain and held a variety of positions including:
 • Staff Nurse and Clinical Instructor, Coronary Care Unit, Naval Regional Medical Center, Portsmouth, Virginia; • Head, Staff Education and Training Department and Interim Director for Administration, Naval Hospital, Naples, Italy; • Director, Health Promotion Division, Headquarters, Naval Medical Command, Washington, DC; • Assistant Director, Nursing Services, Naval Hospital, Pensacola, Florida; • Director, Nursing Services, Naval Hospital, Jacksonville, Florida; • Director, Nursing Services, National Naval Medical Center, Bethesda, Maryland.
 He also served as Specialty Consultant for Nursing Practice to the Surgeon General of the Navy from 1995 to 1998. Retiring from active Naval service in August, 1998, he joined

the Nursing faculty at the University of North Florida in Jacksonville, Florida; where he is currently a Senior Instructor and Director of the Regular Pre-licensure nursing program.

He is certified as a Diplomate of the American Board of Risk Management, and is listed in the National Distinguished Service Registry in Nursing. He is a Wharton School of Business/Johnson & Johnson Nurse Executive Fellow.

He is married to the former Judy Morrison; also a Registered Nurse practicing in geriatric nursing at Taylor Manor in Jacksonville, Florida. They have three children; Allison, a Physical Therapist; Melissa, a secondary school English Literature teacher; and Becky, a Special Education teacher.

Kathy Bloom

Best Researcher/Scholar, Nursing

Dr. Bloom is a certified nurse-midwife and has been in clinical practice in women's health for over 30 years. She teaches research, evidence-based practice, assessment, and women's health to undergraduate and graduate nursing students. Dr. Bloom's research includes investigation of advanced practice nurses' knowledge of and adherence to best practice guidelines and evaluation of innovative educational practice and advanced practice models.

Judy Comeaux

Best Teacher, Nursing

Hubert Gill

Most Helpful to Students, Accounting

Caroline Guardino

Best Researcher/Scholar, Education

Lillia Loriz

Most Helpful to Students, Nursing

Dr Loriz is a Professor in the School of Nursing. She is is the director for the School of Nursing. Dr Loriz is a Nurse Practitioner and maintains a part-time practice at Student Health Services. Her early career started in Washington, DC where she worked closely with HIV/AIDS patients. After completing her PhD, she moved to Florida where she began her employment at the University of North Florida. Dr Loriz is a beaches resident. She is married to Tony Flaris. Her main hobby is scuba diving, particularly cave diving throughout Florida. She is an avid supporter of the environment, particularly water supply in Florida.

Jeffrey Michelman

Most Helpful to Students, Business
Best Overall Faculty Member, Business

Dr. Michelman is Associate Dean for Undergraduate Studies, Director of the Honors Program and a Professor of Accounting at the University of North Florida in Jacksonville, Florida, where he has been on the faculty for the past twenty-four years. He was runner-up for UNF's Distinguished Professor Award, a recipient of the Coggin College of Business' Distinguished Business Leaders Award, the University of North Florida's Outstanding Research, Undergraduate Teaching, Faculty Service and Outstanding International Service Awards. Prior to joining UNF, Jeff spent four years on the faculty at Penn State University and was a Bank Examiner with the FDIC in the Pittsburgh, Pennsylvania Field Office. Dr. Michelman is both a CPA and CMA. Jeff is a 1996 graduate of Leadership Jacksonville. He received his B.S. degree from the University of Delaware in accounting and economics, his M.H.A. degree from Washington University in St. Louis, and both an M.B.A. and a Ph.D. degree in business from the University of Wisconsin-Madison, where his major was Accounting and minor fields were Health Care Fiscal Management and Administrative Medicine. He has published articles in a variety of information systems, accounting, international business and healthcare journals. He has written three textbooks and developed several computer software programs to integrate technology into the accounting curriculum. Dr. Michelman has taught in Paris, France; Guangzhou, China; Warsaw, Poland and Fiji at the University of the South Pacific where he was a visiting scholar in 2005. He is currently a Visiting Professor at Warsaw University where he lectures each year. He has been a member of several visiting committees of the Commission for Academic Accreditation, Ministry of Higher Education and Scientific Research, United Arab Emirates. Dr. Michelman lectures extensively to practitioners on Sarbanes-Oxley, Internal Control, Corporate Governance, Disaster Planning, Fraud, the Internet and the overall use of Accounting and Technology in Healthcare and other service organizations. Dr. Michelman has been a member of the editorial board of Research in Healthcare Financial Management since 1996. Jeff served as a member of the national Healthcare Financial Management Association's Accounting Principles and Practices Board from 1999–2002, where he served on the on the Business Combinations and Corporate Compliance sub-committees. He was President of the Florida Chapter of the Healthcare Financial Management Association for 1998–99, and received the Medal of Honor in 2006. He was a member of the 1997/98 National HFMA Task Force on the Future of the Healthcare Financial Management Profession. Jeff is an active member of the Oceanside Rotary club where he has served as a member of the Board of Directors and currently chairs the New Generations

Avenue as well as chairing the regional Ambassadorial Scholarship Selection Committee; was a member of the Youth Leadership Jacksonville Council where he is a Past Chairman and a member of the Leadership Jacksonville Board of Directors and executive committee where he served as chair of the development committee and treasurer. Dr. Michelman was a member of the inaugural class of the Community Coaches Program which was jointly sponsored by the DuPont and Jacksonville Community Foundations. He currently serves as a consultant to several non-profit organizations in Jacksonville. Dr. Michelman currently serves as the Chair of the Advisory Board for the Career Academies of Marketing and International Business at Ponte Vedra High School.

Karen Patterson

Best Teacher, Education

Dr. Karen Patterson is Associate Professor and Department Chair of Exceptional, Deaf, and Interpreter Education at the University of North Florida. Her research interests include: teacher preparation and classroom management; positive academic and social interventions; learners with intellectual and developmental disabilities; learners with emotional and behavioral disorders; and autism.

Steve Paulson

Best Teacher, Business

Sidney Rosenberg

Best Researcher/Scholar, Business

Diane Tanner

Best Overall Faculty Member, Accounting
Best Researcher/Scholar, Accounting
Best Teacher, Accounting

Diane Tanner has served on the faculty of the University of North Florida for 26 years and is currently a Senior Instructor of Accounting. She is a UNF alum having received both her

BBA and Masters of Accountancy degrees from UNF. She teaches primarily cost accounting and the lower-level financial and managerial accounting courses. She serves as coordinator of both lower-level accounting courses and a mentor to adjuncts who teach those courses.

Tanner is a Certified Public Accountant and has held positions as an internal auditor, cost accountant, controller, and as principal in her own public accounting firm. She is the author of numerous national accounting textbook ancillaries, and has created a plethora of online learning materials, both for national publishers and her own students.

This award is her fifth Outstanding Undergraduate Teaching Award.

Kristine Webb

Most Helpful to Students, Education

Kristine (Kris) Wiest Webb, Ph.D. is a Professor in the Department of Exceptional, Deaf, and Interpreter Education and former director of the Disability Resource Center (DRC) at the University of North Florida (UNF). Under her direction, the DRC experienced tremendous growth in the numbers of students who use accommodations and services and the number of successful graduates increased 10 times the number of students who graduated a decade earlier. Further, the grade point averages of the graduates who were registered with the DRC were equal or exceeded the mean grade point average of UNF graduates from the general population. Kris has written or edited a number of books about the transition process including the most recent, Handbook of Adolescent Transition Education for Youth with Disabilities (2012). Kris is a Past-President of the International Division on Career Development and Transition (DCDT), an organization dedicated to improving life for adolescents and adults with disabilities. She was the 2014 recipient of the Desmond Tutu Peace and Reconciliation Award and in 2007 was awarded the Outstanding Faculty Service Award at the University of North Florida. In addition, Kris was the UNF 2003 CASE Undergraduate Teaching Award nominee and received the Outstanding Undergraduate Teaching Award for 2001–2002. She was the 2012 EVE winner in Education, an award given by the Florida Times Union. Before coming to UNF, Kris served as the Director of the Florida Network: Information and Services for Adults and Adolescents with Disabilities housed at the University of Florida. Prior to that position, she was the coordinator of a collaborative special education intern program at the University of New Mexico. Before her own transition to higher education, Kris was a high school teacher for 17 years in Colorado and New Mexico. In addition to her own service commitments, she has developed several programs for students enrolled in education classes and registered with the Disability Resource Center to highlight the importance of community and professional service. Along with her interest in teacher preparation, Kris has a long-standing passion for promoting successful postsecondary education experiences for individuals with disabilities, family involvement and collaboration, and transition to adult life for individuals with disabilities. "There is something that is much more scarce, something finer far, something rarer than ability. It is the ability to recognize ability." Elbert Green Hubbard

UNIVERSITY OF NORTH TEXAS

Richard Rogers

Best Researcher/Scholar, Psychology
Best Researcher/Scholar, Medicine

Richard Rogers, Ph.D., ABPP, is a Regents Professor of Psychology at the University of North Texas. In 2014, he received the Eminent Faculty Award, one of the university's highest faculty honors, which is bestowed one distinguished professor annually. His previous academic appointments included key positions in the Section on Psychiatry and Law, Rush University, and the Division of Forensic Psychiatry, University of Toronto.

Dr. Rogers is nationally recognized for his contributions to forensic psychology and psychiatry. National awards include (1) the Manfred S. Guttmacher Award from the American Psychiatric Association, (2) Distinguished Contributions to Forensic Psychology Award from the American Academy of Forensic Psychologists, and (3) the Amicus Award for the American Academy of Psychiatry and Law.

His contributions to clinical psychology and the discipline of psychology have also been commended. In 2007, the Society of Clinical Psychology honored Dr. Rogers with Distinguished Professional Contributions to Clinical Psychology. In 2008, American Psychological Association bestowed its prestigious Distinguished Professional Contributions to Applied Research Award. In 2011, American Psychological Association again honored Dr. Rogers with its Distinguished Professional Contributions to Public Policy Award.

As a prolific writer with more than 190 refereed articles, Dr. Rogers has written seven books focused on clinical and forensic practice. He has also developed and validated four psychological measures that are published by Psychological Assessment Resources, Inc. (PAR, Inc.):

Rogers Criminal Responsibility Assessment Scales (R-CRAS) Evaluation of Competency to Stand Trial-Revised (ECST-R) Structured Interview of Reported Symptoms-2nd Edition (SIRS-2) Standardized Assessment of Miranda Abilities (SAMA).

Dr. Rogers just completed a decade of support from National Science Foundation as a principal investigator examining Miranda warnings and waivers. His programmatic research led to the development of Miranda measures (SAMA) and was cited by the American Bar Association in its call for national reform of juvenile Miranda warnings.

Dr. Rogers is well regarded as a teacher, especially at the graduate and post-doctoral levels. He was selected by the UNT Graduate School as the 2004 Toulouse Scholar to honor his outstanding teaching and scholarly achievements.

UNIVERSITY OF NORTHERN COLORADO

Alena Clark

Best Overall Faculty Member, Health Sciences and Nursing
Most Helpful to Students, Health Sciences and Nursing

Alena Clark is currently the Program Coordinator and an Associate Professor in the Nutrition and Dietetics Program at the University of Northern Colorado in Greeley, Colorado. Alena's current research focuses on developing and evaluating programs that support breastfeeding in the community, infant nutrition especially introducing solid foods and preconception health for women of child bearing age. She serves as a lactation counselor for her community. She earned a Ph.D. in Human Nutrition at Colorado State University, a M.P.H. in Public Health Nutrition at the University of Minnesota and a B.A. in Nutrition and Dietetics at Concordia College. In addition, she serves on local and state dietetic associations and is the Colorado Delegate for The Academy of Nutrition and Dietetics. Alena is a registered dietitian and has worked in various settings including hospitals, public health agencies and academia. Alena is involved in many organizations that advocate the need to promote healthy lifestyles for women, children and infants including advisory boards for preconception health and breastfeeding coalitions.

Dave Thomas

Best Teacher, Management

UNIVERSITY OF OKLAHOMA - HEALTH SCIENCES CENTER

Valerie Eschiti

Best Researcher/Scholar, Nursing

"Valerie has a good track record of funded research in care of Native American women."

Mark Fisher

Most Helpful to Students, Nursing

"Mark is student friendly and approachable."

Neil Henderson

Best Overall Faculty Member, Health Sciences and Nursing

"Impressive in many ways"

Patti Landers

Most Helpful to Students, Health Sciences and Nursing

"Very approachable for students"

Rhonda Lawes

Best Teacher, Nursing
Best Teacher, Health Sciences and Nursing

"Rhonda excels in the area of lecture, and making complex concepts more understandable."

Voncella McCleary-Jones

Best Overall Faculty Member, Nursing

"Voncella is a great role model for students as well as faculty."

Janet Wilson

Best Researcher/Scholar, Health Sciences and Nursing

"Janet has expertise working with multiple disciplines in the area of Elder Abuse and Financial Elder Abuse."

UNIVERSITY OF OREGON

Gerald Berk

Best Researcher/Scholar, Political Science

Priscilla Southwell

Best Overall Faculty Member, Public Affairs, Political & Policy Sciences

Francis White

Best Overall Faculty Member, Anthropology

"Excellent department head that department business runs efficiently."

Dr. Frances White is a biological anthropologist who specializes in behavioral ecology. Her research focuses on the evolution of non-human primate sociality and social systems. She has active field projects with wild bonobos in the Democratic Republic of the Congo and free-ranging primates in the US. She directs the Primate Data Lab and the Primate Osteology research laboratory which houses the UO Comparative Osteology Collection. Dr. White's publications have appeared in the American Journal of Primatology, International Journal of Primatology, Folia Primatologica, and American Journal of Physical Anthropology. Her research has been supported by the National Science Foundation, the Leakey Foundation and other sources. She teaches introductory classes on non-human primates and the evolution of human behavior and sexuality and a Freshman Interest Group (FIG) in animal behavior. Her upper level classes focus on primate behavior, conservation, and feeding and nutrition. Dr. White also teaches graduate level biological statistics. Dr. White currently serves as the Department Head of Anthropology.

UNIVERSITY OF PENNSYLVANIA

Jon Merz

Best Researcher/Scholar, Medicine

UNIVERSITY OF PITTSBURGH

Scott Drab

Best Teacher, Pharmacy

Scott R. Drab, PharmD, CDE, BC-ADM Associate Professor, Pharmacy and Therapeutics

Bio: Scott Drab is an Associate Professor of Pharmacy & Therapeutics and Director of University Diabetes Care Associates. He received his Bachelor of Science degree in pharmacy from the University of Pittsburgh and his Doctor of Pharmacy degree from Duquesne University.

Dr. Drab's efforts in contributing to pharmaceutical care led to the creation of one of the first pharmacist run diabetes care centers located in a community independent pharmacy. As a Certified Diabetes Educator and director of the clinic, he is responsible for care plan development, education, and patient follow up. He has managed the care of hundreds of diabetic patients over the years, improving clinical health outcomes. In addition to the direct patient care responsibilities, Dr. Drab is responsible for providing drug information to the surrounding medical community and provides outreach education to benefit the greater good of the community. Today, University Diabetes Care Associates serves as a model and prototype for future care centers.

Dr. Drab's teaching responsibilities at the School of Pharmacy involve clinical instruction as well as classroom teaching. His classes are centered around collaborative and inquiry-based learning while he strives to provide a classroom environment that simulates the practice environment.

Dr. Drab currently serves on several School and University committees including the Experiential Learning Committee, Academic Performance Committee, Pharm D Council, University Senate Community Relations Committee and the Pitt Pathway Committee. He also serves as faculty advisor and liaison between students and the National Community Pharmacists Association. He has received the Roche Preceptor of the Year Award in 2003 and 2008 and was selected as the 2007 Recipient of the Pennsylvania Society of Health System Pharmacist's Joe E. Smith Award. The Joe E. Smith award is awarded to a pharmacist who is a member of PSHP and who demonstrates excellence in practice and is deserving of recognition for service to his institution, the community and the profession.

UNIVERSITY OF PUGET SOUND

Alva Butcher

Best Overall Faculty Member, Business

Lisa Johnson

Best Researcher/Scholar, Business
Best Researcher/Scholar, Business

Lisa's primary areas of interest are the legal and moral status of non-human animals and animal ethics in political theory. For the environmental policy and decision making minor, she teaches Environmental Law and the Environmental Studies Senior Seminar. She also teaches Business and the Natural Environment, which is a freshman seminar. Her students have written grant proposals to the Environmental Protection Agency relating to sustainability issues, and they have completed a preliminary comparative study relating to community gardens and connection to the land. She earned her JD from the Northwestern School of Law of Lewis and Clark College, where she focused on Environmental and Natural Resources Law. Her Master of Public Affairs was earned from Indiana University's School of Public and Environmental Affairs, with a focus on International Environmental Policy.

Lynda Livingston

Best Teacher, Business

Jeff Matthews

Best Overall Faculty Member, Business

Elizabeth Nunn

Most Helpful to Students, Business

Lisa Nunn is Visiting Assistant Professor of International Political Economy and Economics. She earned her BS in Mathematics and Economics from the University of Puget Sound (1985) and her MA (1986) and PhD (1989) in Economics from Washington University in St. Louis. She was a member of the original faculty group that developed the IPE program at Puget Sound and contributed to the first edition of the program's textbook, Introduction to International Political Economy. Her teaching specialties are public finance and American economic history.

Kate Stirling

Best Teacher, Business

Paula Wilson

Most Helpful to Students, Business

UNIVERSITY OF RHODE ISLAND

Bing-Xuan Lin

Best Researcher/Scholar, Finance

Mergers and Acquisitions, Financial Analysts' Forecasts, Corporate Governance and Chinese Capital Market.

Bingxuan Lin is a Professor of Finance at University of Rhode Island. He received his Ph.D. from Georgia State University in 2001. His research focuses on mergers and acquisitions, financial analysts' forecasts, corporate governance and Chinese capital markets. He has published in various academic and practitioner journals including the Journal of Risk and Insurance, Journal of Financial Research, Journal of Accounting and Public Policy, Pacific Basin Finance Journal, Journal of American Taxation Association, and Financial Analyst Journal. He currently teaches both undergraduate and graduate corporate finance courses.

UNIVERSITY OF SAN DIEGO

Dan Rivetti

Best Teacher, Business

"Charismatic Intelligent Easy to Listen to Self Effacing Humor Doesn't Take Himself Seriously, Yet is Obviously On Top of His Game."

Daniel A. Rivetti is an associate professor of Finance at the University of San Diego. As an international finance consultant, with expertise in Real Estate financing, he directed all funding activities for the largest private investment along the U.S. and Mexican border. He established a first and second floor lending arrangement with Banamex and Bancomext respectively for Matrix Aeronautica S.A.de C.V. This project resulted in the largest heavy jet repair facility in Latin America. His consulting experience includes the use of Real Option Theory and has utilized state of the art derivative instruments for one of the largest Crown Corporations in Canada, Ontario Power Generation. He has served on two publicly listed Board of Directors, (BHBC) Beverly Hills Bancorp Inc. (assets sold to Mitsubishi and Sanwa Banks), and (TGRA) Tigera Group, Inc.

UNIVERSITY OF SOUTH CAROLINA

Susan Kuo

Best Teacher, Law

Susan Kuo is a Professor at the University of South Carolina School of Law and an affiliated researcher with the Hazards and Vulnerability Research Institute. Her research focuses on disaster law and policy and has been published in law reviews including the Boston College Law Review, the Washington University Law Review, the Indiana Law Journal, and the U.C. Davis Law Review. She teaches or has taught Criminal Law, Criminal Procedure, Federal Courts, Conflict of Laws, Civil Procedure, Law and Social Justice, and Race and the Law. In 2014, Professor Kuo received the University of South Carolina Michael J. Mungo Graduate Teaching Award. She also received the 2014 Social Justice Award given by the University's Office of Equal Opportunity Programs. At the law school, Professor Kuo was recognized as Best Classroom Teacher in 2014 and received the 2014 Outstanding Faculty Publication Award. In 2008 and 2010, Professor Kuo was voted Outstanding Faculty Member by the student body. In 2013, she was selected for inclusion in a study of "Best Law Teachers" along with 25 other law professors nationwide. The results of this study appear in the book, What the Best Law Teachers Do, published by the Harvard University Press. Professor Kuo served as a Visiting Associate Professor at the University of Alabama School of Law during Fall 2011 and at the University of Iowa College of Law during Spring 2011. Previously, Professor Kuo was an Associate Professor at Northern Illinois University College of Law where she was voted Professor of the Year in 2000, 2002, 2005, and 2006. She also served as a Visiting Assistant Professor at the University of Toledo College of Law. Before entering academia, Professor Kuo was a Special Assistant United States Attorney with the United States Attorney's Office in Atlanta, Georgia. She also completed two federal judicial clerkships, one with Judge Eugene E. Siler, Jr. of the United States Court of Appeals for the Sixth Circuit and the other with Judge Robert H. Hall of the United States District Court for the Northern District of Georgia.

Hayden Smith

Best Overall Faculty Member, Criminal Justice
Best Researcher/Scholar, Criminal Justice
Best Teacher, Criminal Justice
Most Helpful to Students, Criminal Justice

UNIVERSITY OF SOUTH CAROLINA - AIKEN

Arinola Adebayo

Best Teacher, Business
Best Teacher, Business

Leanne McGrath

Most Helpful to Students, Business
Most Helpful to Students, Business

Dr. Leanne McGrath, Professor of Management, brings with her a wealth of knowledge and experience. She challenges her students to apply management concepts to actual situations in the work place.

Bill Shelburn

Best Researcher/Scholar, Business
Best Researcher/Scholar, Business
Best Overall Faculty Member, Business

Professor William Shelburn received his MBA in Marketing from James Madison University in 1973. He obtained his undergraduate degree in history from the College of William and Mary. He has published papers in the areas of international marketing, marketing ethics, marketing research, and sports marketing and management. Professor Shelburn developed and is the advisor for the Golf Course Services Concentration. In fact, he is an editor for one of the texts used in this field. He has done consulting in the areas of banking, retail sales and restaurant location.

Kitty Wates

Best Overall Faculty Member, Business

She uses her experience to integrate classroom knowledge and the business world for students. Professor Wates has several hobbies she enjoys. She is an avid golfer, enjoys snow skiing, and loves to garden in her backyard park.

UNIVERSITY OF SOUTH DAKOTA

Robert Morecraft

Best Teacher, Medicine
Best Teacher, Medicine

"In addition to doing outstanding research for the Sanford School of Medicine, Bob is an amazing teacher."

Steve Waller

Best Researcher/Scholar, Medicine

UNIVERSITY OF SOUTH FLORIDA

Barbara Cruz

Best Overall Faculty Member, Education

Dr. Bárbara C. Cruz is Professor of Secondary Education at the University of South Florida. Her interests include diversity issues in education, teacher preparation, English language learners, and the teaching of Latin America and the Caribbean. Her research on the exclusion and distortions of Hispanic cultures in school books has led to reform efforts in the field. In an effort to correct these misrepresentations, Dr. Cruz has published a number of young adult biographies of inspirational Hispanics (such as Frida Kahlo and Rubén Blades) and developed curricula on Latin America and the Caribbean. She is also the author of César Chávez: A Voice for Farm Workers and Multiethnic Teens and Cultural Identity, for which she received the Carter G. Woodson Book Award. Her current project is a social studies book for English language learners that will be the first of its kind.

In addition to presentations at numerous professional conferences, Dr. Cruz is a frequent presenter at local schools, clubs, and organizations. With the goal of serving as a Latina role model and mentor as well as familiarizing people with the important contributions of Hispanics, Dr. Cruz participates in annual Hispanic Heritage celebrations, is a continuing guest speaker at area schools, and is a frequent presenter at educational conferences and colloquia. In 2000 she was honored as the City of Tampa's Hispanic Woman of the Year in Education.

At the University of South Florida, Dr. Cruz has been recognized for excellence in both research and teaching. She is the recipient of the Jerome Krivanek Distinguished Teacher Award, the Outstanding Undergraduate Teaching Award, and the Honors College Distinguished Service Award. For her nationally recognized research, she has also been honored with the Outstanding Faculty Research Achievement Award and the Faculty Excellence Award. Dr. Cruz also led the Global Schools Project, a collaborative program that prepares teachers and students for an increasingly globalized society. In 2009 she was a lead facilitator on a summer program for Caribbean educators entitled Towards Democracy & Diversity and in 2010 brought 24 teachers from all over the world to Florida for the 6-week program, Interactive Teaching in a Globalizing World.

Robert Dedrick

Best Overall Faculty Member, Education

Dr. Dedrick's research interests include the use of structural equation modeling to examine measurement quality of psychological instruments, the analysis of change using hierarchical linear modeling, and mentoring in doctoral education. His work has been published in Psychological Assessment, Journal of Outcome Measurement, Educational and Psychological Measurement, Sociology of Education, and Mentoring and Tutoring.

Danielle Dennis

Best Overall Faculty Member, Elementary Education
Best Researcher/Scholar, Elementary Education
Best Teacher, Elementary Education
Best Researcher/Scholar, Education
Best Teacher, Education
Best Overall Faculty Member, Education

Dr. Danielle Dennis is an Associate Professor in the Department of Childhood Education and Literacy Studies. She graduated in 2007 with a Ph.D. in Literacy, Language, and ESOL Studies from the University of Tennessee. Dr. Dennis's research agenda focuses on Literacy (1) Assessment, (2) Policy, and (3) Building Teacher Capacity. Currently, Dr. Dennis is collaborating with two partnership teachers on a critical discourse analysis of their language use in literacy instruction. Additionally, Dr. Dennis is conducting research on the impact of Family Literacy Nights on community engagement with schools. Dr. Dennis is the Coordinator of the USF Urban Teacher Residency Partnership Program. She earned the 2009–2010 USF Outstanding Undergraduate Teaching Award, and was recently named a 2013–2014 PDK Emerging Leader.

Deborah Kozdras

Most Helpful to Students, Elementary Education
Most Helpful to Students, Education

Jeff Kromrey

Best Teacher, Education

Dr. Kromrey's specializations are applied statistics and data analysis. He has a general interest in the behavior of numbers, and summaries of numbers, in the context of research. His work has been published in Communications in Statistics, Educational and Psychological Measurement, Journal of Experimental Education, Multivariate Behavioral Research, Journal of Educational Measurement, Psychometrika, and Educational Researcher. He was the associate editor for Review of Educational Research and is currently editor of the Florida Journal of Educational Research and executive editor of Journal of Experimental Education.

UNIVERSITY OF SOUTHERN MAINE - GORHAM

Mark Steege

Best Researcher/Scholar, Education

"Prolific scholar with multiple books and journal articles in school psychology and applied behavior analysis; role model for other scholars."

Mark earned his Educational Specialist (Ed.S.) degree in School Psychology from the University of Iowa in 1982. Mark worked as a school psychologist with the Grant Wood Area Education Agency in Cedar Rapids, Iowa for four years prior to earning his doctorate in school psychology from the University of Iowa in 1986. Mark completed his post-doctoral training as a Pediatric Psychologist within the Department of Pediatrics within the College of Medicine at the University of Iowa. He has written extensively on functional behavioral assessment, single-subject research methods, and the use of empirically-based interventions for students with developmental and behavioral difficulties. He is certified and licensed as a psychologist and is a Board Certified Behavior Analyst-Doctoral.

Jean Whitney

Best Overall Faculty Member, Education

"Outstanding teacher; high standards for accountability for both students and department; excellent administrator; committed scholar; knowledgeable and highly professional; dedicated educator and advocate for social justice and needs of special learners"

UNIVERSITY OF ST. THOMAS

Tonia Bock

Most Helpful to Students, Psychology

Dr. Bock is an educational and developmental psychologist who specializes in moral development. Specifically, she studies how adolescents (late elementary through college-age students) understand moral problems and see themselves as moral beings. She is interested in the developmental trends of their moral understandings as well as what factors are related to their moral cognition and self-perceptions. Dr. Bock teaches Lifespan Development, Psychology of Adolescence, Psychological Testing, General Psychology and History of Psychology in Social Context.

Lauren Braswell

Best Teacher, Psychology

Mary Daugherty

Most Helpful to Students, Finance
Most Helpful to Students, Business

Mary Schmid Daugherty is the founding faculty member of the Aristotle Fund, a $5 million student managed investment fund at the University of St. Thomas, which she developed and has led since 1999. In her role as associate professor of Finance, she has taught in the undergraduate, graduate and Executive UST MBA programs. In addition to teaching, Daugherty does research and consults on corporate organizational structure. She serves on the board of directors for Crescent Electric Supply Company and Mairs & Power Funds, Inc. She chairs the Governance Committee and is a member of the Compensation Committee for the Crescent Electric Board, and chairs the Audit Committee and is a member of the Governance and Nominating committees for Mairs & Power, Inc. As the audit chair, Daugherty is recognized as the Financial Expert as required by the SEC.

Daugherty has been working with private and family businesses in a variety of in-depth financial and corporate governance consulting assignments since 1993. Her experience with boards and C-suite management includes working with transitioning the corporate structure from advisory boards to fiduciary boards, facilitating and developing optimal governance policies and practices, designing and teaching corporate governance educational programs, designing the processes and conducting board and committee evaluations, and advising on other corporate governance issues unique to the client. She has provided consulting services in different industries, serving both small and large companies. Daugherty has published in the area of family and private corporate governance, co-authoring Family Business, 4th Edition, which is the most widely used textbook on family business in university family business programs. She is an active academic researcher in organizational structure and investment funds, and has authored research papers in these areas.

Tom Hamilton

Best Overall Faculty Member, Finance

Tom is an associate professor of Real Estate with the Department of Finance at the University of St. Thomas Opus College of Business. His research specialties include public utility valuation and real estate feasibility studies and investment analysis. He received a Bachelor of Science degree in natural resources from the University of Wisconsin and a Master of Science degree in Finance from the University of Wyoming. He received an M.B.A. and Ph.D in Urban Land Economics from the University of Wisconsin.

Thadavillil Jithendranathan

Best Researcher/Scholar, Finance
Best Teacher, Business

Lorman Lundsten

Best Overall Faculty Member, Business

J. Roxanne Prichard

Best Researcher/Scholar, Psychology

J. Roxanne Prichard, Ph.D. is an Associate Professor of Psychology and Neuroscience at the University of St. Thomas. She earned a B.A. in Biopsychology from Transylvania University in 1998, and her doctorate in Neuroscience from the University of Wisconsin-Madison in 2004. Her graduate work focused on the neuroanatomical systems that coordinate sleep responses to light. She regularly leads continuing educational workshops and public lectures on sleep and health.

Gregory Robinson-Riegler

Best Overall Faculty Member, Psychology

James Shovein

Best Teacher, Finance
Best Researcher/Scholar, Business

UNIVERSITY OF TENNESSEE - CHATTANOOGA

Helen Eigenberg

Best Researcher/Scholar, Education

Dr. Helen Eigenberg is Professor in the Criminal Justice Department at the University of Tennessee at Chattanooga. She was born and raised in Nebraska. She received her B.S. and her M.S. from Eastern Kentucky University and her Ph.D. in Criminal Justice in 1989 from Sam Houston State University. She taught and served as Department Head at Old Dominion University from 1988–1995 and Eastern Kentucky University from 1995–1997. She has been at UTC since 1998. Her research interests include women and crime, victimology, violence against women, institutional corrections, and male rape in prisons. She has published a book on domestic violence: Woman Battering in the United States (2001). She also has published over 25 book chapters and articles in a wide variety of journals including: American Journal of Police, Women and Criminal Justice, Criminal Justice Review, Journal of Criminal Justice Education, Journal of Criminal Justice, Justice Quarterly, and The Prison Journal. She was the Editor of Feminist Criminology for four years, which is an international peer-reviewed journal. She currently sits on their Editorial Board.

She has served on the Tennessee State Coalition Against Domestic and Sexual Violence and the Tennessee Victims of Crime State Coordinating Council (2006–2010). She is the faculty liaison for the Senator Tommy Burks Victim Assistance Academy which provides training for victim advocates throughout the state of Tennessee and is a joint UTC and Tennessee State Coalition Against Domestic and Sexual Violence project. She also worked closely with the UT Law Enforcement Innovation Center (LEIC), Southeast Command and Leadership College to establish that program in 2000 where she played a significant role in developing the curriculum and overall design of the program. She continues to serve as a Curriculum Committee Member: (2000–present) for that program and a Board Member and Chair for LEIC. She wrote and received two National Institute of Justice grants that created the Transformation Project at UTC in 2002. This project provides educational programming and direct services to victims of domestic and sexual violence, and was the impetus for the current Women's Center. She has served as Vice Chair and Executive Counselor for the American Society of Criminology's (ASC) Division on Women and Crime. She has received their New Scholar Award (1995), their Inconvenient Woman Award (which recognizes the scholar/activist who has participated in publicly promoting the ideals of gender equality and women's rights throughout society, particularly as it relates to gender and crime issues), and the Sara Hall Award (for service to the profession). She was selected as the Eastern Kentucky University, College of Law Enforcement, Distinguished Alumnus in 1994. In 2000, she received the UTC College of Health and Human Services' Research Award. In 2003 and 2005 she was selected for the UTC Grant and Program Review Research Award and in 2006 she received the College of Arts and Sciences Service Award.

Tammy Garland

Best Researcher/Scholar, Criminal Justice

Dr. Tammy Garland is Associate Professor and Graduate Coordinator for the UTC Criminal Justice Department. Dr. Garland received her Bachelor's Degree from the University of Kentucky (1997), a Master's Degree from Eastern Kentucky University (1999), and a doctorate from Sam Houston State University (2004). She primarily teaches courses in drugs and crime, victimology, juvenile justice, and crime and popular culture. Her current research interests include juvenile bullying, drug policy, and the victimization of women, children, and the homeless.

Dr. Garland is involved in a number of service activities at both the university and community levels. Currently, she volunteers as a self-defense instructor to prevent violence against women, and works with a number of local agencies to aid the homeless population. Additionally, she has worked with a number of agencies as a program evaluator. In 2011, she was awarded the Outstanding Service for the College of Arts and Sciences.

Kelli Hand

Best Teacher, Nursing

Jamie Harvey

Most Helpful to Students, Physical Ed

Linda Hill

Best Overall Faculty Member, Nursing

Associate Professor, Coordinator, Nurse Anesthesia Concentration

Jenny Holcombe

Best Researcher/Scholar, Nursing

Sarah Knox

Best Overall Faculty Member, Education

Cheryl Lamb

Most Helpful to Students, Nursing

Seong Park

Best Overall Faculty Member, Criminal Justice
Best Teacher, Criminal Justice

Vicky Petzko

Best Teacher, Education

Dr. Petzko came to UTC in 1998 to join the Graduate Studies Division of the College of Education as a faculty member in School Leadership. She had been a middle school principal in suburban Minneapolis (MN) for seven years, where she received a divisional Principal of the Year in 1997 and her school received a National Blue Ribbon School of Excellence award in 1996. She had also been a high school associate principal for seven years, and a teacher for 10 years in the Minneapolis area. Vicki received her Ph.D. in 1990 in Educational Policy and Administration at the University of Minnesota, where her dissertation research focused on the legal limitations and attendance rates associated with excessive absence policies in high schools. Prior to her appointment at UTC, Vicki was awarded a partial sabbatical to study topics related to school law at Hamline University School of Law, and engaged in research that year associated with the identification and recruitment of aspiring principals.

Roger Thompson

Most Helpful to Students, Criminal Justice
Most Helpful to Students, Education

Dr. Thompson is one of the seasoned Criminal Justice faculty having joined UTC in 1976 shortly after completing both Bachelor and Master's degrees in Criminal Justice at Youngstown State University. A doctorate in Education degree was conferred by The University of Tennessee in 1983. He is a UC Foundation Professor.

Dr. Thompson is known for his level and involvement in community service with focus on youth gangs, homelessness, community development, and crime prevention strategies. He was selected by the Tennessee Higher Education Commission among faculty and students across the state of Tennessee to receive the 2011 Harold Love Outstanding Community Service Award.

Current projects include appointment by the Mayor to the Office of Multicultural Affair's Advisory Board, membership on two youth gang committees, and planning committee for an Emergency Shelter for the Homeless. He is also active in challenging the early school start times of many local public schools.

Anne Wilkins

Best Researcher/Scholar, Accounting

UNIVERSITY OF TENNESSEE - KNOXVILLE

Alison Buchan

Best Researcher/Scholar, Microbiology

UNIVERSITY OF TENNESSEE AT MARTIN

Parker Cashdollar

Best Teacher, Economics

Mamhoud Haddad

Best Teacher, Business

Dr. Haddad is a professor of Finance at the College of Business and Public Affairs, the University of Tennessee at Martin. He received a BA and MBA in Accounting from Minnesota State University, and Ph.D. in Finance from the School of Business at The University of Alabama. He taught finance theory, options and futures theory and applications, risk management, international finance, investments, financial institutions and markets, corporate finance, economics, accounting and statistics, at Wayne State University, The University of Tennessee at Martin, University of Alabama, Arab American University-Jenin, American University of Armenia, American University of Beirut, Eastern Michigan University and Minnesota State University.

Dr. Haddad is certified in e-teaching and learning. Mahmoud has been involved with distance and online teaching since 1994 at both the graduate and undergraduate levels. Mahmoud has participated in numerous curriculum development projects that were essential in meet the needs of our students and to improve the quality of the college curriculum.

Dr. Haddad has published over 30 papers in corporate finance, portfolio management, risk measurement and management, international finance and futures and options. His papers have been published in leading refereed journals. Dr. Haddad has presented at and participated in numerous national and international professional business conferences.

Dr. Haddad has broad consultation, managerial and practical experience that includes familiarity with proper staffing techniques, interfacing with all levels of management, and prioritizing national and international banking and investments. Dr. Haddad was the Financial and Accounting Manager at the American Family Insurance Co., the Executive Director of Research at the Palestinian Monetary Authority, the Vice President for Administrative and Financial Affairs and Dean of College of Business and Financial Sciences at Arab American University and a Research Scholar and Head of Economic and Social Studies Division at the Emirates Center for Strategic Studies and Research. Member of the advisory board, Talal Abu-Ghazaleh Organization –ACQM, Dubai, United Arab of Emirates. Dr. Haddad is a Senior Associate Consultant at the William Davidson Institute (WDI), he worked on various USAID educational projects, www.wdi.umich.edu;

http://wdi.umich.edu/news/announcements/wdi-work-in-west-bank-profiled/?searchterm=Mahmoud%20Haddad;

http://wdi.umich.edu/news/announcements/2009/wdi-consultants-stream-into-west-bank/?searchterm=Mahmoud%20Haddad http://microlinks.kdid.org/library/esaf-university-strengthening

He has received numerous honors, awards and grants for his research and teaching, including the Excellence in Research Award, 1997 & 2010, Excellence in Graduate Teaching, 2002, 2005 & 2011 and the Bob Figgins Award for Outstanding Student Service 2011 From the University of Tennessee at Martin, and numerous International Travel Awards for outstanding contributions in International Finance and Business research. Dr. Haddad is listed in Who's Who in American Education and in who's who in America's Teachers. He serves as a Review Board Member for the Journal of Business and Economic perspectives and is the Co-Director of the Tennessee Valley Authority Investment Program. Dr. Haddad is the University of Tennessee at Martin Featured Scholar for the 1998–1999 academic years, and the 1999 Students Government Association Outstanding Faculty Member. Dr. Haddad is a Research Fellow at the Economic Research Forum (ERF) and Vic President of the Organization of Arab Academic Leaders for the Advancement of Business and Economic Knowledge (OAALABEK).

Erik Markin

Most Helpful to Students, Economics

Mike McCullough

Most Helpful to Students, Business

UNIVERSITY OF TEXAS

Juan Dominguez

Most Helpful to Students, Psychology

Juan Dominguez received his Ph.D. in behavioral neuroscience from the University at Buffalo, The State University of New York. His postdoctoral fellowship was conducted with dual appointments in the Department of Psychiatry and the Department of Cell Biology at The University of Cincinnati College of Medicine. Following postdoctoral training, Dr. Dominguez joined the research faculty of the neuroscience program at Florida State University. Before joining the UT faculty, he was an assistant professor of psychology at American University in Washington, DC. The goals of his lab are to elucidate the underlying neural and endocrine mechanisms regulating motivated behaviors, specifically, using the study of sexual behavior as a prototypic model for understanding motivation, its acquisition and associated disorders.

UNIVERSITY OF TEXAS - PAN AMERICAN

Leticia Deleon

Most Helpful to Students, Education

Russell Eisenman

Best Teacher, Psychology

"Terrific, dedicated teacher. Serious and humorous. Makes education fun for students."

Russell Eisenman received his Ph.D. in clinical psychology from University of Georgia in 1966. He has published many books and journal articles and won an award for outstanding teaching.

Luz Murillo

Best Researcher/Scholar, Education

Gregory Sparrow

Best Teacher, Education

Gregory Scott Sparrow is a past President, and a current Board member for the International Association for the Study of Dreams. He also serves as an advisor to the IASD Executive Committee, and is also IASD's site administrator for its new Online Course Center (www.iasdreamcourses.org). He is an Associate Professor in the counseling program at the University of Texas-Pan American, and teaches courses in the LPC track. He is also on the faculty of Atlantic University in Virginia Beach, where he has taught courses in the evolution of consciousness, the yoga of dreaming, and spiritual mentoring.

Scott has maintained a solo private practice in professional counseling since 1982, and currently practices at The Center in Mission, Texas. He specializes in existential, Jungian, systemic, and transpersonal approaches to therapy, and has developed an innovative approach to dream work based on his lucid dream research, called the FiveStar Method. He is founder of DreamStar Institute, which offers mentoring and certification in dream analysis for laypersons and clinicians. He wrote the early classic, Lucid Dreaming: Dawning of the Clear Light.

Scott has lectured and taught courses across the U.S. on such topics as meditation, spiritual experiences, and dream work methods. He is a student of Jungian psychology, Eastern religions, and contemplative Christianity. For several years, he made a study of the "Christ encounter," which is the visionary or dream experience of meeting Christ face-to-face. A book on this research titled, I Am With You Always: True Stories of Encounters With Jesus–a Barnes and

Noble bestseller, Book of the Month selection, and Quality Paperback selection–that was published by Bantam in 1995. Similarly, he subsequently wrote a book about dreams and visions of Mary, titled Blessed Among Women: Encounters with Mary and Her Message (Crown, 1997). He is also author of Sacred Encounters with the Christ (Ave Maria, 2002), Sacred Encounters with Mary (Ave Maria, 2003), and Healing the Fisher King: A Fly Fisher's Grail Quest (BlueMantle, 2009).

Martha Tevis

Best Overall Faculty Member, Education

UNIVERSITY OF TEXAS AT BROWNSVILLE

Charles Chapman

Most Helpful to Students, Government

Charles W. Chapman, Lecturer for the Government Department, received his J.D. and Ph.D. from The University of Texas at Austin. After beginning his practice of law, he served, at various times, as the police adviser, municipal court prosecutor, assistant city attorney, and acting city attorney for the City of San Marcos, Texas. He also served a four-year term as the elected criminal district attorney for Hays County, Texas. He teaches Constitutional Law (Civil Liberties and Federalism), Judicial Process, Legislative Process, the American Presidency, and Public Law.

Rene Corbeil

Most Helpful to Students, Education

"Dr. Corbeil ensures that our online instruction is delivered without problems to our students. Dr. Corbeil is easy to contact when any issues arise and immediately contacts us when we have situations that need assistance or corrections. He makes recommendations on alternatives to improve our courses and that ensures that we stay up-to-date. Always courteous and friendly with staff and students"

My name is Rene Corbeil. I am an instructor in the Online M.Ed. in Educational Technology at the University of Texas at Brownsville. My experience in education spans over a 30-year teaching career in the public schools and in higher education. With over 16 years experience as a middle school teacher of Biology and Computer Literacy, TIF Grant writer and technology coordinator, and teacher trainer, my career in the public schools culminated when my school was awarded a $276,000 TIF grant that I wrote and I received the Brownsville Independent School District's Secondary Teacher of the Year award. At the pinnacle of my teaching career, I set my sight on new challenges in higher education. Being a graduate research assistant for Dr. Badrul Khan, a pioneer in Web-based distance education, allowed me to be instrumental in the design of the first online course at UTB/TSC, and assist with the editing of his book Web-Based Instruction. This was a turning point in my life. Consequently, in my 15+ years of service to the university, I have had the fortuitous opportunity to contribute to the development of courses in many disciplines, such as Educational Technology, Criminal Justice and Allied Health. I have also gotten to know the learners' professional and academic needs first hand. I have used this knowledge to design courses that offer our graduate students what they need to be professionally competitive. Because technology is constantly changing, the courses I design and teach address current needs in the profession and reflect my commitment to life-long learning.

On August 22, 2003, I earned a doctoral degree in Curriculum and Instruction with an emphasis in Instructional Technology from the University of Houston. For my dissertation research, I conducted a study which identified the psychological, environmental and technological factors that limited people's willingness or ability to learn online. It is my desire that knowledge gained from this study will enable this institution to identify its at-risk online learners and to develop intervention strategies to improve retention and increase enrollment in our distance education programs. In May of 2008, I was promoted to Associate Professor with tenure.

I currently teach fully online graduate and undergraduate courses in Educational Technology. In these courses we explore a variety of learning platforms and Web 2.0 technologies to determine their instructional potential. Click on the image to the right to view my Second Life avatar.

Tyler Dial

Best Teacher, Government
Best Researcher/Scholar, Public Affairs, Political & Policy Sciences

Tyler Dial, assistant master technical instructor, received his B.A. in business management from UT Austin and M.P.A. from Texas Tech University. His areas of expertise are public administration, and healthcare policy and management. He has over 25 years of experience as a senior operations manager in the public and private sector, which includes health maintenance organizations, integrated hospital systems, multi-specialty physician groups and a leading academic medical institution. In addition, he is the director of UTB's Certified Public Managers program.

Ronald Lane

Best Researcher/Scholar, Government

Edward G Moore

Best Overall Faculty Member, Public Affairs, Political & Policy Sciences

John Robey

Best Overall Faculty Member, Government
Best Teacher, Public Affairs, Political & Policy Sciences

John S. Robey received his Ph.D. in Political Science from the University of Georgia. Dr. Robey has published work on the war on drugs, civil rights and liberties concerns, and U. S. Mexican/Hispanic issues (e.g., immigration, economic development and environmental problems). He teaches State Government, Public Policy, and American Government and Politics.

UNIVERSITY OF TEXAS AT EL PASO

Irasema Coronado

Best Overall Faculty Member, Political Science
Best Researcher/Scholar, Political Science
Most Helpful to Students, Political Science

Irasema Coronado received her bachelor's degree in political science and a certificate of Latin American Studies from the University of South Florida. She has an M.A. in Latin American Studies and a Ph.D. in Political Science from the University of Arizona. Her area of specialization is comparative politics. Her dissertation topic focused on the role of transboundary political elites on the U.S.-Mexico border.

Irasema Coronado is a professor in the Department of Political Science, a contributing faculty member of the Environmental Science and Engineering Ph.D. program, and associate provost. She is co-author of the book titled "Fronteras No Mas: Toward Social Justice at the U.S.-Mexico Border" and several academic articles "Conflictos Ambientales Internacionales e Intranacionales," "Legal Solutions Vs Environmental Realities: The Case of the United States-Mexico Border Region." She has co-edited "Digame! Policy and Politics on the Texas Border" and the book "Juntos Pero No Revueltos: Estudios sobre la frontera Texas-Chihuahua". She also coauthored "Latinas in Local Government, Políticas: Latina Public Officials in Texas" in 2008.

She was the recipient of a Border Fulbright in 2004 and continues to collaborate at the Universidad Autonoma de Ciudad Juarez. Irasema Coronado was the president of the Association for Borderland Studies 2005–2006. She served as a member of the Environmental

Protection Agency Good Neighbor Environmental Board from 1999–2002 and co-chair of the Coalition Against Violence Toward Women and Children on the Border. She was also part of the National Advisory Committee for the U.S. Environmental Protection Agency from 2003–2006 and member of the academic advisory board for Ms. Magazine. Her research interests continue to evolve around the role of women in politics and cross-border cooperation at the local level on the U.S.-Mexico border region.

Irasema Coronado has had a variety of academic and administrative positions at the University of Texas El Paso. She served as graduate Advisor in the Department of Political Science September 2006–January 2008; Chair, Department of Political Science, University of Texas at El Paso September 1, 2005–September 1, 2006 and associate dean of the college of liberal arts 2007–2008. Presently, she serves as an associate provost at the University of Texas at El Paso.

Tony Payan

Best Teacher, Political Science

UNIVERSITY OF TEXAS AT SAN ANTONIO

Thomas Babcock

Most Helpful to Students, Criminal Justice

Ilna Colemere

Best Researcher/Scholar, Education

Daniel Gelo

Most Helpful to Students, Medicine

Roxanne Henkin

Most Helpful to Students, Education

Dr. Roxanne Henkin, Professor, earned her doctorate from Northern Illinois University and taught 18 years in public schools. She then taught at National-Louis University in Chicago before joining the faculty at The University of Texas at San Antonio. Dr. Henkin's research interests include critical literacy, writing, multiliteracies, multimodal and digital literacies and in-service staff development in literacy. She has also published two books and was the lead co-editor of the journal Voices from the Middle. She received the 2013 Top Education Professors in Texas Award and the 2012 Headline Award for Excellence in Education from The Association for Woman in Communication San Antonio.

The National Council of Teachers of English awarded Dr. Henkin the 2009 Halle Award for significant contributions to literacy at the middle level. She was also awarded the UTSA 2009 President's Distinguished Achievement Award for Community Service. She also received the College of Education and Human Development's 2007 & 2008 Distinguished Achievement Award for Teaching Excellence for Tenured Faculty. In 2011, Dr. Henkin's blog was awarded as one of the Top 50 Blogs for Education Professors.

Dr. Henkin directs the San Antonio Writing Project. The San Antonio Writing Project (SAWP), which Dr. Henkin founded in 2006 is a local, non-profit organization designed to promote the effective teaching of writing in San Antonio Area schools from ages Pre-K through the University level, especially in low income areas and for second language students. The four main activities of the site are the Summer Institute, continued development and leadership opportunities for the Teacher Consultants, in-service faculty development for our local schools, and summer Young Writer's Camps for area youth. The Teacher Consultants complete an intensive Summer Institute and share a belief that writing is essential for learning in all content areas.

SAWP is housed in the Department of Interdisciplinary Studies in the College of Education and Human Development at The University of Texas at San Antonio and is the local affiliate of the National Writing Project. Research projects have been conducted on various aspects of the SAWP Invitational Summer Institute every year since 2007.

Since 2007, there have been eight annual conferences, which are open to all area in-service and pre-service teachers and any other interested parties. SAWP has a strong record of working with area teachers. Since 2006, SAWP has worked with approximately 2,643 teachers in long-term professional in-service in writing and literacy.

Robert Rico

Best Teacher, Criminal Justice

Rob Tillyer

Best Overall Faculty Member, Criminal Justice
Best Researcher/Scholar, Criminal Justice

UNIVERSITY OF TEXAS AT TYLER

Susanne Brians

Best Teacher, Education
Most Helpful to Students, Education
Best Teacher, Education
Most Helpful to Students, Education

John Lamb

Best Overall Faculty Member, Education
Best Researcher/Scholar, Education
Best Overall Faculty Member, Education
Best Researcher/Scholar, Education

UNIVERSITY OF TEXAS HEALTH SCIENCE CENTER

Hacker Carl

Best Overall Faculty Member, Health Science
Best Teacher, Health Science
Best Teacher, Health Sciences and Nursing
Best Overall Faculty Member, Health Sciences and Nursing

Thomas Matney

Best Researcher/Scholar, Health Science
Most Helpful to Students, Health Science
Best Researcher/Scholar, Health Sciences and Nursing
Most Helpful to Students, Health Sciences and Nursing

Dr. Thomas S. Matney was a Houston philanthropist and emeritus professor of genetics and environmental science at The University of Texas Health Science Center and M. D. Anderson Cancer Center Graduate School of Biomedical Sciences. Professor Matney made important contributions to scientific understanding of cancer-causing agents and the genetic mechanisms that underlie the development of cancer. His wide-ranging philanthropic and service activities enhanced the well-being of hundreds of Houston-area children and families.

Thomas Stull Matney was born on September 21, 1928, in Kansas City, Missouri. His family moved to Texas when he was 10 years old. He received bachelor's and master's degrees in biology and chemistry from Trinity University in San Antonio, and the Ph.D. degree in bacteriology from the University of Texas at Austin.

In the 1950s, Matney served as Captain and later civilian Medical Bacteriologist in the U.S. Army Chemical Corps in Fort Detrick, Maryland where he developed protections for chemical and biological weaponry. Dr. Matney moved to Houston to join the Biology Department of the M.D. Anderson Hospital and Tumor Institute in September of 1962. He became the first associate dean of the newly formed UT Graduate School of Biomedical Sciences and a Distinguished Professor Emeritus. He was also a member of M. D. Anderson Steering Committee for Alumni and Faculty.

Dr. Matney was a generous supporter of the University and mentor to many graduate students. He personally financially supported students studying at the UT-Graduate School of Biomedical Sciences. He established endowments of The Thomas Stull Matney Professorship in Cancer Genetics and The Thomas Stull Matney Professorship in Environmental and Genetic Sciences, both to support scientific excellence and service to graduate education.

He was married to Glenda Matney nee Oglesby until her death in 1990 and had three children with her. He remarried to Nancy Lee Matney.

Dr. Matney's community service activities focused primarily on the well-being of at-risk children. He served as a consultant to the City of Houston Parks and Recreation Department about problems concerned with violence prevention in children. He was a Trustee and raised millions of dollars for Hospitality Apartments, which provides affordable housing for those undergoing long term medical treatment in Houston.

Dr. Matney was a past president of Emerson Unitarian Universalist Church, in Houston. He was also a Past President of the Houston Chapter of National Train Passenger Association as well as other national and local rail road organizations.

Dr. Matney died at the age of 82 on November 28, 2010 after an extended illness. (1)

UNIVERSITY OF TEXAS SOUTHWESTERN MEDICAL CENTER

James Richardson

Best Researcher/Scholar, Medicine

James Wagner

Best Overall Faculty Member, Medicine
Best Teacher, Medicine
Most Helpful to Students, Medicine
Best Researcher/Scholar, Medicine
Best Teacher, Medicine
Most Helpful to Students, Medicine
Best Overall Faculty Member, Medicine

UNIVERSITY OF THE INCARNATE WORD

Ramona Parker

Best Overall Faculty Member, Nursing

Dr. Ramona Parker, President of the Faculty Senate, has been at UIW since 2004. She holds the Ph.D. in Nursing from the University of Texas at Austin; a Master of Science in Nursing from the University of the Incarnate Word; and a Bachelor of Science in Nursing from the University of Texas Health Science Center at San Antonio. Dr. Parker, who is an Associate Professor of Nursing and a Registered Nurse, serves as the liaison between the faculty and administration.

UNIVERSITY OF THE PACIFIC

Marilyn Draheim

Best Overall Faculty Member, Education
Most Helpful to Students, Education

"Keeps a close watch on students and gives them feedback so they don't get sidetracked. Sees and keeps up with all students and issues"

From the heartland of Minnesota, where writer Garrison Keillor created Lake Wobegon, I was educated in a small town's school system in which devoted teachers lived and breathed their commitment to the town.

In return, teachers were revered professionals in the community.

The small classes, traditional curriculum, and extra-curricular activities that I adored-concert choir, madrigals, concert and marching band, forensics, Future Teachers of America-helped shape my love for learning in many areas, but particularly in literature, literary criticism, music, theatre, art, and world history.

Reading became my means for opening the treasures of human achievement, and my parents' interest in the arts, history, and hard work supported me from elementary school through the completion of a doctoral degree.

My thoughts about education are eclectic, and, over the past twenty years, I have investigated and thought about the writings of Lev Vygotsky and other social and cultural theorists for their ideas about how the knowledge we learn is woven in our culture, language, and social interactions and how our knowledge is gained through reading, experience, reflection, trial and error, our interests, vicarious learning, and the convivial nature of interaction with others.

Greg Potter

Best Teacher, Education

"Interesting, keeps it moving"

UNIVERSITY OF TOLEDO

Jon Elhai

Best Researcher/Scholar, Psychology

Jon D. Elhai received his Ph.D. in clinical psychology in 2000 from Nova Southeastern University's Center for Psychological Studies. He completed his pre-doctoral internship in 2000 with the Charleston Consortium Psychology Internship Training Program at the Medical University of South Carolina (MUSC) and Charleston Veterans Affairs (VA) Medical Center. He completed a postdoctoral fellowship in posttraumatic stress disorder in 2002 at MUSC and the Charleston VA Medical Center.

Dr. Elhai served as Associate Professor until 2009 in the University of South Dakota's Department of Psychology. He now serves as Professor in the University of Toledo's Department of Psychology (primary appointment) and Department of Psychiatry (joint appointment), in Toledo, Ohio, where he directs the PTSD Research Lab. He is also affiliated with the Anxiety and Stress Lab.

Dr. Elhai's research is on psychological trauma and posttraumatic stress disorder (PTSD), exploring assessment, psychopathology, forensic and treatment issues in PTSD; he has published more than 150 papers, primarily involving empirical studies on psychological trauma and PTSD. He has presented his research at meetings of such organizations as the American Psychological Association, International Society for Traumatic Stress Studies, Association for Behavioral and Cognitive Therapies, and Anxiety Disorders Association of America. He is currently Associate Editor for the Journal of Anxiety Disorders, Academic Editor for PLoS One, and Editorial Board member for Clinical Psychology Review, Psychological Injury and Law, and Psychological Trauma. He has co-authored books on traumatic stress and PTSD.

Dr. Elhai occasionally serves as an expert witness, in his role as a forensic psychological consultant, primarily on PTSD cases.

Dr. Elhai teaches undergraduate courses on such topics as abnormal psychology and research methodology in clinical psychology, and graduate courses such as cognitive-behavioral psychotherapy, research design, and multivariate statistics. He also mentors graduate students on their master's theses and doctoral dissertations, and undergraduate students on their honors' theses.

Bruce Groves

Most Helpful to Students, Health Science

Amy O'Donnell

Best Teacher, Business

Amy O'Donnell joined the College of Business Administration in August 2004. She holds a Master of Science in College Student Personnel Services from Miami University and a Bachelor of Science in Journalism from Bowling Green State University. As the College's Career Development Lecturer, Amy's current emphasis is on Career Development I and II, our two required career courses. In addition, she supports the Business Career Programs Office with its function. Last Updated: 1/3/12

Alex Petkevich

Best Researcher/Scholar, Finance

Francis Pizza

Best Researcher/Scholar, Health Science

My team's research focuses on the inflammatory response in skeletal muscle after injurious and non-injurious exercise. We are particularly interested in determining the contribution of inflammatory cells (neutrophils and macrophages) to exercise-induced muscle injury, repair/regeneration, and hypertrophy. We are also interested in determining the mechanisms by which exercise modulates neutrophil and macrophage activation in skeletal muscle. We use human, rodent, and cell culture models and contemporary laboratory techniques to answer basic and applied questions in these areas.

Kathie Reed

Best Overall Faculty Member, Health & Human Services

"As the director and chairman of the Legal Specialties Department, Professor Reed has impressed me for over twenty years. Her dedication to the Paralegal Studies program is astounding. Her dedication is exceptional."

Barry Scheuermann

Best Overall Faculty Member, Health Science
Best Teacher, Health Science

Don Wedding

Best Overall Faculty Member, Business

"Has a keen interest in students."

UNIVERSITY OF TOLEDO COLLEGE OF LAW

L Strang

Best Researcher/Scholar, Law

Professor Lee J. Strang joined the faculty in 2008, was granted tenure in 2010, and was named Director of Faculty Research in 2014. Before that, he was a visiting Professor at Michigan State University College of Law and an Associate Professor at Ave Maria School of Law. A graduate of the University of Iowa, where he was Articles Editor of the Iowa Law Review and Order of the Coif, Professor Strang also holds an LL.M. degree from Harvard Law School.

Prior to teaching, Professor Strang served as a judicial clerk for Chief Judge Alice M. Batchelder of the U.S. Court of Appeals for the Sixth Circuit. He was also an associate for Jenner & Block LLP in Chicago, where he practiced in general and appellate litigation.

A prolific scholar, Professor Strang has published in the fields of constitutional law and interpretation, property law, and religion and the First Amendment. His most recent article, Originalism and the Aristotelian Tradition: Virtue's Home in Originalism, was published in the Fordham Law Review. Among other scholarly projects, he is currently editing a case book on constitutional law for LexisNexis, drafting a book proposal tentatively titled Originalism: Its Promise and Limits, and writing a book on the history of Catholic legal education.

Professor Strang is a frequent presenter at scholarly conferences. He is also a regular participant in debates at law schools across the country, contributor to the media, and speaker to political, civic, and religious groups.

Professor Strang's course offerings include Constitutional Law, Constitutional Interpretation, Jurisprudence, Property Law, Administrative Law, Business Associations, Federal Courts, and Appellate Practice.

UNIVERSITY OF UTAH

Abe Bakhsheshy

Best Teacher, Business
Best Teacher, Business

"Abe Bakhsheshy is universally respected and admired by students and faculty. Students have told me over and over again that his ethics class changed their lives. His impact on students is huge - both in the classroom and out of the classroom. Students learn how to evaluate situations and act ethically even in the face of opposition. They also learn by observing him how to treat others respectfully and how to listen. We have many outstanding faculty members, but Abe is the one we all aspire to copy."

Dr. Abe Bakhsheshy has more than 35 years of management experience including seven years of executive level position in overseas working with multi national corporations.

Prior to joining the David Eccles School of Business in July of 2002, Dr. Bakhsheshy served the University of Utah Hospitals and Community Clinics as the Director of Quality Resources, Marketing, Employee Recognition, Public Relations and Customer Services. Preceding this position, he had administrative responsibilities for over twelve UUHN Medical centers throughout the state of Utah. Dr. Bakhsheshy is also an organizational development consultant providing human resources and leadership education services for organizations within the state of Utah and nationally.

Jay Barney

Best Researcher/Scholar, Business
Best Researcher/Scholar, Business

"Jay Barney is one of the most influential scholars of all time in the field of strategy. His creativity and energy seemingly have no bounds. Without Jay's work, businesses would not have an understanding of the fundamentals of competitive advantage. Moreover, Jay is no "one hit wonder". The body of his work is outstanding and has had an enormous impact on scholars and practitioners around the world."

Patricia Eisenman

Best Overall Faculty Member, Health Science
Best Researcher/Scholar, Health Science
Best Teacher, Health Science
Most Helpful to Students, Health Science
Best Researcher/Scholar, Health Sciences and Nursing
Best Teacher, Health Sciences and Nursing
Most Helpful to Students, Health Sciences and Nursing
Best Overall Faculty Member, Health Sciences and Nursing

Patricia Eisenman did undergraduate work at Colorado State University and received my doctorate in 1973 from Kent State University. Over the years, I have provided service to a number of professional, sport and commercial organizations. Most recently I am on the certification examination board for the Collegiate Strength and Conditioning Coaches Association and the science advisory board for Turning Point, the manufacturer of core training and assessment devices. While here at the University of Utah for the past 29 years, I have had the good fortune to mentor many undergraduate and graduate students. During my tenure at the U, I have also served as chair of the Department of Exercise and Sport Science and associate dean for academic affairs for the College of Health. I have served on many departmental, college, and university committees and currently I am chair of the university's Diversity Requirement Committee.

Alan Smith

Best Researcher/Scholar, Medicine

Irene Yoon

Best Teacher, Education

"Dr. Yoon is a first year professor in the ELP department. She is an amazing instructor. As a first year doctoral student, I have learned a lot from Dr. Yoon. She attends to both the content on the course as well as the writing and dissertation process. Dr. Yoon provides clear, detailed, and specific constructive feedback, and in a timely manner. She is always available for consultations. She has been by far my best instructor."

UNIVERSITY OF WASHINGTON

Steve Gloyd

Best Overall Faculty Member, Health Sciences and Nursing

Stephen Gloyd serves as director of education and curriculum activities in the Department of Global Health. He is also Executive Director of Health Alliance International, a Seattle-based non-profit organization that provides support to Ministries of Health in Mozambique, East Timor, Ivory Coast, and Sudan. Dr. Gloyd teaches about the history and political economy of global health from a social justice perspective, research methods for developing countries, and management for global health. He has over 30 years experience working in Africa and Latin America, and his work has focused on improving primary health care through implementation science, with an emphasis on maternal-child health, malaria, tuberculosis, and STD/AIDS programs. He speaks Portuguese, Spanish, and French.

Marcos Llobera

Best Overall Faculty Member, Anthropology
Best Researcher/Scholar, Anthropology

My main interest is the development of new computational methods that will hopefully enable us to understand better past landscapes: the significance of certain places; how people may have constructed certain perceptions; how landscape changed and how those changes may be related to changes in social and economic terms, and more subtle experiential, symbolic and ritual ones. Most of this work is inspired in current interpretative approaches to landscapes.

UNIVERSITY OF WEST GEORGIA

Danilo Baylen

Best Researcher/Scholar, Education

Danilo M. Baylen is a tenured Full Professor of Instructional Technology and School Library Media in the Department of Educational Technology and Foundations, College of Education at the University of West Georgia. He received his doctoral degree in Instructional Technology from Northern Illinois University. Also, he has graduate degrees in Elementary Education (Florida Gulf Coast University), Counseling (NIU), and Library and Information Studies (University of Alabama) as well as a graduate certificate in Children's Literature (Pennsylvania State University). His undergraduate degree was in Economics from the University of the Philippines. His research interests include technology integration across the curriculum, media literacy education, visual literacy, online learning and teaching, and instructional design. His recent research projects focused on faculty readiness for online teaching, use of technology for course transformation, and the role of learner characteristics in e-learning among undergraduate and graduate students. Currently, he is exploring the convergence of emerging technologies and children's literature through interactive picture books and digital storytelling. Also, he is investigating the use of digital media in enhancing the educational experience of teacher education students and practitioners.

Jill Drake

Best Overall Faculty Member, Education
Most Helpful to Students, Education
Best Researcher/Scholar, Education

Dr. Jill Drake is a Professor of Mathematics Education in the Department of Learning and Teaching. She teaches graduate and undergraduate courses in early childhood education. Her research focuses on assessment in mathematics education, problem posing, and teaching mathematics to diverse populations. She is internationally recognized for her research on problem posing and she recently published two mathematics activity books for high school and middle school students. Prior to her appointment at UWG, Dr. Drake was an educator at the early childhood, elementary, and middle school levels.

Eileen Muzio

Best Teacher, Education

Eileen Muzio is an Instructor in the Early Learning and Childhood Education Department of the College of Education. She teaches undergraduate education courses and supervises students during the field placements. Her passion is equipping students with the most current research based knowledge in appropriate teaching practices and materials for the PreK–5 classroom. Prior to joining the faculty at the University Of West Georgia, Mrs. Muzio taught 31 years in the Coweta County School System.

Lyn Steed

Most Helpful to Students, Education

Laura Willox

Best Teacher, Education

Lara Willox is the director of the Ed.D. in School Improvement program and an Assistant Professor of Learning and Teaching at the University of West Georgia. Her research is in the area of collaborative action research, elementary social studies education, and teaching for social responsibility. She is active in the College and University Faculty Assembly (CUFA) of the National Council for Social Studies (NCSS); she is the chair of the elementary social studies special interest group. She advocates for the inclusion of social studies curriculum in the elementary grades, and focuses her teaching and research on socially just and responsible education. Prior to completing her doctorate at UNC Chapel Hill, she taught for twelve years in public elementary schools in Charlotte, North Carolina.

UNIVERSITY OF WISCONSIN - EAU CLAIRE

Cheryl Brandt

Best Overall Faculty Member, Nursing

"Cheryl has taught traditional, accelerated, masters and DNP courses. She includes evidence based practice consistently. Cheryl challenges students to apply concepts from pathophysiology to providing care. She has researched a variety of areas in teaching and practice. Her service within the department, college, university, community and profession has been consistent."

Mary LaRue

Best Overall Faculty Member, Kinesiology
Best Teacher, Kinesiology
Most Helpful to Students, Kinesiology

Robert Stow

Best Researcher/Scholar, Kinesiology

Gen Thul

Best Teacher, Nursing

"Gen listens attentively and uses strategies to actively engage students in the classroom."

UNIVERSITY OF WISCONSIN - MADISON

Jeffrey Henriques

Best Overall Faculty Member, Psychology
Best Researcher/Scholar, Psychology
Best Teacher, Psychology
Most Helpful to Students, Psychology
Best Researcher/Scholar, Medicine
Best Teacher, Medicine
Most Helpful to Students, Medicine
Best Overall Faculty Member, Medicine

Larry Nesper

Best Overall Faculty Member, Anthropology

"Above and beyond scholarly excellence, which all the nominees have, Larry has led our Department in building a harmonious relationship with the Wisconsin tribal organizations. US anthropology historically derives from relations with the Native American peoples, and I feel that building this relation has special merit. Also, Larry is just a great guy to work with."

Larry Nesper is a professor in the department of Anthropology and American Indian Studies Program at the University of Wisconsin-Madison. His book, The Walleye War: The Struggle for Ojibwe Spearfishing and Treaty Rights (University of Nebraska Press 2002) exemplifies his interest in the cultural and historical dimensions of contemporary American Indian political and economic projects in the Great Lakes region. His current research is on tribal court development. He has consulted for the Menominee Tribe, the Sokaogon and Bad River Bands of Lake Superior Chippewa and the Great Lakes Indian Fish and Wildlife Commission.

Tina Winston

Best Teacher, Psychology

"She works really hard at conveying complex material in a student-relevant, accessible manner. She's also a great colleague overall."

UNIVERSITY OF WISCONSIN - PLATTEVILLE

David Chellevold

Best Teacher, Education

UNIVERSITY OF WISCONSIN - STOUT

Charles D. Baird

Most Helpful to Students, Business

Xuedong Ding

Most Helpful to Students, Management
Most Helpful to Students, Business

Sally Dresdow

Best Overall Faculty Member, Management
Best Researcher/Scholar, Management
Best Teacher, Management
Best Researcher/Scholar, Business
Best Teacher, Business

Jeanette Kersten

Best Researcher/Scholar, Business

William J. Kryshak

Best Teacher, Business

Adel Mekraz

Most Helpful to Students, Business

Dennis Vanden Bloomen

Best Overall Faculty Member, Business

Scot Vaver

Best Teacher, Management
Best Teacher, Business

I have a diverse work and educational background. My belief is that through the transfer of knowledge, comes strength not only for those you share it with, but also yourself via the free exchange of ideas. I strive at building a sense of belonging by sharing vision, embracing diversity, and acknowledging everyone's value.

My work experience includes:
• CFO and Co-Founder of GolfLip, LLC • CFO/COO of JDSS, LLC • Manager and Project Leader Experience • Domestic and International Management • Forecasting and Accounting Management • Policy and Procedure Development • Extensive training and teaching experience (course room, e-learning, ITV) • Curriculum development and training program development (in person and online delivery)

I have an interest in the educational setting:
• To provide both leadership and assistance to all students to help achieve their goals • Help students excel through processes that help ensure their success both in the academic environment and then transition into the work environment • Work as a team with all staff to garner community support for the programs and pursue innovative solutions to answer the important next generation questions within education

UNIVERSITY OF WISCONSIN - PARKSIDE

Ting He

Most Helpful to Students, Accounting

UTAH STATE UNIVERSITY

Ken Bartkus

Best Overall Faculty Member, Business
Best Researcher/Scholar, Business
Best Teacher, Business
Most Helpful to Students, Business
Best Researcher/Scholar, Business
Best Teacher, Business
Most Helpful to Students, Business
Best Overall Faculty Member, Business

Jan Kelley-King

Best Teacher, Education

Julie Wolter

Best Overall Faculty Member, Comm Disorders/Deaf Education

UTAH VALLEY UNIVERSITY

Genan Anderson

Best Overall Faculty Member, Education
Best Teacher, Education

"Genan Anderson leads out in curriculum issues, causing important changes to happen. She has published in multiple journals and has co-published with many of the faculty. She provides instruction that produces results with her students.

She is willing to help and do cooperative teaching. She is organized and does what she needs to do to stay current and innovative. She is open to her students and helps them out in all situations."

Mary Brown

Best Researcher/Scholar, Health Sciences and Nursing
Best Overall Faculty Member, Health Sciences and Nursing

Hsiu-Chin Chen

Best Researcher/Scholar, Nursing

Matt Flint

Most Helpful to Students, Health Sciences and Nursing

Assistant Professor
    Department of Public & Community Health
    Utah Valley University
    Orem, Utah
    • Teach: Grant Writing, Social Marketing, Nutrition, Stress Management, Personal Health & Wellness
    • Act as the faculty advisor for the student-run Community Health Club at UVU (CHUVU)
• Plan service projects
• Organize professional presentations
• Manage & balance the club's budget
• Participate in scholarly works including being involved in research and grant writing
    • Contribute to various committees including the research committee and the health curriculum committee

Mi Ok Kang

Best Researcher/Scholar, Education
Best Researcher/Scholar, Education

"A faculty member in the School of Education since 2011, Dr. Kang has published and presented a substantial body of research in national and international arenas. Her outstanding scholarship is recognized globally, and it reflects well on the quality of UVU faculty."

Katie Moore

Most Helpful to Students, Nursing

Katie received her BS in nursing from Westminster College. She initially worked on Medical/Surgical, Telemetry and Intensive Care Units and then decided to become a nursing instructor. She received her MS in teaching nursing from the University of Utah in 2009 and is currently enrolled as a PhD student. She has been teaching at UVU for 6 years and recently received tenure.

Troy Nelson

Most Helpful to Students, Nursing
Best Teacher, Health Sciences and Nursing

Floyd Olson

Best Teacher, Marketing

Mike Patch

Best Overall Faculty Member, Education
Most Helpful to Students, Education

Nancy Peterson

Best Teacher, Education

Tracy Sermon

Most Helpful to Students, Education

Allison Swenson

Best Overall Faculty Member, Nursing
Best Teacher, Nursing

Elaine Tuft

Best Overall Faculty Member, Education

"Elaine Tuft is a leader among her colleagues. She is a get it done and get it done right kind of gal. She works hard to make sure her students understand the content she teaches and has helped many faculty members with their statistical research needs."

Ph.D., Michigan State University M.A., Utah State University
B.S., Utah State University
    Specialty: Mathematics Education

Mina Wayman

Best Teacher, Nursing
Best Teacher, Health Sciences and Nursing

VALENCIA COLLEGE

Dianne Gomez

Best Researcher/Scholar, Nursing

"Her experience as well as her knowledge in her area of expertise."

Since the age of seven, and my experience as a patient in a local community hospital, I knew I was going to be a nurse. According to my parents, it is all I would talk about once I was discharged. I grew up on Long Island, New York and attended an Associate degree program at Farmingdale. It was the best experience imaginable. I graduated from that program feeling prepared and ready to face everything. My first job as a new nurse was on the cardiac medical telemetry floor at St. Francis Hospital in Roslyn, NY. In this unit, we cared for patients with a history of cardiac disease or being ruled out for heart problems. Our unit included a small 4 bed post angioplasty unit attached to our 49 other beds.

After a year, I wanted to learn more and took the critical care course that was offered at the hospital and moved onto the Open Heart Unit or CVICU. There I learned to expand my skills and care for a more complex patient type, applying those critical thinking skills I was continually developing. In 1990, I had an opportunity to accept a new challenge in the emergency room of another local hospital. This position was as the assistant nurse manager of the emergency department monitoring unit.

During that time I had been a volunteer fireman for my local fire house and just loved that excitement and wanted to learn as much as I could. It was during that time that I met my husband and we moved to Texas. In Texas, I took a job as the assistant nurse manager of a local community hospital in their MICU/SICU. The nurses would rotate between the 2 units, which allowed everyone to keep up their present skill level and learn more about other patient types.

In 1994, my husband and I moved here to Central Florida where I accepted a position at Florida Hospital Orlando's CVICU. It was at Florida Hospital where my bug for learning took off in a new direction. I had always enjoyed assisting other nurses in developing their skills to care for the ICU level care patient and thought why not learn some more and expand my degree. As a result, I attended my BSN program and merged onto my MSN because I knew that teaching was becoming more and more my focus. Eventually I wanted to teach at the academic level. In preparation for my eventual goal of professor, I accepted a position as a clinical education specialist in Maine. There I learned what a simulation lab could do and how it benefited the nursing staff I worked with.

Due to budget cuts, my husband had to relocate and I with him. We moved back to Central Florida were I took a position as clinical nurse specialist at Osceola Regional Medical Center. It was a great experience and challenge. During this time I noticed that Osceola did not have students and I questioned the administration as to why and learned that there was a need for educators and would I be interested in looking at the program at Valencia. I accepted a position as adjunct faculty in the nursing arts lab, then went on to function as clinical instructor for Nursing IV, V, & VI. I was able to accept a full time faculty position here at Valencia and I know that it is the culmination of what I have been working for. VCC's philosophy and dedication to the nursing profession is everything I could hope for and I feel privileged to be included as a Valencia faculty member. I strongly believe that a Community College level entrance into nursing provides a strong foundation to grow as a practicing professional nurse and I am proud that I can help students fulfill their dreams of success.

Susan Gosnell

Best Overall Faculty Member, Health Science

Lisa Gray

Best Overall Faculty Member, Accounting

Sean Jennings

Best Researcher/Scholar, Psychology

Laurinda Lott

Best Overall Faculty Member, Education

"Lauri Lott, is an adjunct faculty but she goes beyond and above her time and effort towards helping build our Teacher Education Program. Lauri makes herself available to all her students and is working hard to infuse technology in our curriculum thus helping prepare teachers who will be well prepared to teach a high tech generation!"

Eric Model

Best Teacher, Psychology

Dr. Model is a nationally and internationally recognized sport psychology consultant and an expert in the role of motivation, relaxation and pre-performance routines.

His academic background includes an undergraduate degree from the University of British Columbia with majors in both Human Kinetics and Psychology, a Master of Science in Psychology from Arizona State University and a Doctoral Degree in Psychology (Sport Specialization) from the University of Florida.

He is a former national athlete for Canada and currently lives with his wife and two children in Celebration, Florida.

Jaime Shipley

Most Helpful to Students, Business

David Skinner

Best Overall Faculty Member, Psychology
Most Helpful to Students, Psychology

VALPARAISO UNIVERSITY

Nola Schmidt

Best Researcher/Scholar, Nursing

Joining the faculty in 1992, Professor Schmidt teaches in the undergraduate and graduate programs. Her area of expertise is nursing research and evidence-based practice. Professor Schmidt's areas of scholarly interest include evidence-based practice, children's pain, Orem's nursing theory, nursing education, ethics, service learning, and spirituality. She co-edited an undergraduate research textbook, Evidence-Based Practice for nurses: Appraisal and Application of Research with Dean Janet Brown, and co-edited Spiritual Care in Nursing Practice with Professor Kris Mauk. Professor Schmidt also enjoys her volunteer work teaching "Safe Sitter," a nationally recognized babysitting class for children ages 11–13.

VANDERBILT UNIVERSITY

Jon Kaas

Best Researcher/Scholar, Psychology
Best Researcher/Scholar, Medicine

Jon Kaas

Best Researcher/Scholar, Psychology

"Recommendation for Dr. Jon Kaas for a 2014 faculty award
I recommend Jon Kaas for a faculty award for his long term productivity and visibility, and his outstanding and continuing neuroscience research. His recent, cutting edge research has been on parietal-frontal sensorimotor networks: that is, how external and/or internal sensory messages are converted into organized motor acts. Jon and his laboratory group have shown that across primate species, the posterior parietal cortex is divided into 7 or more regions they call domains where micro-stimulation evokes a very specific movement, such as reaching to pick something up, or moving hand to mouth, or grasping an object. These movements are mediated though connections with functionally matched domains in pre-motor cortex (part of the frontal lobe) and from there to primary motor cortex. In the latter area, components of the matched movements can also be evoked. Stimulating two domains at once shows that they interact at each level to produce a combined movement or to prevent a competing movement. Their conclusion in a very recent paper is that movement selection or decision making occurs at each level based on different activating inputs: sensory for posterior parietal cortex and pre-frontal cognitive for pre-motor cortex. These results are presented in detail in their most recent paper: Stepniewska et al. J. Neuroscience 111:1100–1119, 2014. So Jon Kaas' research continues to be at the front edge of research on the function of the cerebral cortex and he would be a deserving recipient of a faculty award from Vanderbilt University this year."

Olubunmi (Bunmi) Olatunji

Best Overall Faculty Member, Psychology

Olatunji's primary research interest lies in cognitive behavioral theory, assessment, and therapy for anxiety disorders. He is particularly interested in the role of basic emotions other than fear in the etiology of anxiety pathology. His current research employs basic descriptive and experimental psychopathology methodology to examine the relationship between the experience of disgust and specific anxiety disorder symptoms.

Leslie Smith

Best Teacher, Psychology
Best Teacher, Medicine

Andy Tomarken

Most Helpful to Students, Psychology
Most Helpful to Students, Medicine
Best Overall Faculty Member, Medicine

VERNON COLLEGE

Beth Arnold

Best Teacher, Nursing
Most Helpful to Students, Health Sciences and Nursing

VIRGINIA COMMONWEALTH UNIVERSITY

David Downs

Best Researcher/Scholar, Business

Walter Griggs

Most Helpful to Students, Business

Graduate degree in history and JD in law from the Univ. of Richmond. Ed.D. from the College of William and Mary. Author of 7 books on topics ranging from moose to the history of Richmond. Served as Associate Dean and Department Chair.

Manu Gupta

Best Overall Faculty Member, Finance

"Voted best teacher, is an able researcher and contributes to the department - both strategically and tactically."

Philip Olds

Most Helpful to Students, Accounting

Laura Razzolini

Best Teacher, Business

Robert Reilly

Best Overall Faculty Member, Business

Roxanne Spindle

Best Teacher, Accounting

Jayaraman Vijayakumar

Best Overall Faculty Member, Accounting
Best Researcher/Scholar, Accounting

Dr. Vijayakumar researches primarily in the areas of State and Local Government Finances and Accounting, Municipal Bonds, and Capital Market issues relating to Financial Reporting and Insider Trading. His publications have appeared in journals such as the Journal of Accounting and Public Policy, Journal of Accounting, Auditing, and Finance, Journal of Corporate Finance, the Journal of Financial Services Research, The Financial Review, and the Municipal Finance journal. He serves on the editorial board of the Journal of Public Budgeting, Accounting, and Financial Management. He teaches Managerial Accounting (for graduates and undergraduates), Management Control Systems (for graduates and undergraduates), and has also offered doctoral seminars in Financial Reporting and Accounting Research at VCU. Dr. Vijayakumar is a past President of the Government and NonProfit section of the American Accounting Association having served that section in numerous other capacities such as Director of Research, and Secretary/Treasurer. Dr. Vijayakumar has held several senior positions in the Government of India as a member of the Indian Revenue Service, a prestigious branch of the civil service in India. In addition, his work experience includes service as an Industrial Engineer in the Indian Telephone Industries, Bangalore, India. Education: Dr. Vijayakumar has a Bachelor's degree in Mechanical Engineering from Bangalore University, India, a Post-Graduate Diploma in Industrial Management from the Indian Institute of Science, Bangalore, and a Ph.D in Accounting from the University of Pittsburgh, USA.

Isaac Wood

Best Overall Faculty Member, Medicine
Best Teacher, Medicine

"really leading the medical school in updating teaching standards"

Have received the award for Educational Innovation and the Faculty Teaching Excellence Award given by the School of Medicine though my work as the M2 Behavioral Sciences Course Director and M3 Psychiatry Clerkship Director. Have received the award for the outstanding teacher in the Behavioral Sciences consecutively for over 10 years. As Senior Associate Dean for Medical Education, I have led a total revamping of the curriculum and design of a medical education center to create a state-of-the-art educational experience in undergraduate medical education.

VITERBO UNIVERSITY

Bill Bakalars

Best Teacher, Psychology

Deb Murray

Best Researcher/Scholar, Psychology

WAGNER COLLEGE

Richard Larocca

Most Helpful to Students, Finance
Best Teacher, Business

B.S. in Finance and Economics, Boston College Carroll School of Management, 1989; M.B.A. Finance, St. John's University; D.P.S. in Finance and International Economics, Pace University Lubin School of Business, 2012.

Mary Lo Re

Best Researcher/Scholar, Business
Most Helpful to Students, Business
Best Overall Faculty Member, Business

Dr. Mary L. Lo Re is a Professor of Finance and the Dean of Adult Education & Extension Programs at Wagner College having previously held the following leadership positions: Interim Associate Provost, Chair of the Department of Business Administration, Director of MBA Programs and Director of Assessment. She obtained her Ph.D. from CUNY Graduate

Center specializing in Monetary Theories & Policies and International Trade. She has: written/obtained several external/internal grants; numerous newspaper/media community and student involvement mentions; presented and published journal articles in the areas of: civic engagement/service-learning, best practices such as writing across the curriculum, EU/EMU convergence & the economic financial crisis; recently authored 2 book chapters—"Service-Learning Community: A Win for Students, Community, Institution & Faculty" in Service-Learning Pedagogy: How Does It Measure Up? and "Balancing Management & Leadership" in The Resource Handbook for Academic Deans; and is the author of the book, "Experiential Civic Learning: Construction of Models & Assessment" which can be found at: http://www.na-businesspress.com/estore/ExperientialCivicLearning.html. Prior to academia, Dr. Lo Re obtained 17 years of practical business experience—including 9 years in computer technology and 5 years in executive management—working at American Express, the United Nations and the Board of Education.

WAKE TECHNICAL COMMUNITY COLLEGE

John Annis

Most Helpful to Students, Criminal Justice

Amin Asfari

Best Teacher, Criminal Justice

WALSH COLLEGE

James McHann

Best Overall Faculty Member, Business

"A true scholar with stellar credentials, yet welcoming to students and fellow staff. A guiding star for all who know him."

James McHann serves as Professor of Strategy and Management at Walsh College in Troy, Michigan, where he teaches courses in the areas of strategic management, organizational development and change, the evolution of economic and management thought, and building organizational value through knowledge and intellectual capital. Recognizing the impact of his teaching, Dr. McHann was nominated for a U.S. Professor of the Year award in 2013.

In 2014, Dr. McHann served as a Fulbright Scholar in India where he gave a series of lectures on strategic management and human capital development in the College of Commerce and Management Sciences at Berhampur University, Odisha, India, and at the University of Calcutta and at Globsyn Business School in West Bengal, India. In 2011, he served as Visiting Scholar in strategy and organization in the School of Business, Economics, and Law at the University of Gothenburg, Sweden.

Dr. McHann completed Postdoctoral studies in management at the A.B. Freeman School of Business at Tulane University. He earned a PhD from King's College, University of Aberdeen, Scotland in the field of philosophical hermeneutics (the theory of how human beings understand meaning communicated across time, languages and cultures). He holds an MBA from the Kellogg School of Management at Northwestern University with emphases in strategy and global marketing.

Dr. McHann has over fifteen years of general management experience at the president & CEO level, and he has done management consulting work for organizations as diverse as General Motors and Cincinnati Children's Hospital & Medical Center. He has lived and worked in Europe, Africa, and Asia.

WALTERS STATE COMMUNITY COLLEGE

Cheryl McCall

Most Helpful to Students, Health Sciences and Nursing

"Cheryl has a warm personality that makes her approachable by even the most shy student. Dr. McCall is a wonderful person who is respected by not only the students but by the faculty she works with as well."

Donna McGaha

Best Overall Faculty Member, Nursing

"Donna wants each student to excell in their studies and takes the extra time to make sure they are on the right track."

WASHBURN UNIVERSITY

Becky Dodge

Most Helpful to Students, Health Sciences and Nursing

"Becky goes the extra mile for her students every day!"

Jim Martin

Most Helpful to Students, Business

WASHINGTON STATE UNIVERSITY

Ana Espinola-Arredondo

Best Overall Faculty Member, Economics
Best Researcher/Scholar, Economics

Ana Espinola-Arredondo is an Associate Professor in the School of Economic Sciences at Washington State University. She specializes in the areas of environmental economics and industrial organization. She received a Ph.D in Economics from the University of Pittsburgh and was awarded the Research Medal Award of the Global Development Network. She has recently published, among others, in the Journal of Environmental Economics and Management, Ecological Economics and International Journal of Industrial Organization.

Dogan Gursoy

Best Researcher/Scholar, Business

Dr. Dogan Gursoy is the Taco Bell Distinguished Professor in Hospitality Business Management at Washington State University in the School of Hospitality Business Management and the editor of Journal of Hospitality Marketing & Management. He developed and designed the "Hotel Business Management Training Simulation" (http://www.hotelsimulation.com/), a virtual management training game where participants are divided into teams and assigned the task of running 500-room hotels in a competitive virtual marketplace. The Hotel Business Management Training Simulation has been used for both revenue management and hospitality management purposes by several institutions.

Dr. Gursoy received his Ph.D. degree from Virginia Tech. His research goal is to advance the theoretical and practical knowledge by meaningfully contributing to the state of the knowledge in hospitality and tourism. This process may involve investigation of an issue that has been neglected, re-examination of an existing area to advance the existing body of knowledge by integrating other theoretical constructs and theoretical structures, or re-modeling or improving of existing theoretical models.

Dr. Gursoy is recognized as one of the leading researchers in the hospitality and tourism area. His area of research includes services management, hospitality and tourism marketing, tourist behavior, travelers' information search behavior, community support for tourism development, cross-cultural studies, consumer behavior, involvement and generational leadership. His research has been published broadly in refereed Tier I journals such as Annals of Tourism Research, Journal of Travel Research, Tourism Management, International Journal of Hospitality Management, Journal of Hospitality and Tourism Research. His research has also been presented at numerous hospitality and tourism conferences and received numerous research awards. Dr. Gursoy has recently been recognized as one of the top 10 authors in the world in terms of publications in the top six hospitality/tourism journals during the past decade (Journal of Hospitality and Tourism Research (2011), 35(3), 381–416).

Dr. Gursoy serves on the editorial board of several journals including Annals of Tourism Research (resource editor), Journal of Social Inquiry (associate editor), International Journal of Hospitality Management, Journal of Hospitality and Tourism Research, Tourism Analysis, Journal of Travel & Tourism Marketing, International Journal of Contemporary Hospitality Management, etc. Dr. Dogan Gursoy also receives frequent invitations to give keynote speeches at international hospitality and tourism conferences. He is also an active member of International Council of Hotel, Restaurant and Institutional Education (ICHRIE) and Travel and Tourism Research Association (TTRA).

Pat Kuzyk

Best Teacher, Economics
Most Helpful to Students, Economics

Felix Munoz

Best Researcher/Scholar, Economics
Best Teacher, Economics

WASHINGTON STATE UNIVERSITY - VANCOUVER

Brian McTier

Best Researcher/Scholar, Business

Tahira Probst

Best Overall Faculty Member, Psychology
Most Helpful to Students, Psychology

WASHINGTON UNIVERSITY IN ST. LOUIS

Rebecca Dresser

Best Researcher/Scholar, Law

"Most insightful and relevant researcher and scholar."

Professor Rebecca Dresser is an expert in biomedical ethics. She holds a joint appointment with Washington University School of Medicine, teaching law and medical students about legal and ethical issues in end-of-life care, biomedical research, genetics, assisted reproduction, and related topics. She has written extensively in her field and is the co-author of a casebook on bioethics and law and a book on the ethical treatment of animals. She is also the author of a book on patient advocacy and research ethics. In addition to her teaching, research, and scholarship, she is on the advisory or editorial boards of several prestigious journals devoted to bioethics. A past member of the President's Council on Bioethics, she currently sits on the Washington University Medical Center Institutional Review Board, as well as its Embryonic Stem Cell Research Oversight Committee. Professor Dresser is a prolific speaker and panelist at national and international symposia, conferences, and workshops on such topics as bioethics and cancer; advance treatment directives; stem cell research; biomedical research policy; and human cloning. Before becoming a law professor, she clerked for the Hon. James E. Doyle, U.S. District Court for the Western District of Wisconsin, and held a postdoctoral fellowship in the Psychiatry Department at the University of Wisconsin–Madison.

Daniel Keating

Best Teacher, Law

"Most committed teacher and loves teaching."

Dean Daniel L. Keating is a nationally known expert in bankruptcy, commercial law, and UCC Article 2. The author of two casebooks on commercial law, as well as a treatise on the employment law implications of bankruptcy, he has written extensively on such issues as bankruptcy reform, the implication of bankruptcy on collective bargaining agreements, pension insurance, and the Pension Benefit Guaranty Corporation (PBGC). His scholarship also has covered the subject of sales law and practice. He is an elected member of the American Law Institute and a Fellow of the American College of Bankruptcy. Throughout his career, he has given extensive service to the Association of American Law Schools, including chairing several committees. At the law school, Dean Keating has served three times as interim dean, as well as vice dean and associate dean. He is the recipient of a Founder's Day Distinguished Faculty Award and the law school's Outstanding Professor Award. In addition to his scholarship, service, and teaching, Dean Keating is a frequent speaker and panelist, making presentations to the American Bankruptcy Institute, National Conference of Bankruptcy Judges, the UCC Institute, and various academic conferences. Before joining the faculty, he was a John Olin Fellow in Law and Economics at the University of Chicago and practiced law for two years as a commercial attorney with The First National Bank of Chicago.

William Lowry

Best Teacher, Political Science

"He has taught more students in the last 20 years than any other professor at Washington University and he has won numerous teaching awards while doing so."

Bill Lowry is a Professor of Political Science at Washington University. Before going into political science, he served in the U.S. Navy, drove a taxi cab, worked for the National Park Service, and managed drug stores. He received his PhD in Political Science from Stanford University in 1988. He studies American politics, environmental and energy policy, and natural resource issues. He is the author of five books, the most recent being Repairing Paradise, as well as numerous articles and chapters in edited volumes. His recent work focuses on American energy policy. In 2010, he received a Distinguished Faculty Award from Washington University.

Gary Miller

Best Overall Faculty Member, Political Science
Best Researcher/Scholar, Political Science

"His work has changed the way we think about topics such as organizations and parties. He has mentored countless junior faculty and graduate students. He's the best on numerous dimensions."

Gary Miller received his Ph.D. from the University of Texas at Austin in 1976. He taught previously at California Institute of Technology, Michigan State University, and Washington University's John M. Olin School of Business. His primary substantive interest has been decision-making in bureaucracies, committees, and small groups. He has done a number of laboratory experiments testing hypotheses about group decision-making, derived from game theory and organizational economics. His teaching assignments include a graduate seminar in the politics of the U.S. bureaucracy, an undergraduate course on the civil rights movement, and an undergraduate course on the U.S. Presidency.

Andrew Sobel

Most Helpful to Students, Political Science

"Sobel has worked with countless undergraduate and graduate students over the years."

Andrew Sobel is a political scientist in the program in International and Area Studies (IAS) at Washington University in St. Louis. He specializes in the politics of global finance with a focus upon domestic explanations of international behavior. He is the Director of Undergraduate Studies in IAS and the Program Director of the M.A. Program in International Affairs in University College at Washington University. He currently serves on the Faculty Advisory Councils to the Center for the Interdisciplinary Study of Work and Social Capital, the Whitney R. Harris World Law Institute, and the Doctor of Liberal Arts Program in University College at Washington University.

He is the author or editor of six books and numerous articles. His first book, Domestic Choices, International Markets, examines the politics underpinning the liberalization and globalization of national securities markets in Japan, the United Kingdom, and the United States. His second book, State Institutions, Private Incentives, Global Capital, explores the extraordinary transformation and reawakening of global financial markets, systematic differences in access for borrowers in the global capital pool, and the effects of national political institutions in explaining the metamorphosis and the differential access. Congressional Quarterly Press published his third book, Political Economy and Global Affairs. In his fourth book, The Challenges of Globalization, he edited a volume of papers from a conference on Globalization, State and Society. His fifth book, Birth of Hegemony: Crisis, Financial Revolution, and Emerging Global Networks, came out in the summer of 2012 from the University of Chicago Press. This book explores the public and private financial foundations of liberal hegemonic leadership by examining the three cases of such leadership over the past 400 years—the Dutch Netherlands, England, and now the United States. A sixth book, International Political Economy in Context: Individual Choices, Global Effects, was released by Sage/CQ Press in September 2012.

WAUBONSEE COMMUNITY COLLEGE

Paula Hladik

Best Overall Faculty Member, Business
Best Teacher, Business
Most Helpful to Students, Business

Steven Skaggs

Best Teacher, Business
Most Helpful to Students, Business
Best Overall Faculty Member, Business

Leatha Ware

Best Researcher/Scholar, Business
Best Researcher/Scholar, Business

Dr. Ware received her Ed.D. from Northern Illinois University with an emphasis in human resource development and a cognate in instructional technology. She also holds a master's degree in management from National-Louis University and a bachelor's degree in mathematics from Tougaloo College.

WAYLAND BAPTIST UNIVERSITY

Stephen Horton

Best Researcher/Scholar, Business

John McClusky

Most Helpful to Students, Business

Kelly Warren

Best Researcher/Scholar, Business
Best Overall Faculty Member, Business

WAYNE STATE UNIVERSITY

Suzanne Billingsley

Most Helpful to Students, Nursing

Karen Feathers

Most Helpful to Students, Education

Maria Ferreira

Best Overall Faculty Member, Education
Best Teacher, Education
Best Researcher/Scholar, Education
Best Overall Faculty Member, Education

"Maria Ferreira goes beyond expectations to teach and mentor her students."

Professor Ferreira joined the faculty of the College of Education at Wayne State University in the fall, 1996. Before coming to Wayne State University, and while working on her doctorate, she worked as a research and teaching assistant at Indiana University (Bloomington). Previous to that she taught high school biology and algebra. Originally from Portugal, she immigrated to the USA in 1980 and became an American Citizen in 1986. Besides Portugal and the United States, she lived in Angola, France, Poland and Russia. She is fluent in four languages and enjoys traveling, the outdoors and the arts.

Nancy George

Best Teacher, Nursing

Janet Harden

Best Overall Faculty Member, Nursing

Mary Anne McCoy

Best Overall Faculty Member, Nursing

Mary Anne McCoy

Most Helpful to Students, Nursing

"She has been an advisor for graduate students since 2008. Program Director for the AG-ACNP program. Her door is always open and she goes the extra mile to help students far beyond just with their academic issues."

James Moseley

Most Helpful to Students, Education

Sally Roberts

Best Researcher/Scholar, Education

"Sally Roberts GOGirls project has lasting significance to develop life skills in the young people she works with."

Dr. Sally Roberts was a mathematics specialist and elementary school teacher in the Detroit Public Schools for 23 years prior to coming to Wayne State University. Dr. Roberts has a special interest in elementary and middle school mathematics education. Her research interests include mathematical problem solving; the teaching and learning of geometry; and the development of the mathematical understandings of teachers. Dr. Roberts has made conference presentations on these topics a the national, state and local level and has published articles on these topics in Teaching Children Mathematics and Mathematics Teaching in the Middle School.

Dr. Roberts is the Director of the GO-GIRL (Gaining Options: Girls Investigate Real Life) program at Wayne State University. The program serves adolescent girls from the community and provides an alternative pre-student field experience for students in the College of Education.

Janna Roop

Best Teacher, Nursing
Most Helpful to Students, Nursing

John Strate

Best Teacher, Political Science

"Dr. Strate is very helpful and friendly. He devotes his time to students and never complains of their visits to his office. He deserves to be honored by the the whole university and acknowledged by President Wilson."

Professor Strate teaches courses in statistics, policy analysis, biopolitics, introductory political science, and American government. Areas of research interest currently include administrative ethics and end-of-life issues and policies including physician-assisted suicide and pain management. Recent publications include articles in Public Integrity, Mortality, American Review of Politics, and Politics and the Life Sciences.

Karen Tonso

Best Teacher, Education

April Vallerand

Best Researcher/Scholar, Nursing
Best Researcher/Scholar, Health Sciences and Nursing

"Has been consistently funded as a researcher. Works well with PhD students in helping them become nurse researchers."

WEBER STATE UNIVERSITY

Melina Alexander

Most Helpful to Students, Education

Melina Alexander is an associate professor at Weber State University in the Department of Teacher Education. She completed her PhD at Utah State University. She currently teaches a variety of courses in teacher education including introduction to teaching, introduction to special education, and a variety of special education methods courses. Her primary areas of research interest include instructional methods for students with and without special needs, and teacher preparation.

Dan Fuller

Most Helpful to Students, Economics

Doris Geide-Stephenson

Best Teacher, Business
Best Overall Faculty Member, Business

Therese Grijalva

Best Overall Faculty Member, Economics
Best Teacher, Economics

Therese Grijalva received her Ph.D. in Economics from the University of New Mexico. Her research focuses on the efficient allocation of scarce environmental resources, and on using market incentives to improve the stock and quality of environmental resources.

Thom Kuehls

Best Overall Faculty Member, Political Science
Best Researcher/Scholar, Political Science

Thom Kuehls' area of specialization in political science is political theory. He is the author of Beyond Sovereign Territory: The Space of Ecopolitics, as well as several essays in edited collections. He is constantly working on his American Government, American Ideals (unpublished), which he uses in his American National Government courses, and he serves as the academic advisor for political science majors and minors.

Alex Lawrence

Most Helpful to Students, Business

Hi, I'm Alex Lawrence, the founder of Startup Flavor. I work at Weber State University. Before that, I was an entrepreneur. I am a teacher, writer, and tech enthusiast, with other interests as well. I bet I'm like many of you; the kid who started out with the lemonade stand and then went on to start larger companies. I'd like to think I'm known as a passionate, energetic, fiery competitor that is enthusiastic about work, helping other entrepreneurs and seeing people and processes get better. At least that is how I'd describe myself (for the most part). I've been launching and growing successful companies for more than 20 years. I'm pretty well-versed in business strategy, sales, Internet marketing, social media, venture and debt capital, mobile and web application development, franchising, mergers and acquisitions, commercial real estate, and operations. Prior to coming to Weber State, my most recent partnership, Lendio, ranked #34 on the Inc. 500 list a few years back. You can read about the exciting things going on with Lendio in an article on TechCrunch. As I mentioned, I am currently working at Weber State University. My official title is Vice Provost of Innovation and Economic Development. I get to do a lot of fun things at Weber State. I'm a Visiting Professor of Entrepreneurship at the John B. Goddard School of Business and Economics, where I earned my MBA. If you are at or near WSU, consider taking one of my classes. I teach real-world strategies that are based on cutting edge, proven approaches that are actually used in business today. As Vice Provost, I work with faculty, staff, students and business members in the community to help start and grow businesses. I also serve as the Executive Director of the Weber State University Research Foundation (WSURF), a wholly owned WSU entity that fosters and supports on campus innovation and commercialization with students, faculty and the surrounding business community. Finally, I enjoy being Chairman of Startup Ogden, an exciting new partnership between Ogden City and Weber State to help area entrepreneurs create and grow strong and exciting businesses (read more about it here). Teams I am responsible for work together in Technology Commercialization (TCO), Contracts (OSP), Small Business (SBDC), Innovation (UCAID) and State of Utah initiatives (USTAR). I'm a contributor on Entrepreneurship for

Forbes Magazine. I also write a regular column for Utah Business Magazine. I've been featured on TechCrunch, 40 under 40 Top Entrepreneurs, V100 Top 100 Technology Entrepreneurs, E-50 Top 50 Social Entrepreneurs, WSU Entrepreneur Alumni of the Year, and more. I am on the advisory board for the Women Tech Council, The NorthFront Entrepreneur Alliance, the WSU Entrepreneur Association, and I'm a Trustee for Launchup.org. I've been married for 14 years and no one makes life more enjoyable than my wife and daughters. My family and I live a quiet life in the mountains of Utah, where we enjoy golf, fly fishing and other outdoor activities like biking and hiking. When I'm not having fun with my family, I'd love to be playing golf, riding my bike or upgrading Apple products. You might see me getting a speeding ticket now and again as well. If you'd like to connect with me+ on various social networks, you'll find me on Facebook, LinkedIn, and Google+. The easiest way to catch me though is either to send me a tweet @_AlexLawrence or get out with me on a bike ride or a golf course!

John Mbaku

Best Researcher/Scholar, Economics
Best Researcher/Scholar, Business

Louise Moulding

Best Researcher/Scholar, Education
Most Helpful to Students, Education
Best Overall Faculty Member, Education

Leah Murray

Best Teacher, Political Science
Best Researcher/Scholar, Public Affairs, Political & Policy Sciences
Best Teacher, Public Affairs, Political & Policy Sciences
Best Overall Faculty Member, Public Affairs, Political & Policy Sciences

Stephanie Wolfe

Most Helpful to Students, Public Affairs, Political & Policy Sciences

"Stephanie this year alone published a academic book on Rwanda and mentored students on a summer trip to Rwanda. To recognize the 25th year after the genocide, she organized an excellent conference on campus in the spring of 2014 that brought experts on the Rwandan tragedy to the entire university and community. She coordinates her research, teaching, and service and does so exceptionally."

WEBSTER UNIVERSITY

Paul Steinmann

Best Researcher/Scholar, Education

Debbie Stiles

Best Teacher, Education

WEST COAST UNIVERSITY ORANGE COUNTY

Maria Gonzalez

Best Researcher/Scholar, Nursing

Doctorate University of La Verne Public Administration MS Seton Hall University, New Jersey Nursing BS University of San Tomas, Philippines Nursing Licensure: RN

Afsaneh Helali

Most Helpful to Students, Nursing

"Is Faculty advisor for PALS, a STARS counselor, and is a support to our students."

Esther Montoya

Best Overall Faculty Member, Nursing
Best Teacher, Nursing
Most Helpful to Students, Nursing
Best Overall Faculty Member, Health Sciences and Nursing
Best Teacher, Health Sciences and Nursing

Cheryl Rojas

Best Overall Faculty Member, Nursing
Best Researcher/Scholar, Nursing
Best Teacher, Nursing
Best Researcher/Scholar, Health Sciences and Nursing
Most Helpful to Students, Health Sciences and Nursing

"For many reasons, she has led our LC group for years and is consistent and dependable, is innovative in her teaching methods and always considers what is best for the students. She began review sessions when she saw a need for success of the students. She now has totally flipped her course and has all lectures video taped so the students can have the best of both worlds. She was invaluable to me as a new faculty member. I do not think I would have survived without her. She is just supportive of all students, staff, and the organization."

WEST LOS ANGELES COLLEGE

Arvie Malik

Best Overall Faculty Member, Dental Hygiene

Arvie Malik received her Bachelors of Science degree in Dental Hygiene with a minor in General Science from Texas Woman's University. She has over 23 years of clinical experience in both Periodontal and Prosthodontic practices in Texas and Califorinia.

Currently, Arvie practices full time in a private Prosthodontic practice in Newport Beach, California. In addition to being a periodontal therapist and an oral health educator, she facilitates in the role of patient treatment coordinator and educates patients in receiving multidisciplinary care and full mouth restorative care and maintenance. Furthermore, she is a doctor and patient liaison in refining patients esthetic desires for treatment outcomes.

In conjunction with private practice, Arvie has been teaching part time for over 15 years at West Los Angeles College in the Department of Dental Hygiene. Her responsibilities include clinical instruction and teaching "Introduction to Dental Hygiene: a pre-clinic course" for incoming dental hygiene students. Additionally, she serves as a lecturer in the section of Periodontics at the University of California, Los Angeles (UCLA). She also instructs hygienists' in the use of magnification and other technological advances as a clinical instructor at the Newport Coast Oral Facial Institute (NCOFI), a non-profit teaching facility. Formerly, she was a Clinical Educator Consultant for Dentsply Corporation providing educational seminars to dental and dental hygiene schools on the use of ultrasonics.

Arvie is a member of the American Dental Hygienists' Association, the California Dental Hygienists' Association, the Orange County Dental Hygienists' Society, the California Dental Educators' Association and the American Dental Education Association.

WEST VALLEY COLLEGE

Sondra Ricar

Best Overall Faculty Member, Public Affairs, Political & Policy Sciences

WEST VIRGINIA UNIVERSITY

Jane Cardi

Best Teacher, Education
Most Helpful to Students, Education
Best Overall Faculty Member, Education

Helen Hazi

Best Researcher/Scholar, Education

Barbara Ludlow

Best Overall Faculty Member, Special Education
Best Researcher/Scholar, Special Education
Best Teacher, Special Education
Most Helpful to Students, Special Education

Andrea Taliaferro

Best Overall Faculty Member, Physical Education
Best Researcher/Scholar, Physical Education
Best Teacher, Physical Education
Most Helpful to Students, Physical Education

Dr. Andrea Taliaferro joined Coaching and Teaching Studies faculty in the College of Physical Activity and Sport Sciences in the fall of 2010. Taliaferro earned her Ph.D. (2010) and M.Ed. (2002) from the University of Virginia in the area of Adapted Physical Education. She holds an B.S. in Kinesiology with a concentration in Physical Education from James Madison University.

Prior to obtaining her Ph.D., Taliaferro was as an Adapted Physical Education Specialist in three public school systems in Virginia for seven years. She also has experience as a general physical educator and coach. Taliaferro is a Certified Adapted Physical Educator through the National Consortium for Physical Education and Recreation for Individuals with Disabilities (NCPERID), and is also an American Red Cross Instructor Trainer in Water Safety and Lifeguarding. Taliaferro has reviewed manuscripts for five different journals including Research Quarterly for Exercise and Sport and Adapted Physical Activity Quarterly. Taliaferro is a member of the American Alliance for Health, Physical Education, Recreation, and Dance (AAHPERD), the National Association for Kinesiology in Higher Education (NAKHE), NCPERID, and the North American Federation of Adapted Physical Activity (NAFAPA), and has served on the Board for WVAHPERD and NCPEID.

WESTERN CONNECTICUT STATE UNIVERSITY

Monica Sousa

Most Helpful to Students, Nursing

WESTERN ILLINOIS UNIVERSITY

Bart Jennings

Best Teacher, Management

Mary Jensen

Most Helpful to Students, Education

Mary Jensen has been a professor at WIU since 1990, teaching in both the department's undergraduate and graduate programs. She received her BA in Psychology from the University of Utah in 1976. In 1977, she received her Masters degree in Special Education from Kent State University. Dr. Jensen taught students with learning and behavioral disorders in public school and residential treatment for ten years. In 1990 she received her Ph.D from the University of Wisconsin-Madison.

Dr. Jensen has authored two text books, Gangs: Straight Talk, Straight Up (Sopris West, 1997) and Introduction to Emotional and Behavioral Disorders: Recognizing and Managing Problems in the Classroom (2005, Pearson/Merrill/Prentice Hall). In addition, Pearson Custom Publishing has published specialized course handbooks created and produced by Dr. Jensen for her classes at WIU.

Dr. Jensen is a 2-time recipient of the Teacher of the Year Award (1995 and 2006) for the COEHS at WIU. She has also received a Golden Apple Award from the WIU Student Alumni, a Most Inspiring Teacher Award from the Student Education Association, an Outstanding Faculty Mentor Award, and two Faculty Excellence Awards.

Dr. Jensen's primary research interests and consulting services revolve around the topics of proactive behavior management, innovative technology in teaching, social skills, bully behavior, and gangs and school violence. She provides numerous professional development programs for school districts and conferences.

Jim Kenny

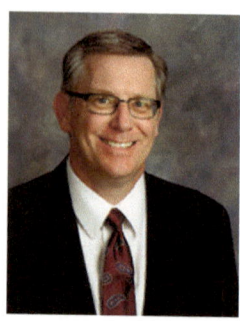

Best Teacher, Marketing

Education Ph.D., Marketing, Oklahoma State University, 1993 MBA, Oklahoma State University, 1987 BS, Marketing, Oklahoma State University, 1983 Teaching Interests Personal Selling Marketing Strategy Services Marketing Research Interests Personal Selling Services Marketing Distribution

Carla Paciotto

Best Researcher/Scholar, Education

Sara Simonson

Best Teacher, Education

Dr. Simonson earned her Ph.D. from the University of Iowa, Iowa City. She teaches graduate courses in secondary reading methods, decoding, vocabulary, assessment, literacy techniques and strategies, and leadership in reading. She has presented at the state, national, and international conference levels. Dr. Simonson is very active with in-service and workshops at the local level. She works with the Illinois State Board of Education, writing and presenting modules across the state to share best practice in reading instruction at a variety of grade levels. Dr. Simonson also writes a column for professional development in literacy for the Illinois Reading Council Journal.

WESTERN KENTUCKY UNIVERSITY

Elizabeth Main

Best Teacher, Health Sciences and Nursing

Beverly Siegrist

Best Overall Faculty Member, Nursing

Beverly Siegrist, EdD, MS, RN, CNE Distinguished University Professor College of Health & Human Services, School of Nursing Coordinator Graduate Nursing Program Member of Kappa Theta Chapter of Sigma Theta Tau International, American Nurses Association, Faith Community Nurses International. Editor of International Journal of Faith Community Nurses Education: University of Louisville, Virginia Commonwealth University/MCV, Medical College of Georgia, and WKU.

WESTERN MICHIGAN UNIVERSITY

Onur Arugaslan

Best Researcher/Scholar, Finance

Dr. Onur Arugaslan's primary teaching interests are in the areas of personal financial planning, corporate finance, financial strategy and international finance. His research includes the separation of cash flow rights and voting rights, the endogeneity of liquidity, the risk-adjusted performance of American Depository Receipts and the market reaction to the acquisitions by unified dual class firms. Arugaslan has published articles in the Journal of Finance, the Journal of Corporate Finance, Managerial Finance, the Journal of Global Business, Management Research News, International Journal of Banking and Finance, International Journal of Commerce and Management, International Journal of Management Theory and Practice, Business Quest, The Business Review, Cambridge and Lahore Journal of Economics. His research received the Best Paper Award at the 2006 Association for Global Business Conference and a Highly Commended Paper Award at the Emerald Literati Network Awards for Excellence 2008. Arugaslan holds a B.S. degree in electrical engineering from Bilkent University in Ankara, Turkey.

Marianne Fraunknecht

Best Teacher, Education

WICHITA STATE UNIVERSITY

Robert Ross

Best Researcher/Scholar, Marketing
Best Teacher, Marketing

Received his Ph.D. Marketing, University of Oklahoma; BA Economics and M.B.A. Cornell University. Joined Wichita State's faculty in 1977, Director of Graduate Studies in Business in 1980–1991. A past President of the Wichita Chapter of the American Marketing Association, Dr. Ross is active in the Marketing Research Association and the Case Research Association Prior to entering academia, and held a variety of positions in the marketing department of the Scott Paper Company, rising to the rank of Associate Marketing Product Manager.

WILKES UNIVERSITY

Harvey Jacobs

Most Helpful to Students, Pharmacy

Harvey Jacobs, Ph.D. graduated in 1972 with a B.A. in Psychology from Wilkes College, in 1976 with a B.S. in Chemistry, in 1982 with a B.S. in Pharmacy, and in 1987 with a Ph.D. in Pharmaceutics all from the University of Utah. He joined the faculty of the School of Pharmacy in 1996 as an Assistant Professor, was promoted to Associate Professor with tenure in 2000, and was appointed as Assistant Dean of Student Affairs in 2001; the position he currently holds. Dr. Jacobs has a successful career in teaching, assessment, pharmaceutical research, and administration during his 18 years at Wilkes University.

Arthur Kibbe

Best Overall Faculty Member, Pharmacy
Best Researcher/Scholar, Pharmacy

Dr. Kibbe has a Ph.D. in Pharmacokinetics and Biopharmaceutics and a score of years working directly or indirectly with the Pharmaceutical Industry, FDA, Professional and Trade Associations and Academia. Spoke on behalf of the profession of pharmacy before Congress and the regulatory agencies. Organized and conducted successful marketing programs for two contract research organizations. He has and is currently training professional pharmacists and pharmaceutical scientists. He has investigated the operations of the FDA on behalf of the commissioners and testified in court on formulation patents as an expert witness in litigation.

He spoke on behalf of the profession of pharmacy . . . before Congress and the regulatory agencies. He also chaired the Pharmaceutical Advisory Committee of the Federal Drug Administration (FDA) and trained professional pharmacists and pharmaceutical scientists at the University of Mississippi.

Dr. Kibbe is an expert in pharmaceutical excipients and was the editor-in-chief of the internationally recognized reference text Handbook of Pharmaceutical Excipients, 3rd Edition.

He investigated the operations of the FDA . . . on behalf of the commissioner, is currently the chair of the Governor's Renal Disease Advisory Committee, and is active as a consultant to the pharmaceutical industry.

Adam Vanwert

Best Teacher, Pharmacy

Dr. VanWert returned to Wilkes in the fall of 2008 after receiving his Ph.D. from the Medical University of South Carolina (MUSC). His doctoral research was in the area of drug transport in the kidney. His didactic training was broad, with a focus on toxicology. His current research involves investigation of the interactions between antibiotics and kidney transporter proteins (e.g., OAT3). Techniques used include HPLC, Western blotting, and cell culture. He works with students in the lab during both the school year and summer; the latter being supported by a Wilkes Mentoring Grant. He was introduced to research at Wilkes while working with Dr. James Culhane from 2000–2003. He was offered jobs in both the FDA and biotech sectors, but chose to return to Wilkes because of his passion for teaching, his positive experience with the faculty, and the family-like environment on campus.

On the teaching front, Dr. VanWert has been recognized as Teacher of the Year (2009–2010). He teaches pharmacology of several disease states, e.g., cardiovascular, neoplastic, and gastrointestinal. He has also developed a toxicology elective in which guest lecturer, Jeanna Marraffa, Pharm.D., DABAT (Upstate Medical University Poison Center, SUNY), introduces clinical toxicology to pharmacy students.

Most recently, Adam has been invited to become a Graduate Affiliate Faculty member in the Pharmaceutics Department at Virginia Commonwealth University (VCU). He will serve on dissertation committees in this role. In addition, he was invited to present his research at VCU in the Pharmaceutics Seminar Series on October 23, 2012.

WILLIAM PATERSON UNIVERSITY

Nadine Aktan

Best Overall Faculty Member, Nursing
Most Helpful to Students, Nursing
Best Researcher/Scholar, Health Sciences and Nursing
Most Helpful to Students, Health Sciences and Nursing

"Nadine is attentive to faculty and students' needs and steers the program to good advantage."

Heejung An

Best Researcher/Scholar, Elementary Education

Steve Betts

Best Teacher, Business

Giuliana Campanelli

Best Teacher, Economics

"great knowledge and personal skills"

Jean Levitan

Best Teacher, Nursing

"Very receptive to students and faculty who need support."

General Education Curriculum Reform Chair, Faculty Senate 2013–1015 Member of the Board of Directors, Society for the Scientific Study of Sexuality, 2008–2014, President 2013 Member of the Board of Directors, Foundation for the Scientific Study of Sexuality, 2014–present Co-author Healthy Sexuality (Blonna and Levitan) Courses Taught Human Sexuality, Women's Health, Reporductive Rights, Health Care in the US, Healthy Living after 30, HealthyU

Kem Louie

Best Researcher/Scholar, Nursing
Best Overall Faculty Member, Nursing
Best Teacher, Nursing
Best Researcher/Scholar, Health Sciences and Nursing
Best Teacher, Health Sciences and Nursing
Best Overall Faculty Member, Health Sciences and Nursing

"Kem has had much experience in nursing education longitudinally and uses that information to fulfill program needs

Dr Louie is the best overall faculty member. She is tough but has a compassionate side to her. She understands her students thoroughly and supports them as needed. She tells you outright, what's good or bad, very straightforward."

Renee Pevour

Most Helpful to Students, Nursing
Most Helpful to Students, Health Sciences and Nursing

"Renee coordinates the Learning Center in Nursing and assists many student with their difficulties"

Karen Phillips

Best Teacher, Health Sciences and Nursing

Holly Seplocha

Best Teacher, Education

Janis Strasser

Best Overall Faculty Member, Education

Department: Elementary and Early Childhood Education Position: Professor, Tenured Area Specialization: Early Childhood Education Academic Degrees and Certifications:- Ed. D. Columbia University, Teachers College - Curriculum and Instruction M. Ed. Columbia University, Teachers College - Early Childhood M.S. Bank Street College of Education - Educational Leadership B.S. State University of NY, College of New Paltz - Sociology/Elementary Education Professional Experience:- Public and Preschool Teacher Education Coordinator, Head Start Curriculum Writer Consulting Editor Educational Consultant Current Teaching and Administrative Responsibilities:- CIEC-365-01 M 2–3:40pm CIEC-300-01 & CIEC300-02 R 10:30am–12:30pm ELCL629 M 5:00–7:30pm (spring semesters) ELCL630 M 5:00–7:30pm (fall semesters) CIEC634 W 5:00–7:30pm (fall semesters) CIEC-635-60 W 5:00–7:30 pm (spring semesters) Coordinator of M.Ed. in Curriculum & Learning, Early Childhood Concentration Supervise Student Teachers & Practicum Students Other Collegiate Assignments:- Facilities Advisory Committee Teach Pre-Practicum Seminar on Multiculturalism/Diversity Student Relations Committee Tenure/Promotion Committee Teach Online Module 7 (GSPTQ) "Culturally Responsive Teaching" Professional Affiliations and Activities:- Consulting Editor, Young Children Write Teacher/Researcher Column "Studying Our Own Practice" in Teaching Young Children Journal Past President, New Jersey Association of Early Childhood Teacher Educators (NJAECTE) Preschool Curriculum Consultant for New Jersey State Museum Current Research Interests:- Diversity/Multiculturalism Curriculum in Early Childhood Arts Education Awards and Honors:- National Association of Early Childhood Teacher Educators Pearson Teacher Educator of the Year, 2010. New Jersey Association of Early Childhood Teacher Educators Outstanding Teacher Educator Award 2005. New Jersey Association of Early Childhood Teacher Educators Outstanding Researcher Award 2005. Faculty Excellence Award for Service, William Paterson University, May 2004. Bergen/Passaic Association for the Education of Young Children Special Member Recognition Award for 2004. Bank Street College of Education Alumni Award, 2001

Janet Tracy

Best Overall Faculty Member, Health Sciences and Nursing

Linnea Weiland

Most Helpful to Students, Education

WILLIAMS BAPTIST COLLEGE

Summer Deprow

Best Overall Faculty Member, Business

B.S. & M.B.A. - Arkansas State University Ph.D. - University of Mississippi Specialist in Community College Teaching - Arkansas State University Further study at Arkansas State University

Teaching at Williams since 1997

WILMINGTON UNIVERSITY

Bonnie Kirkpatrick

Best Teacher, Education

"Bonnie is very knowledgeable in the field of online education. Not only is she is exceptional educator, but she is also an exemplary leader."

WINSTON-SALEM STATE UNIVERSITY

Nelson Adams

Most Helpful to Students, Psychology

Amber Debono

Best Overall Faculty Member, Psychology
Best Researcher/Scholar, Psychology
Best Teacher, Psychology
Most Helpful to Students, Psychology

I am an Assistant Professor at Winston-Salem State University in the Department of Psychological Sciences. The primary courses I teach are: Social Psychology, Research Methods and Statistics 1 & 2, and Introduction to Psychological Sciences. I love getting students excited about research in psychology!

David Kump

Best Teacher, Medicine

"Anatomy and Physiology is a very hard class, but Dr. Kump really helps you to understand and remember the material. He's the best!"

WOFFORD COLLEGE

John MacArthur

Best Overall Faculty Member, Economics
Best Teacher, Economics

"Dedicated to the success of all students, the "A" students as well as the "C" students. Reads widely and relates exceptionally well to students. Also great in his leadership of the department."

Richard Wallace

Best Researcher/Scholar, Economics

"The breadth and depth of his scholarship is incredible. Such scholarship and mentorship to his colleagues and students rarely come together."

WOODBURY UNIVERSITY

Angelo Camillo

Best Researcher/Scholar, Business
Best Researcher/Scholar, Business

"I have worked with Dr. Camillo for many years, and have been impressed with the breadth and depth of his publications and presentations in peer reviewed journals and conferences. His accomplishments are summarized at http://woodbury.libguides.com/aacamillo"

Angelo A. Camillo, PhD, is Associate Professor of Strategic Management. He has more than 35 years of international hospitality management experience and has worked and lived in ten countries on four continents. He holds an undergraduate degree from the Heidelberg (Germany) School of Hotel Management, an MBA degree from San Francisco State University, and a Ph.D. from Oklahoma State University. He teaches courses in Strategy, Global Enterprise Management, Business Ethics, Organizational Behavior, and special topics in Hospitality Entrepreneurship and Business Development. His research encompasses three streams of interest that reflect his specializations and courses he teaches: Strategic Management (Mainstream), Industry Specific, and Cross-disciplinary. Research topics include Strategic and International Management, Food and Wine Business, Marketing Best Practices, Social Responsibility, and Ethics.

E. B. Shaya Gendel

Best Researcher/Scholar, Business

"The late Professor Gendel deserves recognition for what he achieved."

E.B. "Shaya" died in a fatal car accident January 4 2015 in Long Beach, CA. Born Eugene Bernard Gendel on February 27, 1948 in Hartford, CT, E.B. taught economics at Woodbury University in Burbank and was a leader of progressive causes in Long Beach, where he had lived since 1985. The founder of the local chapter of Occupy Wall Street, E.B. had recently filed to run for mayor of Long Beach. He was also active in IKAR, a progressive Jewish community based in West Los Angeles, spearheading or taking part in many social action and environmental projects. E.B. was raised in Manchester and West Hartford, CT, earning a bachelor's degree from the University of Connecticut and a doctorate in economics from Boston University. He originally moved west to assume the directorship of the California Museum of Science and Industry's Mark Taper Hall of Economics.

Robert Jinkens

Best Overall Faculty Member, Accounting
Best Teacher, Accounting
Most Helpful to Students, Accounting
Best Teacher, Business
Most Helpful to Students, Business
Best Overall Faculty Member, Business

"Over the past 5 years, Dr. Bob Jinkens has proven to be a skilled, innovative, and knowledgeable teacher-scholar who is dedicated to enhancing learning, advising students, and leading pedagogical change. He has served key roles in the Department of Accounting, where he teaches. High among them is that he has run our boot camp classes where we start to move students' perceptions of (and desires for) accounting classes as mechanical, rote, experiences where teachers pour knowledge into students' heads into what students really need: student based learning experiences which focus on critical analyses of open-ended facts.

When I was his Department Chair, I was copied on all of his standard student evaluations, visited his classes, and fielded comments by his students. These were consistently positive. Indeed, he has proven to be an excellent colleague, a hard-working and effective teacher, and willing to help the program out in every way. He has taught a variety of classes, and taught them all quite well, even when he has had to step in at the last minute.

In the 5 years I have worked with Bob, both as his department chair and as a teaching colleague, he has time and again impressed me – as well as students who I advised – with his astute advising and generous efforts to help students get good jobs. Indeed, as a reflection of this, he was overwhelmingly recruited by students to serve as the faculty advisor to the Accounting Society.

I have worked with Bob over the past 5 years. Students have always told me that he is a great teacher, something which I observed myself when I visited his classrooms in my capacity as Department Chair. They also have commended him to me for his astute advising, and generous investment of time in helping students get jobs in their filed, and transition into the profession. Indeed, students invited him to serve as the faculty advisor to the Accounting Society. He also has worked hard and successfully to publish and present useful scholarship in peer reviewed journals and conferences."

Dr. Jinkens began his professional career with four years of public accounting after completing a BS in Accounting and a BA in Mathematics at the University of Southern California. During this time he also obtained a California CPA and a California Real Estate Broker's License. He then managed a grocery store for 10 years while simultaneously teaching accounting at a community college, as well as obtaining an MBA from the University of California at Irvine; managing commercial real estate, syndicating commercial real estate investments; building two homes; and because of his love for education, starting a PhD at the University of Southern California. Dr. Jinkens was not able to complete this PhD because of family demands following the death of his father, and instead earned a master's degree in Finance, but he continued to teach, and finally completed a PhD in Education with a concentration in accounting education at the University of Hawaii at Manoa, and a Post Doctorate in Accounting and Finance at the University of Florida at Gainesville. In total, he has over 35 years of experience in education and over 15 years of applied experience in business, as well as 6 degrees from prestigious universities, 10 teaching credentials, a California CPA, a Hawaii CPA, and a California Real Estate Broker's License.

John Karayan

Best Teacher, Business
Most Helpful to Students, Business
Best Overall Faculty Member, Business

WORCESTER STATE UNIVERSITY

Mariana Cache

Best Teacher, Health Science

"Dr. Cache is devoted to finding different strategies to help all students understand that nutrition literacy is an essential base from which to create persuasive and creative interventions that influence behavior change. She spends countless hours, beyond the classroom, advising, guiding and encouraging students to do health research projects."

Shelley (Michelle) White

Best Researcher/Scholar, Health Sciences and Nursing

"Shelley has published numerous articles in health and sociology journals. She is a health activist, especially with Save the Children of which she is a leading officer, and has traveled, and currently travels, internationally to work to improve the health of children in developing countries."

WRIGHT STATE UNIVERSITY

Marie Bashaw

Most Helpful to Students, Nursing

Anita Dempsey

Best Researcher/Scholar, Nursing
Best Teacher, Nursing
Best Researcher/Scholar, Health Sciences and Nursing

Colleen Finegan

Best Overall Faculty Member, Education

"Colleen is involved in every aspect of education and looks out for the rights of the students."

Suzanne Franco

Best Researcher/Scholar, Education
Best Researcher/Scholar, Education

"Suzanne is a an outstanding researcher and has done some amazing research."

Scott Graham

Most Helpful to Students, Education

"Scott works well with his students. He is always willing to work with them."

Bobbe Gray

Best Overall Faculty Member, Nursing
Best Teacher, Health Sciences and Nursing
Most Helpful to Students, Health Sciences and Nursing
Best Overall Faculty Member, Health Sciences and Nursing

Tracy Kramer

Most Helpful to Students, Education
Best Overall Faculty Member, Education

"Tracy works closely with her students and spends a great deal of time with them when they are in the field.

Tracy is diligent in her interaction with students. She works closely with them to establish a feeling of belonging to the dicipline."

Dan Noel

Best Teacher, Education
Best Teacher, Education

"Dan earns the respect of his students and uses current technologies to keep them engaged."

XAVIER UNIVERSITY

Delane Bender-Slack

Best Researcher/Scholar, Education

"Laney is the most rigorous and disciplined researcher I have ever had contact with. She works towards the greater good and social justice issues are at the heart of all of her research. She is as inspiring as she is brilliant!"

Delane Bender-Slack was an English Language Arts teacher in Mason City Schools in Mason, Ohio before coming to Xavier University where she is on faculty in the Department of Childhood Education and Literacy. She earned her graduate degree in English at Miami University in Oxford, a graduate certificate in Women's Studies at the University of Cincinnati, and then a doctorate in Literacy at the University of Cincinnati. She teaches literacy courses to graduate and undergraduate students preparing to be K–12 teachers, supervises field experiences, and conducts professional development for area schools. More recently, she has co-developed and implemented study abroad in Peru. Her research on teacher induction, preservice teachers' classroom observations, teaching for social justice, critical literacy, study abroad, adolescent literacy, and curriculum studies has been published in such journals as English Education, Teaching Education, The Reading Professor, Feminist Teacher, Teacher Education and Practice, and the Midwestern Educational Researcher.

Mack Mariani

Most Helpful to Students, Political Science

Mack Mariani is a Buffalo native and the 2008 chicken wing eating champion of Xavier University. He also earned his BA at Canisius College (1991) and his MA and PHD at the Maxwell School of Citizenship and Public Affairs at Syracuse University (1992, 2006). Professor Mariani's teaching and research interests include campaigns and elections, Congress and the legislative process, women and politics, and political internships/experiential learning. Mariani is co-author of Diverging Parties (Westview Press, 2003) and co-editor of the Insider's Guide to Political Internships (Westview Press, 2002). His research has appeared in Legislative Studies Quarterly, Political Science Quarterly, PS: Political Science and Politics, Politics and Gender, Comparative State Politics and the Journal of Terrorism and Political Violence.

Courses taught: POLI 140 - Introduction to American Politics and Government POLI 236 - Presidential/Congressional Campaign Internship POLI 244 - Congress and the Legislative Process POLI 361 - Campaign Strategy and Persuasion POLI 367 - Gender and Politics: Women, Elections, and Representation POLI 391 - Senior Seminar: American Government POLI 397 - Political Internship

Teresa Young

Best Overall Faculty Member, Education

"Teresa is just the perfect role model for a professional and for an educator. Her teaching is clear, organized and engaging. She works tirelessly to meet the needs of her students including extra time at work or actually going into the classrooms and observing and coaching. A true professional with a heart of gold. She lives the words James COmer coined "there is no significant learning without significant relationships."

YOUNGSTOWN STATE UNIVERSITY

Molly Roche

Best Teacher, Health Sciences and Nursing

"Outstanding instructor both in the clinical area as well as the classroom! Always has time for ALL of her students, bright, caring, genuine (one-of-a-kind) individual!!!"

Index

A

Abate, M., 23
Abba, K., 193
Abdollahian, M., 71
Abela, A., 372
Abramowicz, K., 389
Abramowicz, M., 167
Ackert, L., 221
Adamich, J., 355
Adams, N., 552
Adebayo, A., 474
Adkins, M., 26
Adkins, R., 447
Agesa, J., 251
Aguilera, N., 193
Ahrens, W., 459
Akcinaroglu, S., 23
Akpan, J., 207
Aksan, A.-M., 137
Aktan, N., 548
Alberto, P., 175
Alexander, M., 536
Allen, M., 367
Aloia, S., 45
Alpe, T., 265
Alter, M., 289
Altman, M., 396
Altsech, M., 132
Alvarez, Y., 255
Amundson, R., 395
An, H., 548
Anaya, J., 60
Anderson, A., 410
Anderson, G., 511
Andre, J., 46
Andrews, C., 145
Annis, J., 523
Anthony, S., 80

Aranella, P., 396
Arca-Contreras, K., 82
Archer, K., 411
Ariail, J., 360
Arndt, A., 309
Arnold, B., 518
Arnott-Hill, E., 70
Arugaslan, O., 545
Arya, A., 301
Asfari, A., 523
Athaide, G., 245
Atkins, J., 318
Attaran, M., 40
Auzenne, L., 348
Aviles, F., 255
Aymard, L., 5

B

Babcock, J., 416
Babcock, T., 491
Bach, G., 293
Bagley, J., 352
Bagnardi, M., 211
Baird, C. D., 508
Baiyee, M., 126
Bakalars, B., 522
Baker, H., 221
Baker, P., 381
Bakhsheshy, A., 500
Balboni, A., 80
Baldwin, N., 387
Baltodano, M., 241
Baltz, D., 301
Banyard, V., 450
Barclay, M., 155
Barkley, D., 28
Barnard, M., 132
Barnes, A., 225

Barney, J., 501
Barnhart, D., 197
Barnhill, K., 30
Barrow, R., 255
Barth, J., 13
Bartkus, K., 510
Bartos, J., 191
Bashaw, M., 556
Batkay, W., 274
Bavonese, J., 207
Baylen, D., 503
Beals, K., 46
Beaubien, E., 133
Bechtol, B., 5
Beck, J., 111
Bedard, J., 19
Begum, N., 109
Bejtlich, M., 56
Bekar, C., 234
Bell, D., 221
Bellner, B., 302
Bender-Slack, D., 558
Bennardo, G., 297
Benton, J., 380
Berk, G., 467
Berman, G., 331
Bernard, V., 32
Berzett, J., 12
Betsill, M., 88
Betts, S., 548
Bianchi, C., 333
Biglow, B., 150
Billingsley, S., 532
Binder, M., 453
Binford, A., 82
Birdsong, L., 15
Birrittella, T., 147
Biscoglio, J., 78
Bishop, J., 126
Blackburn-Harris, S., 367
Blackman, M., 46
Blackmon, V., 346
Blake Jones, M., 77
Blanchette, L., 329
Blondy, L. C., 126
Bloom, K., 460

Blouch, W., 245
Blum, J., 91
Blundell, P., 61
Bochain, S., 58
Bock, T., 478
Boerio, R., 153
Bolsen, T., 176
Borelli, S., 387
Boschini, D., 40
Bosfield, S., 135
Bounds, L., 295
Bourke, S., 408
Boury, T., 153
Bower, J., 131
Bower-Russa, M., 184
Bowers, A., 348
Bowman, J., 93
Boyd-Batstone, P., 53
Boyer, M., 278
Bracey, E., 81
Brady, D., 453
Bragger, J., 274
Brandel, M., 293
Brandt, C., 505
Branham, S., 313
Braswell, L., 479
Bravo, M., 20
Brazel, J., 291
Bresser, P., 352
Brewer, J., 308
Brians, S., 493
Brigham, J., 429
Brody, S., 214
Brooks, K., 331
Broom, E., 368
Brown, D., 233
Brown, K., 405
Brown, L., 359
Brown, M., 511
Brubaker, R., 111
Bruenger, A., 400
Buchan, A., 484
Buchbinder, S., 376
Budryte, D., 173
Bumpus, S., 126
Bundy, M. B., 111

Burer, S., 423
Burke, J., 228
Burnett, M. S., 266
Burnett, P., 136
Bushouse, B., 429
Bussard, D., 360
Bussiere, E., 431
Butcher, A., 469
Butterfield, S., 427
Butto, N., 33
Buzan, B., 46
Buzawa, E., 434
Byrd, C., 33
Byrns, R., 455

C

Cacciatore, R., 315
Cache, M., 556
Cain, M., 412
Calabria, K., 153
Calais, G., 252
Calhoun, J., 151
Callahan, C., 112
Camillo, A., 553
Campanelli, G., 548
Campbell, B., 156
Campbell, S., 221
Canham, D., 341
Cannon, R., 360
Capel, M., 445
Caplis, L., 377
Carbrey, N. D., 383
Cardaciotto, L., 228
Cardi, J., 541
Carey, H., 176
Carl, H., 494
Carlin, R., 177
Carlson, C., 390
Caron, S., 427
Carpenter, M., 309
Carpenter, R., 127
Carter, T., 135
Cascarano, M., 275
Cashdollar, P., 485

Casteel, D., 382
Cavalaris, J., 256
Cerrato, L., 90
Cervone, D., 420
Chanbonpin, K., 215
Chang, E., 267
Chaplin, L., 103
Chapman, C., 488
Charlton, R., 186
Charlton, S., 400
Chatterji, M., 23
Chavez, V., 336
Chavis, B., 47
Cheak-Zamora, N., 445
Chellevold, D., 507
Chen, C.-D., 362
Chen, H.-C., 511
Chenven, J., 59
Chiritescu, A., 110
Chizever, C., 425
Choudhary, A., 233
Choy, R., 256
Christiansen, E., 103
Christianson, K., 24
Christie, N., 321
Chrosniak, L., 161
Chung, E. B., 302
Cirace, J., 233
Clark, A., 465
Clark, J., 147
Clarke, J., 186
Claver, M., 53
Clelland, I., 324
Cleveland, G., 254
Cloward, D., 31
Codling, C., 264
Cody, D., 72
Coleman, Y., 263
Colemere, I., 492
Colley, K., 324
Collins, D., 133
Collins, E. (Ed), 405
Collins, E. (Elicia), 72
Combs, J., 159
Comeaux, J., 460

Comparini, L., 364
Compean-Garcia, N., 368
Conant, J., 197
Condra, D., 278
Connelly, K., 435
Coogan-Pushner, D., 93
Cook, D., 309
Cook, S., 267
Cook, V., 420
Coons, W., 427
Coppedge, R., 92
Corbeil, R., 488
Coronado, I., 490
Corprew, C., 243
Corrales, A., 256
Corzine, E., 12
Coscia, C., 432
Costa, V., 47
Costello, J., 329
Cothran, M., 207
Couts, P., 405
Cox, M., 344
Cox, P., 252
Craig, L., 291
Crain, S., 267
Crain, W., 71
Cramer, A., 321
Crane, D., 47
Craven, W., 284
Crawford-Jones, K., 196
Creamer, C., 329
Creel, E., 348
Crews, D., 302
Crockett, D., 295
Crowley, J., 284
Cruikshank, B., 429
Cruz, B., 475
Cruz-Zuniga, M., 372
Cumiskey, K., 82
Cundiff, N., 390
Cunningham, M., 333
Curcio, D., 83
Curtis-Boles, H., 3
Curwen, M., 66
Custin, R., 57

D

Dancer, T., 361
Dancey, W., 302
Daniel, L., 325
Daniels, A., 193
Daniels, L., 183
Daramola, C., 145
Darling, S., 143
Darling, T., 351
Dastoor, B., 299
Daugherty, M., 479
Davis, E., 81
Davis, M., 390
Davis, S., 292
Dawson, J., 292
De La Cruz, R., 256
Deak, E., 138
Debono, A., 552
DeBoskey, D., 334
Declouette, N., 173
Dedrick, R., 476
Degrandpre, B., 360
Delacruz, J., 234
Delaney, B., 307
Deleon, L., 487
Delgado, C., 74
Delpy Neirotti, L., 162
Demchak, M. A., 448
Dempsey, A., 556
Dennis, D., 476
DeNoble, A., 334
Deprow, S., 551
Desmarais, P., 402
Devine, C., 141
Dial, T., 489
Dicataldo, F., 331
DiCicco-Bloom, B., 83
Dickerson, C., 62
Dicus, S., 31
Diederichs, M., 279
Dilts, D., 204
Dimatteo, H., 285
Dinan, D., 161
Dinardo, B., 316
Ding, X., 508

Dingman, S., 249
Dodge, B., 525
Dodge, T., 163
Dominguez, J., 486
Dong, H., 72
Donlan, R., 198
Dooley, K., 59
Dosch, R., 458
Doughty, C., 293
Downey, M., 354
Downs, D., 519
Doyle, T., 281
Drab, S., 468
Draheim, M., 496
Drake, J., 503
Dresdow, S., 508
Dresser, R., 528
Drew, M., 298
Drezner, Z., 48
Drumming, S., 142
DU, J., 265
Dua, A., 279
Duffey, M., 383
Duke, B., 405
Dukes, C., 143
Duncan, D., 436
Dunlop, S., 266
Dunn, I., 70
Dunne, T., 399
Dunworth, P., 60
Durham, B., 188
Duschinski, H., 306
Dutcher, C., 222
Dutra, D., 44
Dyo, M., 54

E
Eccles, D., 256
Edelman, P., 171
Edison, D., 320
Edrington, M., 104
Edwards, K., 450
Edwards, R., 371
Eigenberg, H., 481
Eisenman, P., 501

Eisenman, R., 487
Eley, S., 198
Elhai, J., 497
Eliason, M., 336
Elizabeth, P., 331
Elmore, A., 181
Elshahat, I., 378
Emenaker, R., 86
Ensher, E., 242
Epping, E., 400
Esara, P., 373
Eschiti, V., 466
Escoto, R., 293
Esinhart, E., 310
Eskola, K., 400
Espelin, J., 58
Espinola-Arredondo, A., 526
Esser, J., 356
Essig, M., 106
Estape, D., 257
Ethiraj, S., 442
Etienne, E., 142
Ettinger, M., 284
Evans, M., 177
Evans, S., 87
Ezell, S., 416

F
Fabrizi, S., 145
Fagan, M., 326
Fairchild, R., 198
Faircloth, S., 448
Falcone, D., 229
Falls, D., 131
Faoye, O., 257
Farmer, M. J., 350
Farnum, J., 275
Farren, A., 83
Farrington, M., 232
Fassetta, M., 271
Favre, D., 292
Feathers, K., 533
Feden, P., 230
Fees, B., 218
Fennig, B., 350

Ferguson, J., 112
Ferholt, B., 83
Fernandez, M., 365
Ferng-Kuo, S.-F., 198
Ferraro, R., 458
Ferreira, M., 533
Ferrell, P., 339
Fetter, G., 325
Finegan, C., 557
Finn, T., 141
Finnegan, L., 81
Firat, I., 194
Fischer, L., 380
Fisher, A., 402
Fisher, M., 466
Flattely, R., 321
Fleisher, S., 42
Fleming, D., 335
Flint, M., 512
Flores, A., 60
Ford, A., 425
Ford, J., 310
Foreman, S., 231
Forsberg, I. N., 78
Forsgren, B., 351
Forster, R., 6
Forth, N., 420
Foster, V., 72
Fountain, J., 300
Franco, S., 557
Franklin, K., 3
Fraser, L., 48
Fraunknecht, M., 545
Frazier, Y., 2
Fred-Mensah, B., 195
Freedman, M., 378
Fried, B., 79
Friedberg, A., 270
Friedburg, A., 270
Friedman, P., 63
Friman, H. R., 250
Frost, W., 314
Fruchtman, J., 378
Frush, K., 406
Fuller, D., 536

Fuller, M., 355
Furlong, C., 381

G

Gackenbach, E., 191
Gaedke, B., 130
Gaines, C., 81
Galen, L., 184
Gales, D., 235
Gallaher, S., 253
Galletta, A., 74
Gamba, M., 413
Ganiban, J., 164
Garcia, J. (Jesse), 361
Garcia, J. (Justin), 262
Gardner, L., 373
Garland, T., 482
Garman, J. F., 227
Garner, Y., 257
Garon, M., 48
Gaskins, S., 387
Geide-Stephenson, D., 536
Geisler, K., 448
Gelo, D., 492
Gendel, E. B. S., 554
Gentry, K., 380
George, J., 94
George, N., 533
Gerber, B., 358
Gerencser, S., 203
Gerhardt, M. J., 57
Germaine, R., 282
Gershon, S., 177
Gewirtzman, D., 287
Ghazzawi, I., 426
Ghent, L., 110
Giannetta, T., 44
Giannetti, V., 106
Giardino, V., 361
Gibson, E., 393
Gibson, S., 253
Gilbert, D. G., 349
Gilbert, J., 44
Gilbert, W., 45
Gilchrist, K., 40

Gilders, R., 306
Gilger, J., 397
Gill, H., 460
Gillespie, M., 451
Gilman, F., 375
Girocco, C., 236
Glaeser, B., 48
Glanc, G., 365
Glenn Paul, D., 275
Glezakos, C., 54
Gloyd, S., 502
Goeller, W., 54
Golden, C., 300
Goldfarb, K., 276
Goldman, S., 286
Goldstein, P., 144
Gollub, E., 148
Gomez, D., 514
Gonzalez, M., 539
Gonzalez, R., 81
Goodwin, M., 232
Goot, J., 356
Gopinath, M., 310
Gore, J., 112
Gornicki, J., 248
Gosnell, S., 515
Gottdiener, W., 213
Gottesman, M., 172
Gottfried, P., 136
Gottfried, R., 340
Gotts, K., 49
Gould, C., 127
Grafstein, B., 414
Graham, B., 185
Graham, C., 253
Graham, S., 557
Grant, M. L., 364
Gray, B., 557
Gray, D., 343
Gray, L., 515
Gray, N., 113
Grebner, L., 261
Green, B., 407
Green, M. (Martin), 33
Green, M. (Michael), 236

Green, M. (Michael), 237
Green, Y., 69
Greenberg, L., 276
Gregitis, S., 146
Gregory, O., 97
Griggs, W., 519
Grijalva, T., 536
Gritzmacher, D., 72
Groebner, J., 183
Grossman, D., 383
Groves, B., 498
Gryczman, A., 60
Guardino, C., 460
Guell, R., 198
Guerrero, F., 449
Gupta, M., 519
Gurian, P.-H., 415
Gursoy, D., 526
Gutmann, J., 426
Gutshall-Seidman, K., 326

H

Hachey, A., 27
Hackney, D., 182
Haddad, M., 485
Hagan, A., 319
Hagler, R., 38
Hagstrom, P., 140
Hahs-Vaughn, D., 403
Hain, C., 281
Hall, M., 96
Hamilton, T., 479
Hamilton, W., 283
Hammond, T., 336
Hand, K., 482
Handorf, W., 164
Haney, W., 30
Hankla, C., 178
Hanna, D. R., 271
Harden, J., 533
Harmon, M., 199
Harms, C., 219
Harper, B., 75
Harper, J., 199

Harris, A., 434
Harris, J., 237
Harris, P. (Pamela), 212
Harris, P. (Paula), 278
Harrison, D., 263
Harvey, J., 482
Hasaad, R., 252
Haun, K., 361
Hawes, D., 226
Hawk, T., 158
Hawkins DeBose, C., 358
Haynes-Mays, I., 369
Haze, D., 264
Hazi, H., 541
He, T., 510
Heafner, T., 456
Heard, L., 279
Heck, T. (Terri), 134
Heck, T. (Theresa), 352
Hedman, A., 263
Hedtke, J., 37
Hefner, A., 308
Hefter, W., 336
Heggans, D., 164
Heide, F., 3
Heintz, J., 425
Helali, A., 539
Heller, W., 421
Hemingway, L., 12
Hendersen, R., 185
Henderson, N., 466
Hendrickson, R., 131
Hendrix, J., 160
Henkin, R., 492
Henriques, J., 506
Henry-Tett, D., 297
Hensley, T., 226
Hentges, B., 417
Herm, S., 353
Hershey, J., 232
Hervey, J., 49
Hetzel, K., 329
Heyman, M., 216
Hibbard, S., 99
Hickman, K., 182
Hill, D., 197

Hill, J., 398
Hill, K., 348
Hill, L., 482
Hill, P., 36
Hill, R., 282
Himberg, C., 42
Hindenlang, B., 356
Hines, K., 340
Hines, S., 127
Hinton, S., 113
Hirsch, J., 49
Hirsch, R., 342
Hladik, P., 531
Hoffman, A., 20
Hoffman, D., 89
Holcombe, J., 483
Holihan, A., 365
Holland, L., 69
Hollensbe, E., 408
Holloway, J., 4
Holly, J., 261
Honig, L., 356
Hooten, M. A., 381
Hope, A., 272
Hopla, J., 31
Hoppe, R., 98
Horace-Moore, M., 128
Horn, D., 26
Horton, S., 532
Hoshino-Browne, E., 283
Houge, T., 423
Housand, A., 457
Housenick, C., 7
Howard, S., 306
Hsieh, B., 54
Hsieh, S.-J., 337
Huang, G., 75
Hubbard, B., 8
Huckaby, R., 292
Hudgins, S., 310
Huggins, J., 38
Hughes, C., 201
Hughes, S., 12
Hull, H., 16
Hunn, L., 61
Hunt-Brown, L., 62

Hurford, D., 322
Hurrell, M., 78
Hurst, T., 414

I

Ibrahim, M., 7
Ingram, Y., 235
Irwin, D., 421
Isidro, M., 371
Islam, M., 143
Isoldi, K., 238
Izzo, F., 245

J

Jabbour, G., 165
Jack, D., 376
Jacobs, H., 546
Jaeger, D., 184
Jain, A., 384
James, F., 388
James, S., 193
Jefferson, P., 279
Jenkins, B., 317
Jenkins, N., 100
Jennings, B., 543
Jennings, S., 515
Jensen, M., 543
Jinkens, R., 554
Jithendranathan, T., 480
John, D., 87
Johns, K., 208
Johnson, D., 31
Johnson, G., 415
Johnson, L. (Linda), 344
Johnson, L. (Lisa), 470
Johnson, M. (Miguel), 91
Johnson, M. (Molly), 308
Johnson, O., 362
Johnson, P., 388
Johnson-Leslie, N., 6
Johnston, A., 417
Jolly-Smith, S., 265
Jones, L., 240
Jones, M., 32
Jones, M., 439
Jones, R. M., 418

Jones, S., 128
Jorgensen, R., 322
Joris, K., 239
Joyce, T., 17
Jozefowicz, J., 202
Julius, E., 38

K

Kass, J. (Jon), 517
Kass, J. (Jon), 517
Kaiser, M. Y., 137
Kalyagin, D., 61
Kanet, R., 440
Kang, M. O., 512
Kappeler, S., 123
Kappeler, V., 123
Karakatsanis, N., 203
Karayan, J., 555
Kasavana, M., 259
Kashlak, R., 245
Katz, A., 11
Kauneckis, D., 449
Kausler, A., 212
Kaya, A., 423
Kayes, A., 359
Kearns, D., 407
Keating, D., 529
Keenan, A., 289
Keister, C., 151
Keller, S., 283
Kelley-King, J., 510
Kelly, L., 272
Kemp, T., 361
Kenny, J., 544
Kent, J., 267
Kerner, J., 13
Kerper, R., 262
Kerr, G., 445
Kerr, P., 346
Kersten, J., 508
Kessler, J., 204
Khislavsky, A., 397
Kibbe, A., 546
Kibler, M., 320
Kido, R., 64
Kiel, J., 107

Kilgore, S., 69
Kimsey-Ortega, M., 257
Kinder, M., 41
King, L., 109
King, N., 208
Kirkpatrick, B., 551
Klasa, S., 391
Kleiman, L., 26
Kleindienst, C., 49
Klingner, J., 443
Klink, R., 247
Knackendoffel, A., 219
Knapp, M., 257
Knopp, A., 211
Knox, S., 483
Koch, R., 437
Kochanek, R., 410
Koffel, L., 194
Kolenko, T., 222
Kolodner, R., 398
Konieczny, L., 58
Konow, J., 242
Kostadinova, T., 148
Kovacic, W., 168
Kozdras, D., 477
Kramer, T., 557
Kromrey, J., 477
Kryshak, W. J., 509
Kuehls, T., 537
Kump, D., 552
Kundu, S., 148
Kuo, S., 473
Kuzyk, P., 527
Kyriakos, M., 78
Kyser, J., 100

L

La Raja, R., 430
Ladores, S., 403
Lakey, B., 185
Lamb, C., 483
Lamb, J., 494
Lambert, P., 27
Lammers, W., 401
Landers, P., 466
Landsman, J., 20
Lane, B., 73
Lane, R., 489
Lane, W., 455
Lange, E., 382
Lange, I., 49
Langendorfer, S., 29
Langford, B., 178
Langmead, J., 247
Lapides, P., 223
Larkin, L., 123
Larocca, R., 522
Larson, E. (Ellen), 295
Larson, E. (Erik), 316
Larue, K., 206
LaRue, M., 505
Lauter, N., 276
Lavoy, L., 133
Lawes, R., 466
Lawhead, T., 264
Lawrence, A., 537
Lea, B.-R., 269
Leasure, J. L., 416
Lederman, M., 172
Lee, S., 437
Lee, Y., 337
Legare, M., 87
Lektzian, D., 371
Leli, P., 403
Lemberger-Truelove, M., 453
Levitan, J., 549
Lewinsky, K., 230
Lewis, M., 109
Lin, B.-X., 472
Lindeblom, G. J., 34
Lippa, R., 50
Lipsitz, K., 94
Little, M., 24
Liu-Thompkins, Y., 311
Livingston, L., 470
Llobera, M., 502
Lo Re, M., 522
Lockett, Y., 330
Loman, N., 42
Long, C., 303
Longo, P., 447
Looby, A., 458

Loriz, L., 460
Lott, L., 515
Louie, K., 549
Lousin, A., 216
Lowe, P., 21
Lowey, S., 364
Lowry, P., 208
Lowry, W., 529
Lucas, D., 299
Ludlow, B., 541
Luna, D., 17
Lundsten, L., 480
Lutwak, N., 17
Lyons, P., 158
Lytle, R., 43

M
Ma, D., 105
MacArthur, J., 553
MacDonald, L., 440
Mack, R., 558
Madalinski, B., 19
Maddox, N., 358
Maddy, M., 407
Magaddino, J., 54
Magenheim, B., 65
Maggs, G., 169
Mahoney, E., 196
Main, E., 544
Makris, E., 294
Malcolm, R., 425
Malik, A., 540
Malin, D., 418
Maloney, E., 340
Maltese, J., 415
Mameli, P., 213
Mancini, N., 298
Manning, C., 178
Manos, N., 286
Mansfield, H., 188
Mapp, P., 320
Maratea, R. J., 285
Marcoulides, L., 50
Mari, M., 257

Mariani, M., 558
Markin, E., 486
Markowski, E., 311
Marotti, H., 135
Marshall, B., 290
Martelli, J., 413
Martin, J. (Jennifer), 446
Martin, J. (Jim), 525
Martinie, S., 219
Marwaha, A., 416
Masoomian, R., 279
Massie, E., 156
Mathews, J., 353
Mathis, D., 174
Matney, T., 494
Matthews, B., 123
Matthews, J., 470
Mattson, M., 175
Maxey, C., 39
May, B., 277
May, C., 99
Mayo, P., 90
Mazariegos, W., 138
Mbaku, J., 538
McAloon, J., 141
McCall, C., 524
McCarty, C., 208
McCleary-Jones, V., 466
McClendon, J., 91
McCloskey, D., 84
McClusky, J., 532
McConachie, M., 87
McConnell, L., 308
McCord, H., 29
McCormick, D., 296
McCoy, M. A. (Mary Anne), 533
McCoy, M. A. (Mary Anne), 534
McCray, A., 290
McCrory, D., 206
McCullough, M., 486
McDaniel, T., 199
McDiarmid, J., 397
McDonald, H., 237
McDonough, C., 284
McFadden, C., 379

McGaha, D., 525
McGinnis, J., 276
McGrath, L., 474
McGuire, T., 54
McHann, J., 524
McHugh, M., 92
McIntosh, J., 455
McKinley, D., 186
McLean, J., 89
McMahan, R., 200
McMahon, B., 235
McMillan, H., 346
McQuaid, B., 319
McQuillen, C., 181
McTier, B., 528
McVey, M., 154
Mearns, J. D., 50
Meaux, J., 401
Mechling, L., 457
Medeiros, G., 331
Medina, K., 236
Mee, S., 84
Meehan, N., 73
Meguerditchian, S., 196
Mehta, A., 326
Meiners, E., 294
Meize-Grochowski, R., 454
Mekraz, A., 509
Melendez, R., 205
Melillo, K. D., 435
Mendez, M., 50
Meredith, D., 280
Merz, J., 468
Mesler, D., 343
Meyer, D., 32
Meyers, M., 349
Meyers, S., 332
Michelman, J., 461
Mickles, A., 91
Middlebrook, M., 380
Milich, M., 94
Miller, C., 257
Miller, D., 254
Miller, G., 530
Miller, J., 306
Miller, R., 131

Milliman, J., 409
Milostan Egus, K., 205
Miniard, P. W., 149
Mirabella, J., 210
Mitchell, M., 104
Mitchell, R., 124
Mitchell, W., 268
Moallem, M., 457
Model, E., 515
Moeller, L., 424
Momeyer, M. A., 303
Monahan, B. J., 200
Mongomery, C., 99
Monserrate, J., 258
Montgomery, J., 66
Montgomery, M., 185
Montoya, E., 539
Monzo, L., 67
Moore, E. G., 490
Moore, K., 512
Moore, P., 34
Moos, J. C., 266
Morales, M., 258
Morand, O., 411
Morecraft, R., 475
Moreno, M., 365
Morgan, A., 315
Moritz, M., 304
Moroney, S., 231
Morris, D., 146
Morris, L., 195
Morris, W., 135
Morrison, M., 101
Morse, D., 265
Morton, N., 454
Mosca, J., 272
Moseley, J., 534
Moses, L. M., 151
Moulding, L., 538
Moulton, S., 304
Mudge, S., 368
Mullins, 344
Munoz, E., 2
Munoz, F., 527
Murillo, L., 487
Murphy, C., 98

Murphy, K., 67
Murray, D. (Deb), 307
Murray, D. (Deb), 522
Murray, L., 538
Mutti, J. J., 185
Muzio, E., 504

N

Nagel, M., 216
Napolitano, B., 301
Napshin, S., 224
Narkiewicz, V., 210
Nassar, M., 391
Neal, J., 98
Neathery-Castro, J., 448
Nelson, G., 96
Nelson, T., 512
Nemmer, J., 186
Nero, D., 232
Nesper, K., 384
Nesper, L., 506
Neubauer, J., 322
Newman, D., 422
Newport, S., 15
Nguyen, H., 54
Niblette, C., 187
Nichols, C., 243
Nichols, J., 158
Niederriter, J., 75
Nientimp, M., 134
Nieves, E., 258
Nightingale, C., 428
Nitti, Y., 258
Njai, M., 373
Noel, D., 558
Noelke, W., 5
Northrop, A., 51
Novak, J., 61
Nteta, T., 430
Nudelman, B., 187
Nunn, E., 471

O

Oblak, M., 394
O'Brien, E., 432
O'Brien, M. P., 298
O'Brien, P., 317
O'Connor, M., 452
O'Donnell, A., 498
Ofek, H., 25
Offenberg, J., 242
Offstein, E., 159
Okada, M., 51
Okon, S., 146
Olatunji, O. (B.), 517
Olds, P., 520
O'Leary, S., 260
Olson, F., 513
Oneal, F., 389
O'Neill, C., 107
Onyeozili, E., 428
Opp, S., 90
Orme, I., 90
Orndorff, H., 97
Orr, J., 175
Ortiz-Walters, R., 323
Osborne, J., 68
Osterheld, K., 21
Otogun, P., 231
Otten, R., 51
Ousley-Exum, D., 458
Owen, S., 325
Owens, L., 208

P

Paarlberg, R., 189
Paciotto, C., 544
Packard, C., 422
Padula, M., 27
Palit, M., 28
Palmer, D., 208
Pan, M.-S., 344
Parent, J., 440
Parent, N., 411
Parent, P., 370
Park, E., 270
Park, J., 200
Park, K. H., 347
Park, S., 483
Parke, S., 234

Parker, J., 297
Parker, R., 496
Parshall, M., 454
Parsons, P., 11
Partovi, F., 105
Patch, M., 513
Patterson, E., 326
Patterson, K., 462
Paul, D., 272
Paulson, S., 462
Pavloski, R., 203
Payan, T., 491
Payne, M., 43
Pearce, A., 7
Pearce, D., 366
Pedro, J., 41
Peer, N., 59
Peissig, J., 51
Peltz, R., 436
Peng, L., 18
Peoples, P., 63
Perez, M., 45
Perkes, L., 31
Perrine, R., 124
Perry, P., 160
Peters, J., 84
Peters, J.-L., 213
Peters, L., 55
Peterson, M. J., 430
Peterson, N., 513
Peterson, S., 391
Peterson, T., 169
Peterson-Murphy, A., 174
Petkevich, A., 498
Petratos, V., 84
Pettay, R., 220
Petzko, V., 483
Pevour, R., 549
Phelps, T., 321
Phillips, K., 549
Pierson, M., 52
Pinzon-Perez, H., 45
Pizza, F., 498
Platania, J., 331
Plesetz, S., 280
Plumm, K., 458

Plunkett, S., 382
Pluta, A., 70
Pokay, P., 128
Pons, I. B., 403
Pope, D., 1
Porcelain, S., 440
Poston, R., 438
Potter, G. (Gary), 124
Potter, G. (Greg), 497
Powell, C. O., 28
Powell, O., 201
Powers, B., 141
Powers, T., 6
Poyo, S., 154
Prasad, A., 452
Prichard, J. R., 480
Primuth, E., 96
Pringle, G., 192
Probst, T., 528
Prochaska, N., 224
Proffitt, S., 361
Pryse, Y., 409
Pullis, J., 241
Purcell, E., 287
Purcell, L., 196
Putnam, R., 189
Pye, A., 73

Q
Quigley, P., 330

R
Rabin, R., 385
Rabon, K., 379
Rachal, P., 95
Rahimi, R., 11
Rahman, S., 194
Rakes, V., 36
Ramesy, B., 157
Ramos, R., 91
Ramsay, J., 374
Rana, S., 307
Rank, R., 285
Ratcliff, J., 374
Ratliff, D., 211

Rauseo, N., 149
Ravenell, W. B., 143
Rawlings, T., 375
Rawlins, T., 124
Ray, A., 362
Razzolini, L., 520
Redmond, W., 347
Reed, K., 499
Reeder, 142
Rees, M., 426
Reese, N., 328
Rehman, S., 166
Reiber, C., 25
Reichl, A., 95
Reid, M., 393
Reif, G., 428
Reilly, R., 520
Reis, E., 18
Rezvani, F., 276
Rhoades, K., 249
Rhodes, J., 431
Ricar, S., 541
Ricard, R., 366
Rich, C., 318
Richard, T., 14
Richardson, J., 495
Richardson, P., 268
Richey, S., 179
Rickey, B., 98
Rico, R., 493
Riddle, K., 349
Rider, C., 250
Rietz, T., 424
Riley, G., 208
Ripley, D., 237
Rivetti, D., 472
Roberts, B. W., 422
Roberts, D., 322
Roberts, K., 13
Roberts, S., 534
Roberts, T., 240
Robey, J., 490
Robinson, D., 200
Robinson-Riegler, G., 480
Rocco, M., 34
Roche, M., 559

Rock, R., 39
Rockey, E., 319
Roederer, C., 145
Roepke, T., 2
Rogers, R., 464
Rojas, C., 540
Romig, N., 55
Roop, J., 535
Roque, E., 342
Rosa, T., 159
Roscoe, A., 435
Rose, K., 69
Rosemond, L., 363
Rosenberg, S., 462
Ross, C., 201
Ross, L., 390
Ross, R. (Ratchneewan), 227
Ross, R. (Robert), 546
Ross, S., 330
Rotruck, K., 159
Roweton, W., 62
Rubin, C., 108
Rudolph, B., 356
Ruebner, R., 217
Ruff, T., 281
Ruth, S., 60
Rutkowski, E., 52

S

Sagar, P. L., 278
Salinas, C., 345
Salitore, R., 101
Salsbury, J., 22
Sammut, S., 155
Sample, J., 192
Sanabria, R., 160
Sanborn, R. H., 385
Sanchez, E., 337
Sanchez, R., 259
Sandel, M., 190
Sanderson, P., 296
Sanso, L., 359
Santora, J., 1
Santos, G., 259
Saunders, M., 174

Sausen, R., 255
Savin, G., 424
Scagnoli, N., 422
Scanlan, T., 244
Schaefer, M., 187
Schaeffer, S., 62
Schaffner, B., 431
Schecter, R., 271
Schept, J., 125
Scheraga, C., 138
Scheuermann, B., 499
Schiavone, L., 218
Schiff, A., 379
Schiff, M., 144
Schirner, K., 327
Schmidt, N., 516
Schneller, P., 446
Schrumpf, D., 15
Schultz, J., 444
Schumann, T., 317
Schwager, J., 330
Schwartz, N., 43
Schwieger, D., 347
Scoullis, M., 277
Scullin, B., 134
Scully, M., 432
Sedgley, N., 247
Segal, N., 52
Seidel, S., 366
Seplocha, H., 550
Serdyukov, P., 282
Sergent, V., 52
Sermon, T., 513
Sessum, B., 402
Sevilla-Dela Cruz, J., 345
Sewell, M., 411
Sexton, C., 183
Shafer, C., 317
Shah, D., 179
Shahbahrami, P., 260
Shanmugam, R., 371
Sharkes, N., 92
Sharp, C., 417
Sharp, T., 195
Shaw, S., 207
Sheets, D., 304

Shelburn, B., 474
Sherrod, R., 389
Sheu, C., 220
Shewfelt, B., 152
Shinn, P., 359
Shipe, N., 345
Shipley, J., 516
Shockley, K., 18
Short, M., 418
Shorter, B., 238
Shovein, J., 480
Shumake, F., 320
Siegel, E., 290
Siegfried, W., 456
Siegrist, B., 544
Sieple, R., 409
Sigrist, G. L., 157
Simko, G., 273
Simmers, C., 268
Simon, A., 441
Simon, C., 280
Simon, J., 288
Simones, J., 353
Simonson, S., 544
Simonton, D., 395
Simpson, J., 361
Simpson-Wood, T., 16
Singhapakdi, A., 311
Singletary, W., 151
Singleton, R., 242
Sirikul, B., 349
Sisco, M., 332
Skaggs, S., 531
Skiba, M., 273
Skinner, D., 516
Skolnikoff, J., 332
Slaubaugh, M., 204
Slaymaker, A., 255
Smith, A. F., 76
Smith, A., 501
Smith, H., 473
Smith, J. G., 237
Smith, J., 203
Smith, K., 280
Smith, L., 518
Smith, N., 59

Smith, P., 129
Smith, R., 81
Snyder, P., 160
Sobel, A., 530
Sobel, J., 398
Soderstrom, I., 125
Solomowitz, N., 341
Sommer, K., 19
Song, H., 369
Soroosh, J., 248
Sousa, M., 542
Southwell, P., 467
Spalding, T., 40
Spanbauer, J., 217
Sparks, P., 392
Sparrow, G., 487
Spatig, L., 251
Sperling, J., 386
Spielmans, G., 255
Spiker, K., 322
Spindle, R., 520
Spires, E., 305
Spivey, M., 397
Spruill, J., 318
Srinivasan, S., 105
Stacher, J., 227
Stack, J., 150
Stalley, P., 100
Stanaczyk, D., 96
Stark, J., 41
Starn, H., 39
Staubs, M., 209
Steed, L., 504
Steege, M., 477
Steigerwalt, A., 180
Steimetz, S., 54
Stein, H., 357
Steines, J., 289
Steinmann, P., 539
Stiles, D., 539
Still, K., 76
Stirling, K., 471
Stoker, R., 166
Stone, D., 142
Stoner, M., 55
Stout, E., 161

Stow, R., 505
Strader, T., 104
Strang, L., 499
Strasser, J., 550
Strate, J., 535
Strong, G., 439
Stuenkel, D., 341
Stupianski, N., 134
Su, Y., 338
Suk, I., 385
Suneye, H., 361
Suntornpithug, N., 204
Suri, R., 106
Susi, J., 231
Sussman, G., 312
Swafford, J., 458
Swaine, E. T., 170
Swanson, P., 197
Swenson, A., 513

T

Tai, C.-S., 370
Talbert, S., 404
Taliaferro, A., 542
Tanna, W., 64
Tanner, D., 462
Tanner, M., 157
Tarantino, K., 53
Tatarek, N., 307
Tatum, H., 240
Taylor, A., 139
Taylor, M., 195
Taylor, S., 251
Teed, E., 360
Tellefsen, T., 85
Terrance, C., 459
Terrell, H., 459
Tevis, M., 488
Thakur, R., 426
Tharp, T., 260
Thibodeau, J., 22
Thiry, M., 187
Thoman, J., 76
Thomas, C., 59
Thomas, D., 465

Thomas, J., 209
Thomas, P., 349
Thompson, A., 251
Thompson, G., 273
Thompson, P., 299
Thompson, R., 484
Thul, G., 506
Tiggs, A., 195
Tillyer, R., 493
Tolman, D., 32
Tomarken, A., 518
Tomassini, L., 305
Tonso, K., 535
Toole, J., 205
Tournaki, N., 85
Tracy, J., 550
Trevitt, J., 53
Trussell, T., 262
Tuft, E., 513
Tumposky, N., 277

U

Udeoganlaya, V., 252
Ugrin, J., 220
Ullah, S., 34
Ulmschneider, G., 205
Umebayashi, T., 362
Ungson, G., 338
Upadhyay, M., 110
Upshaw, M., 60
Urban-Bollis, J., 197
Uscinski, J., 441
Uttarro, R., 412

V

Vaiman, V., 39
Valerio, J., 345
Vallerand, A., 535
Van Hise, J., 139
Vanden Bloomen, D., 509
Vanderkooi, G., 140
Vanderlaan, S., 98
Vanwert, A., 547
Vasilou, M., 101
Vaughn, S., 53

Vaver, S., 509
Veney, K., 323
Vielhaber, M., 129
Vijayakumar, J., 521
Villegas, A. M., 277
Vincent, D., 200

W

Wagner, J., 495
Wakefield, D., 282
Walencik, V., 277
Walker, J. C., 316
Wall, A., 241
Wall, C., 318
Wallace, R., 553
Wallace, S., 108
Waller, S., 475
Walters, S., 381
Ware, L., 531
Warner, B., 451
Warren, K., 532
Warren, M., 414
Warren, T., 254
Wates, K., 475
Watson, J., 399
Watson, T., 32
Watts, D., 5
Wayman, M., 513
Weatherly, J., 459
Webb, C., 69
Webb, K., 463
Wedding, D., 499
Wegge, S., 85
Weiland, L., 551
Weiner, J., 106
Weinstein, H., 225
Welch, C., 386
Wellinski, S., 129
Wendel, J., 449
Wendt, J., 363
Wesley, W., 134
West, J., 441
West, R., 172
Whelan, M., 271
White, D., 381

White, F., 467
White, K., 382
White, S. (M.), 556
White, S., 433
Whitmore, J., 188
Whitney, J., 478
Wiener, S., 399
Wildman, L., 41
Wiley, J., 180
Wiley, S., 167
Wilkins, A., 484
Williams, C., 240
Williams, J., 260
Williams, W. E., 162
Willis, J., 419
Willoughby, D., 73
Willox, L., 504
Wilson, C., 144
Wilson, D., 41
Wilson, J., 467
Wilson, K., 181
Wilson, P., 471
Winn, A., 68
Winston, T., 507
Wisenbaker, J., 320
Wismont, J., 443
Wolf, B., 79
Wolf, M., 205
Wolfe, S., 538
Wollan, M., 110
Wolter, J., 510
Wong, Y.-Y., 339
Wood, I., 521

Woodiel, D. (K.), 130
Woodward Kaupert, C., 333
Wray, L., 125
Wright, N., 419
Wu, Q., 328
Wu, T.-Y., 130
Wyatt, R., 4
Wynn, S., 389

Y

Yankey, K., 135
Yarensky, P., 451
Yeaman, J., 351
Yeoh, P.-L., 22
Yesihak, D., 324
Yetiv, S., 312
Yoon, I., 502
Yost, D., 230
Young, E., 64
Young, R., 305
Young, T. (Teresa), 559
Young, T. (Tyler), 87
Yuen, S., 81

Z

Zakrzewski, J., 60
Zeind, C., 251
Zenanko, M., 209
Zieff, S., 339
Ziolkowski, T., 281
Zmaj, B., 69
Zola, J., 96